Springer Undergraduate Texts in Mathematics and Technology

Springer Undergraduate Texts in Mathematics and Technology (SUMAT) publishes textbooks aimed primarily at the undergraduate. Each text is designed principally for students who are considering careers either in the mathematical sciences or in technology-based areas such as engineering, finance, information technology and computer science, bioscience and medicine, optimization or industry. Texts aim to be accessible introductions to a wide range of core mathematical disciplines and their practical, real-world applications; and are fashioned both for course use and for independent study.

Cónall Kelly

Computation and Simulation for Finance

An Introduction with Python

 Springer

Cónall Kelly
School of Mathematical Sciences
University College Cork
Cork, Ireland

ISSN 1867-5506 ISSN 1867-5514 (electronic)
Springer Undergraduate Texts in Mathematics and Technology
ISBN 978-3-031-60577-2 ISBN 978-3-031-60575-8 (eBook)
https://doi.org/10.1007/978-3-031-60575-8

Mathematics Subject Classification: 35, 60, 65, 91

This Springer imprint is published by the registered company Springer Nature Switzerland AG
The registered company address is: Gewerbestrasse 11, 6330 Cham, Switzerland

If disposing of this product, please recycle the paper.

For Tazuko, Tadhg, and Ailbhe.

Preface

The increase in available computational power in the past three decades has led to the widespread adoption of sophisticated computational methods by the financial industry. These methods are rooted in the mathematics of finance, and their responsible application requires a decent understanding of the underlying mathematical framework, the principles of numerical analysis, and the kinds of issues that can crop up when implementing an algorithm.

This textbook offers an up-to-date introductory treatment of numerical techniques applied to problems in computational finance, placing particular emphasis on issues such as numerical stability, convergence, and error analysis in both deterministic and stochastic settings. It is aimed at advanced undergraduate or postgraduate students pursuing studies in mathematical finance with an emphasis on simulation and computational techniques. We assume some exposure to core mathematical topics such as linear algebra, ordinary differential equations, multivariate calculus, probability, and statistics at the undergraduate level.

We choose Python as our language of instruction, reflecting its growing usage among quantitative analysts. Experience with Python is not assumed, though readers should be familiar with basic programming constructs such as variables, loops, and conditional statements. Nonetheless, this is not a Python reference: after a tutorial in Chap. 1 we will introduce functionality as we go. Where possible we make use only of core Python libraries, and we avoid black-box approaches where it is sensible to do so. The result is intended to be an accessible introduction to computational techniques with enough guidance on the programming side to ensure that readers can easily get practical experience.

The book is divided into three parts. Part I (Modelling Assets and Markets) is an introduction to the mathematics of finance and the pricing and hedging of derivative securities in the Black-Scholes framework, as well as a tutorial introducing the reader to Python as a programming language. Part II (Computational Pricing Methods in the Black-Scholes Framework) covers the three main computational methods for pricing options and their associated hedge parameters: binomial trees, Monte Carlo methods, and finite difference methods, and demonstrates their

---

The original version of the book has been revised. A correction to this book can be found at
https://doi.org/10.1007/978-3-031-60575-8_9

application to the valuation of European, American, and exotic options written on a single underlying asset.

Part III (Simulation Methods Beyond the Black-Scholes Framework) treats a set of more advanced topics and techniques, introducing Python methods for data analysis (providing a point of entry for students interested in machine learning), and modelling with stochastic differential equations. The financial context includes the modelling of several correlated assets, stochastic models of interest rates, and asset models with local or stochastic volatility.

Throughout Parts I and II, the reader is encouraged to build up a library of functions for reuse. At the beginning of each chapter is a table showing which of these functions from earlier in the book are used there. Part III relies less on the user-defined functions specified in Parts I and II, since readers will by this stage have developed some familiarity with Python, and this should give instructors flexibility in how they present the material.

For students encountering the mathematical theory of derivatives and option pricing for the first time, a one-semester introductory course with a computational flavour can be constructed by covering all of Part I and a selection of topics from Part II. For students who have already studied this theory, the book can be covered in a two-semester sequence, with Parts II and III each corresponding to a single-semester course, and Part I providing a background reference. A solutions manual is available for instructors, containing worked solutions for theoretical exercises and guidance on coding exercises.

Cork, Ireland   Cónall Kelly
March 2024

Acknowledgements

This book is based upon material delivered to undergraduate and postgraduate students pursuing degrees in the School of Mathematical Sciences at University College Cork in Ireland. Elements of Parts I and II have also been delivered in a condensed form at the African Institute for Mathematical Sciences (AIMS) in Mbour, Senegal.

During the writing process, I have benefited from the guidance of a supportive editor, Remi Lodh, and constructive feedback provided by anonymous reviewers. Heartfelt thanks are due to my students, who have served as patient test subjects, and to AIMS Senegal for inviting me to contribute to their graduate programme. I'm also grateful to my colleague Tom Carroll at UCC for encouraging me to take on this project, and for his valuable comments on various drafts of the manuscript.

Any errors that remain are my responsibility.

Contents

Part I

Modelling Assets and Markets

Introduction 1

In this chapter, we build up the stochastic model for a tradable financial asset used in the Black-Scholes framework for the pricing of derivatives, and examine its properties. We also demonstrate how some basic programming principles are implemented in Python and use them to explore the properties of the asset model computationally.

Specifically, in Sect. 1.1 we will look at the basics of NumPy and Python, including setting up a working environment, importing code from a Python library or package, using vectorised arithmetic, and creating and manipulating arrays. In Sect. 1.2 we set up the mathematical framework for modelling tradable financial assets as stochastic processes. We review the notion of compound interest and the time value of money, set up an asset model as a stochastic differential equation (SDE) and review the Itô calculus for working with such equations. We then solve the model equation explicitly, yielding the Black-Scholes asset model with Gaussian log-returns. In Sect. 1.3 we explore pseudo-random number generation in NumPy and show how one can use the `random` module to sample from the distribution of the asset prices and their returns, and to simulate an ensemble of asset price trajectories.

1.1 Getting Started with Python and NumPy

Python is a flexible general purpose programming language widely used in academia and industry. NumPy is a package for numerical computation with Python. Python with NumPy, SciPy, and Matplotlib, the latter for graphics and visualisation, are together a powerful toolset for numerical computing and simulation.

A Python distribution gives access to a vast collection of pre-written code organised into a hierarchical structure. A *package* (such as NumPy or SciPy) is a

© The Author(s), under exclusive license to Springer Nature Switzerland AG 2024
C. Kelly, *Computation and Simulation for Finance*,
Springer Undergraduate Texts in Mathematics and Technology,
https://doi.org/10.1007/978-3-031-60575-8_1

directory that organises collections of pre-written code called *modules*. A *library* (such as Matplotlib) is a collection of packages, though the two terms are often used interchangeably and a package need not be contained within a library.

To get started, we recommend to use the Anaconda distribution, available at https://www.anaconda.com/distribution. This distribution includes Python, NumPy, and other useful tools for scientific computing and data analysis. The NumPy project page may be found at https://numpy.org/, and in particular documentation with a full API reference with details of the NumPy functionality encountered in this book as well as tutorials and a beginner guide are available at https://numpy.org/doc/stable/.

The Anaconda distribution also contains Jupyter, a browser-based application for constructing computational documents that include cells of executable code and discursive markup. The project page is at https://jupyter.org.

1.1.1   Jupyter Notebooks

Once you have installed Anaconda, open Anaconda navigator, launch Jupyter, and open a new notebook. A Jupyter notebook is composed of individual cells, each taking one or more lines of code which can be executed as a block. Each input cell is labelled `In []:`. The `return` key inserts a line break within the current cell of the notebook. To evaluate the cell as a whole you must press `shift-return`. Any output will appear immediately below the input cell and is labelled `Out[]:`. The order of cell execution is reflected in the numbers that appear in the square brackets; we omit these numbers in our sample code. Fig. 1.1 illustrates some typical output.

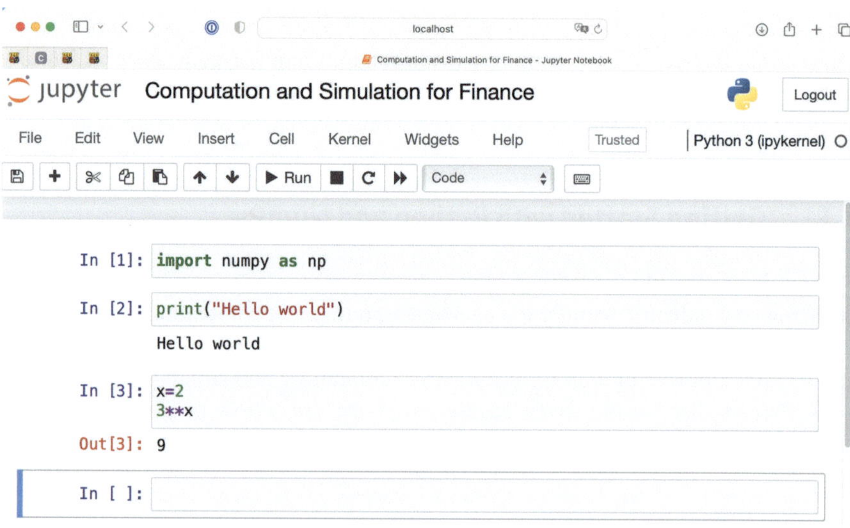

Fig. 1.1 A typical view of a Jupyter notebook

To import code we refer to it by its place in the hierarchy using the syntax *library.package.module*. When importing libraries, packages, or modules, we often give them a short nickname (e.g. np instead of numpy) for convenience. For now, import the complete NumPy package:

```
In []: import numpy as np
```

1.1.2 Variables and Arithmetic

As with any programming language, Python stores data in structures called variables. For example:

```
In []: x=2
```

assigns the value 2 to the variable x. Recover the information contained in x by typing x. However, since we haven't told Python how to interpret this value when creating the variable x, we should check its type:

```
In []: type(x)
```

```
Out[]: int
```

We see that the variable x is of integer type. Python uses dynamic type allocation. This means that whenever data is entered, Python must decide how to interpret it in order to construct a variable to contain it. If we wanted to create a double precision floating point number we would indicate this by including a decimal point:

```
In []: x=2.0
```

or simply

```
In []: x=2.
       type(x)
```

```
Out[]: float
```

Arithmetic with double precision floating point numbers is carried out to 16 digits of precision, and although these are stored they are not always all displayed. For example NumPy arrays display 10 significant digits. As we can see in Fig. 1.1, character strings are indicated by the use of quotation marks, as in print("Hello world").

Once we have created a variable, we can do simple arithmetic. Try each of the following

```
In []:  x+2
        3*x
        3**x
```

```
Out[]: 9.0
```

Notice that only the result of the last line is provided as output. To print the results of all three lines when evaluated in a single cell, we could use the `print()` method:

```
In []: print(x+2); print(3*x); print(3**x)
```

```
Out[]: 4.0
       6.0
       9.0
```

When writing multiple commands on the same line, we use semicolons to separate them. However if we use line breaks, as in

```
In []: print(x+2)
       print(3*x)
       print(3**x)
```

then semicolons are not needed, and the output will be the same.

Alternatively, we can create a `list` variable with all three operations:

```
In []:  [x+2,3*x,3**x]
```

```
Out[]: [4.0, 6.0, 9.0]
```

Other useful arithmetic operators are contained in the NumPy package. Try

```
In []: [np.exp(x), np.log(x), np.sqrt(x)]
```

```
Out[]: [7.38905609893065,0.6931471805599453,1.4142135623730951]
```

The logarithm computed here is the natural logarithm (base e). If we need to compute logarithms with base 2 or 10 we can use

```
In []: [np.log2(x), np.log10(x)]
```

```
Out[]: [1.0, 0.3010299956639812]
```

The constants π and Euler's number e may be invoked as follows:

```
In []: [np.pi, np.e]
```

```
Out[]: [3.141592653589793, 2.718281828459045]
```

1.1.3   Numerical Arrays

While lists are useful, in order to effectively manipulate vectors and matrices we will need to use NumPy arrays. These can be constructed by passing a list of values to the `np.array()` call:

```
In []: a=np.array([2,4,6])
       a
```

```
Out[]: array([2, 4, 6])
```

Individual elements of an array can be accessed by invoking the appropriate index in square brackets. Indexing starts at zero. Let's output a list containing the first (index 0) and third (index 2) terms of a:

```
In []: [a[0], a[2]]
```

```
Out[]: [2, 6]
```

What happens if we try to multiply arrays? Construct a new vector b and multiply using the * operator:

```
In []: b=np.array([1,2,3])
       a * b
```

```
Out[]: array([2,  8, 18])
```

We can see that, using *, multiplication is carried out entry-by-entry. To treat a and b as vectors and to take an inner product, we should use the `dot()` call:

```
In []: a.dot(b)
```

```
Out[]: 28
```

Suppose we wish to carry out a matrix-vector multiplication of the form $C\underline{a}$ where

$$C = \begin{pmatrix} 1 & 2 & 3 \\ 4 & 5 & 6 \\ 7 & 8 & 9 \end{pmatrix} ; \quad \underline{a} = \begin{pmatrix} 2 \\ 4 \\ 6 \end{pmatrix}$$

We will use the already defined array a to represent the vector $\underline{a}$. For the matrix C, define a 2-dimensional square array as follows:

```
In []: C=np.array([[1,2,3],[4,5,6],[7,8,9]])

Out[]: array([[1, 2, 3],
              [4, 5, 6],
              [7, 8, 9]])
```

Notice that we are passing the `np.array()` call a list of rows; each row itself is constructed as a list, and successive rows are separated by commas. Now we can do matrix-vector multiplication using the `dot()` call:

```
In []: C.dot(a)

Out[]: array([ 28,  64, 100])
```

The matrix transpose of C is given by

```
In []: C.T

Out[]: array([[1, 4, 7],
              [2, 5, 8],
              [3, 6, 9]])
```

We can extract individual entries or subarrays. For example the entry in the first row and second column of C is given by (note again that indexing starts at zero):

```
In []: C[0,1]

Out[]: 2
```

The : symbol can be used to represent a full row or column. So if we want to extract the first row of C only, we type

```
In []: C[0,:]

Out[]: array([1, 2, 3])
```

and the second column is obtained by

```
In []: C[:,1]

Out[]: array([2, 5, 8])
```

It will be useful to be able to efficiently generate certain types of large array. For example we can construct a vector of zeros as

```
In []: np.zeros(10)

Out[]: [0. 0. 0. 0. 0. 0. 0. 0. 0. 0.]
```

A 2-dimensional array of zeros can be generated by passing the row and column dimensions as a list:

```
In []: np.zeros([3,4])

Out[]: array([[0., 0., 0., 0.],
              [0., 0., 0., 0.],
              [0., 0., 0., 0.]])
```

We can do the same for an array of ones:

```
In []: np.ones([3,4])

Out[]: array([[1., 1., 1., 1.],
              [1., 1., 1., 1.],
              [1., 1., 1., 1.]])
```

or an identity matrix, which being square only requires that we specify a single dimension:

```
In []: np.eye(3)

Out[]: array([[1., 0., 0.],
              [0., 1., 0.],
              [0., 0., 1.]])
```

We can create ordered arrays of evenly spaced values starting at 0, ending right before 10, and taking steps of arbitrary length. For example we could take steps of length 1 or 2:

```
In []: count1=np.arange(0,10,1); count2=np.arange(0,10,2)
       print(count1); print(count2)

Out[]: [0 1 2 3 4 5 6 7 8 9]
       [0 2 4 6 8]
```

Or we can step backwards:

```
In []: count3=np.arange(9,-1,-1); count4=np.arange(8,-1,-2)
       print(count3); print(count4)
```

```
Out[]: [9 8 7 6 5 4 3 2 1 0]
       [8 6 4 2 0]
```

The `range` function allows us to select subarrays from an array:

```
In []: [count1[range(6)],count2[range(1,3)]]
```

```
Out[]: [array([0, 1, 2, 3, 4, 5]), array([2, 4])]
```

Notice that if we don't specify a starting index, it defaults to 0.

NumPy supports vectorised arithmetic, whereby operations can be applied across each element of an array. Indeed writing efficient code in Python requires that we make use of vectorisation. Say we wanted to raise 2 to the power of every value in the array a. Then we would simply type

```
In []: 2**a
```

```
Out[]: array([ 4, 16, 64])
```

If we want to take the natural logarithm of each term in a, we can simply apply `np.log()` directly to it:

```
In []: np.log(a)
```

```
Out[]: array([0.69314718, 1.38629436, 1.79175947])
```

Although we have demonstrated vectorised operations on the 1-dimensional array variable a, this works equally well for the 2-dimensional array C or indeed any NumPy array.

1.2   Financial Assets as Stochastic Processes

An *asset* is a tradable financial object whose value is currently known but is liable to change unpredictably in the future. For example the prices of a share of stock, a commodity, or a currency pair are all assets. Figure 1.2 shows the time plot produced by the closing prices in USD of one share of stock in Apple Inc. over a period of 34 months.

Financial assets are traded in markets, for example shares of stock are bought and sold on exchanges. The nature of a trade involving such an asset may be simple, for example buying and selling a quantity of the asset, or somewhat more complex, as

Fig. 1.2 A time plot showing daily closing prices in USD for one share of stock in Apple Inc. over a period of just under 34 months

when a derivative security is written on an asset and sold. In this section we discuss how asset prices inform the value of the derivative contracts written on them, and then set up a mathematical framework by which they may be modelled as SDEs with a standard Brownian motion injecting the uncertainty associated with asset price risk. We then solve the resulting SDE leading to the classic Black-Scholes model of a risky asset with lognormal distribution.

1.2.1   Two Examples of Derivatives Written on a Single Underlying Asset

Consider the case of a *European call option*: a contract which gives its holder the right (but not the obligation) to buy from the writer a prescribed asset for a prescribed price (the strike price E) at a prescribed time (the expiry time T) in the future. This is an example of a tradable derivative, and its value $V(S, t)$ depends on the time $t \in [0, T]$ as well as the value the asset happens to take at that time, S. The value of the option at expiry, $V(S, T)$, is determined by the payoff. For example the payoff for a European call option with strike price E is

$$V(S, T) = \max(S(T) - E, 0) = (S(T) - E)^+.$$

Derivatives may depend on the history of the asset price over the period $[0, T]$ as well as the value of the asset at expiry. For example the payoff for an *up-and-out barrier call option* with barrier B and strike price $E < B$ is given by

$$V\left((S(t))_{t \in [0,T]}, T\right) = \begin{cases} (S(T) - E)^+, & \max_{t \in [0,T]} S(t) < B; \\ 0, & \text{otherwise} \end{cases}$$

Notice that the payoff depends on the history of the asset price over $[0, T]$, rather than just on $S(T)$. Notationally, we often suppress any dependence on the asset price trajectory when discussing V in a general setting, writing $V(t)$ to denote the value at time $t \in [0, T]$, and denoting the final value $V_T := V(T)$.

Our first task is to identify and characterise an appropriate stochastic model for S which can be used to compute the option value $V(S, t)$ at time $t \in [0, T]$, approximately if necessary. We can then learn to sample from the distribution of $S(t)$ at a future time $t \in (0, T]$, and to simulate an ensemble of trajectories.

1.2.2   The Stochastic Framework for an Asset Price Model

Let $S(t)$ be the price of such an asset at time t. At time $t = 0$, we have no specific and certain information about future asset values $S(t), t \in (0, T]$, though we can often say something about their distribution. Therefore each $S(t), t \in (0, T]$, may be represented mathematically as a random variable, and over time the evolving asset price $\{S(t) : t \in [0, T]\}$ is modelled as a stochastic process.

Definition 1.1 (Stochastic Process) A *stochastic process* $S = \{S(t) : t \in \mathcal{T}\}$ is a collection of random variables parameterised over the time set $\mathcal{T} = [0, T]$, each defined on the same probability space $(\Omega, \mathcal{F}, \mathbb{P})$.

Each outcome ω in the sample space Ω is associated with a specific *trajectory*, or *sample path*, of the price process S, and when it is convenient to do so we will specify individual trajectories by including an additional argument, writing $(S(t, \omega))_{t \in [0, T]}$ for each $\omega \in \Omega$. $\mathcal{F}$ is the σ-algebra of events: sets of ω to which we wish to assign a probability determined by the probability measure $\mathbb{P} : \mathcal{F} \to [0, 1]$. An event $A \in \mathcal{F}$ is called *almost sure (a.s.)* if $\mathbb{P}[A] = 1$. A stochastic process is *continuous* if for almost all $\omega \in \Omega$ the function $S(t, \omega)$ is continuous on $t \geq 0$. Left and right continuity can be defined similarly.

The increasing set of information available to a market observer regarding the evolution of the price process may be modelled via the events in $\mathcal{F}$ that, at each time $t \in [0, T]$, are decidable by the observer; they are known to have either occurred or not occurred. To this end, we equip the probability space $(\Omega, \mathcal{F}, \mathbb{P})$ with a filtration.

Definition 1.2 (Filtration) A *filtration* is a family $(\mathcal{F}_t)_{t \in [0, T]}$ of increasing sub-σ-algebras of $\mathcal{F}$ such that, for all $0 \leq s < t < T, \mathcal{F}_s \subset \mathcal{F}_t \subset \mathcal{F}$. A stochastic process $(S(t))_{t \in [0, T]}$ is adapted to the filtration $(\mathcal{F}_t)_{t \in [0, T]}$ if, for every $t \in [0, T]$, $S(t)$ is $\mathcal{F}_t$-measurable.

The quadruple $(\Omega, \mathcal{F}, (\mathcal{F}_t)_{t \geq 0}, \mathbb{P})$ is known as a *filtered probability space*. Let $\mathcal{P}$ denote the smallest σ-algebra on $\mathbb{R}^+ \times \Omega$ with respect to which every left continuous process is a measurable function of (t, ω). Then a stochastic process S is *predictable*, or *previsible*, if it is $\mathcal{P}$-measurable.

The canonical example of a continuous stochastic process, and the process we will use to model the effect of market uncertainty on the evolution of risky assets, is the standard Brownian motion.

Definition 1.3 (Standard Brownian Motion) Let $(\Omega, \mathcal{F}, (\mathcal{F}_t)_{t \geq 0}, \mathbb{P})$ be a filtered probability space. A *standard Brownian motion with respect to* $\mathbb{P}$, denoted $W^{\mathbb{P}} = \{W^{\mathbb{P}}(t) \ : \ t \geq 0\}$, is a real-valued, $\mathcal{F}_t$-adapted stochastic process with $W_0^{\mathbb{P}} = 0$, trajectories that are $\mathbb{P}$-a.s. continuous, increments $W^{\mathbb{P}}(t) - W^{\mathbb{P}}(s)$ that are independent of $\mathcal{F}_s$ for $0 \leq s < t$, and which satisfy, for any $x \in \mathbb{R}$,

$$\mathbb{P}\left[W^{\mathbb{P}}(t) - W^{\mathbb{P}}(s) \leq x\right] = \int_{-\infty}^{x} \varphi(y; 0, t - s)dy, \quad 0 < s < t, \tag{1.1}$$

where φ is the density function of a normal random variable with mean m and variance v^2,

$$\varphi(x; m, v) = \frac{1}{\sqrt{2\pi}\,v} e^{-\frac{(x-m)^2}{2v^2}}, \quad x \in \mathbb{R}. \tag{1.2}$$

The increments of $W^{\mathbb{P}}$ are sometimes referred to as $\mathbb{P}$-Normal. This is to place emphasis on the probability measure under which (1.1) holds, where it might not already be clear from context. They are also referred to as *centred normal*, since we can see from (1.1) that the mean of the distribution of each increment is zero.

Definition 1.3 requires that $W^{\mathbb{P}}$ be adapted to the filtration $(\mathcal{F}_t)_{t \geq 0}$. Consider for example the *natural filtration generated by* $W^{\mathbb{P}}$ which may be constructed as

$$\mathcal{F}_t = \sigma\left(W^{\mathbb{P}}(s) \ : \ 0 \leq s \leq t\right), \quad t \geq 0,$$

so that each $\mathcal{F}_t$ is the σ-algebra generated by $\{W^{\mathbb{P}}(s) \ : \ 0 \leq s \leq t\}$. $W^{\mathbb{P}}$ is automatically adapted to its natural filtration.

The following class of (potentially infinite) random times allow us to model the occurrence of events that can be determined by an observer who is unable to see into the future.

Definition 1.4 Let $(\Omega, \mathcal{F}, (\mathcal{F}_t)_{t \geq 0}, \mathbb{P})$ be a filtered probability space. A random variable $\tau \ : \ \Omega \to [0, \infty]$ is a *stopping time with respect to the filtration* $(\mathcal{F}_t)_{t \geq 0}$ if $\{\omega \ : \ \tau(\omega) \leq t\} \in \mathcal{F}_t$ for any $t \geq 0$.

For example, consider a standard Brownian motion $W^{\mathbb{P}}$ satisfying the conditions of Definition 1.3. The first hitting time of $a \in \mathbb{R}$ by $W^{\mathbb{P}}$, defined as the random variable $\theta = \inf\{t \geq 0 \ : \ W^{\mathbb{P}}(t) = a\}$, is a stopping time with respect to the natural filtration, since an observer of $W^{\mathbb{P}}$ can identify the realised value of θ at the moment it occurs. By contrast, the time at which $W^{\mathbb{P}}$ attains its maximum over an interval $t \in [0, T]$, defined as the random variable $\{t \in [0, T] \ : \ W^{\mathbb{P}}(t) = \max_{s \in [0, T]} W^{\mathbb{P}}(s)\}$, is not a stopping time with respect to that filtration, since an observer would have to wait until the end of the interval $[0, T]$ to pick out the maximum of $W^{\mathbb{P}}$.

We also make the following common assumption about large liquid financial markets:

Assumption (The Efficient Market Hypothesis) The past history of an asset is fully reflected in the present price, which does not hold further information. Furthermore, markets respond immediately to any new information about an asset.

As a result, any mathematical process that describes the change in asset price from time s to time t should depend only on the price at time s and not on the price at earlier times. Mathematically we would interpret this to mean that S should be modelled as a Markov process.

Definition 1.5 (The Markov Property) Suppose that the asset price process S is adapted to the filtration $(\mathcal{F}_t)_{t \in [0,T]}$. S has the *Markov property* if, for each bounded and measurable function $g : \mathbb{R} \to \mathbb{R}$,

$$\mathbb{E}_{\mathbb{P}}[g(S(t))|\mathcal{F}_s] = \mathbb{E}_{\mathbb{P}}[g(S(t))|S(s)], \quad a.s, \quad 0 \le s < t \le T.$$

1.2.3 The Time Value of Money

The Black-Scholes framework for the pricing of financial derivatives assumes that investors can borrow or lend any amount of money at a risk-free interest rate that is a known function of time. At least to start with, we will suppose that there is a single constant rate denoted $r > 0$. This is reasonable when valuing stock options with a time to expiry of less than one year, since interest-rate fluctuations over this timescale are unlikely to affect the option value significantly. In Chap. 7 we will also investigate models for stochastically varying interest rates; these are necessary for valuing options on interest-rate dependent assets such as bonds and longer-term derivatives.

1.2.3.1 Risk-Free Investment: Savings Accounts and Zero-Coupon Bonds

Suppose we deposit S_0 with the bank in a savings account with a constant and continuously compounded interest rate $r > 0$. The value of the account will grow exponentially according to the relation

$$S(t) = S_0 e^{rt}, \quad t \in [0, T]. \tag{1.3}$$

Such growth is guaranteed and so we view this as a risk-free investment. If, at the end of the deposit period $[0, T]$, we wish to receive an amount M, we can determine how large our initial deposit S_0 should be by setting $t = T$ and $S(T) = M$ in (1.3) and multiplying through by e^{-rT}:

$$S_0 = e^{-rT} M. \tag{1.4}$$

Thus the initial deposit S_0 is understood as the final amount M discounted back to time $t = 0$ using rate r. We can express a similar relation between M and the value in the account at any intermediate time $t \in [0, T]$ by multiplying both sides of (1.4) by the compound interest factor e^{rt} and applying (1.3) to the LHS, yielding the relation

$$S(t) = e^{-r(T-t)} M, \quad t \in [0, T].$$

In general, we refer to $e^{-r(T-t)} M$ as the *value of M discounted to time $t \in [0, T]$*, and to $e^{-r(T-t)}$ as the *discount factor*.

$S(t)$ defined in (1.3) is the unique solution of the first-order ordinary differential equation given by

$$\frac{dS(t)}{dt} = rS(t); \quad S(0) = S_0, \quad t \in [0, T],$$

which may be written in integral form as

$$S(t) = S_0 + \int_0^t rS(u)\,du, \quad t \in [0, T]. \tag{1.5}$$

We can generalise this idea to one where interest is continuously compounded at a rate $r(t)$ that may vary in time, either deterministically or stochastically. The simulation of stochastic models for r, particularly where it represents the rate at which short-term lending and borrowing is conducted between financial institutions, are considered in Chaps. 7 and 8. Now the amount in the savings account is given by

$$S(t) = S_0 e^{\int_0^t r(s)ds}, \quad t \in [0, T]. \tag{1.6}$$

and we can discount a future value M from T to $t \in [0, T]$ if we multiply by the potentially random discount factor $e^{-\int_t^T r(s)ds}$.

Mathematically, making a deposit in a savings account is indistinguishable from a loan made to the bank. Suppose that, instead, we buy a *zero-coupon bond* with issue price S_0 and maturity T where interest is continuously compounded at a variable rate $r(t)$ (the *T-year zero-rate of interest*). This is a fixed-income instrument that represents a loan made by an investor (the bondholder) to a borrower (the issuer) where all interest and principal is realised at maturity. It also has value given by Eq. (1.6).

Table 1.1 Example term
structure of interest rates

T (years)	r_T (%)
0.5	2.0
1.0	2.2
1.5	2.5
2.0	3.0

1.2.3.2 The Term Structure of Interest Rates

For a zero-coupon bond, the rate at which interest is paid or charged may vary according to the length of time to maturity T at the time of the initial investment $t = 0$. The mapping from the set of possible maturities to the corresponding interest rates is referred to as the *term structure*. Consider as an example the term structure given in Table 1.1. We can represent this as the NumPy array `termRates`.

```
In []: termRates=np.array([[0.5,1.0,1.5,2.0],
                           [0.02,0.022,0.025,0.03]])
```

The array of times to maturity can now be accessed as `termRates[0,:]`, and the array of interest rates is `termRates[1,:]`.

1.2.4   Python: User-Defined Functions

It will be convenient to define a function in Python that will return the discount factor $e^{-r(T-t)}$ when given the arguments r and $dt = T - t$. We can do this using the `def` statement, followed by our chosen name for the function and the arguments in round brackets.

```
In []: def dFac(r,dt):
           return(np.exp(-r*dt))
```

Notice the use of indentation to mark out the code that is inside the function definition, and the : symbol after the arguments. We have also represented the time remaining $T - t$ as a single parameter `dt`. We could ask for the single discount factor associated with rate $r = 0.02$ over a 6 month period:

```
In []: dFac(0.02,0.5)
```

```
Out[]: 0.9900498337491681
```

However, since the `np.exp()` call operates termwise on NumPy arrays, we can pass an entire term structure to dFac and receive back the corresponding array of discount factors. Using the example `termRates` constructed in Sect. 1.2.3.2, we would get

```
In []: dFac(termRates[0,:],termRates[1,:])
```

```
Out[]: array([0.99004983, 0.97824024, 0.96319442, 0.94176453])
```

We treat theory and computational methods for bonds and interest rates in more detail in Chap. 7.

1.2.5 Incorporating Asset Price Risk

We wish to describe the real-world dynamics of an asset such as the price of a share in a company which has a "natural" intrinsic growth rate of μ, but which is also subject to random influences. Since the growth of the asset in the absence of uncertainty is now coming from the intrinsic growth of the company rather than interest paid on deposited money, we start by replacing r with μ in (1.5). So we start with the integral equation

$$S(t) = S_0 + \int_0^t \mu S(u)du, \quad t \in [0, T]. \tag{1.7}$$

Now consider how we might develop (1.7) to incorporate asset-price risk. According to the Efficient Market Hypothesis the price of this asset should react instantaneously to new information as it becomes available to traders in the market. To this end we assume that trading of the asset may take place at any time $t \in [0, T]$ and in any amount.

We need to include a source of randomness, or noise, to model the unpredictable nature of market movements, and for this we use the process $W^{\mathbb{P}}$, a standard Brownian motion under the measure $\mathbb{P}$, satisfying Definition 1.3. Now we can include an integral term on the right hand side of (1.5) to account for the stochastic effect of asset price risk.

$$S(t) = S_0 + \int_0^t \mu S(u)du + \int_0^t \sigma S(u)dW^{\mathbb{P}}(u), \quad t \in [0, T]. \tag{1.8}$$

The new parameter $\sigma > 0$ allows us to control the intensity of the noise input and is called the *volatility* of the asset price model. The stochastic integral on the RHS of (1.8) is known as an Itô integral and it is taken with respect to the process $W^{\mathbb{P}}$. We will not go into the details of its construction here except to comment that it is defined in such a way that if $g : [s, t] \to \mathbb{R}$ is a deterministic function such that $\int_s^t |g(s)|^2 ds < \infty$, then

$$\int_s^t g(s)dW^{\mathbb{P}}(s) \sim \mathcal{N}\left(0, \int_s^t |g(s)|^2 ds\right), \quad 0 \le s < t \le T. \tag{1.9}$$

Moreover, for any constant $c \in \mathbb{R}$

$$\int_s^t cdW^{\mathbb{P}}(u) = c(W^{\mathbb{P}}(t) - W^{\mathbb{P}}(s)), \quad 0 \le s < t \le T.$$

Since (1.8) is intended to describe the real-world dynamics of the asset, we refer to $\mathbb{P}$ as the *real-world probability measure*.

Equation (1.8) describes an example of a scalar Itô process.

Definition 1.6 Let $(\Omega, \mathcal{F}, (\mathcal{F}_t)_{t\ge 0}, \mathbb{P})$ be a filtered probability space and suppose that $W^{\mathbb{P}}$ is adapted to $(\mathcal{F}_t)_{t\ge 0}$. A *scalar Itô process* is a continuous $\mathcal{F}_t$-adapted process $X(t)$ on $t \ge 0$ of the form

$$X(t) = X(0) + \int_0^t f(s)ds + \int_0^t g(s)dW^{\mathbb{P}}(s), \tag{1.10}$$

where $\int_0^T |f(s)|ds < \infty$, and $\int_0^T g(s)^2 ds < \infty$ a.s. for all $T < \infty$. $X(t)$ is said to have *stochastic differential* $dX(t)$ on $t \ge 0$ given by

$$dX(t) = f(t)dt + g(t)dW^{\mathbb{P}}(t).$$

The stochastic differential form of the asset model (1.8) is

$$dS(t) = \mu S(t)dt + \sigma S(t)dW^{\mathbb{P}}(t), \quad t \in [0, T], \quad S(0) = S_0 > 0. \tag{1.11}$$

Here, $f(t) = \mu S(t)$ and $g(t) = \sigma S(t)$. It is important to understand that this is a notational convenience: the trajectories of $W^{\mathbb{P}}$ and hence S are non-differentiable almost everywhere (a.e.). We call (1.11) a stochastic differential equation (SDE) of Itô type.

1.2.6   The Itô Calculus

Our next step is to find the unique solution for the SDE (1.11) explicitly in terms of S_0, μ, σ, and $W^{\mathbb{P}}$. To do this we recall the stochastic version of the change of variables formula, known as Itô's formula, in the standard differential notation.

Theorem 1.7 (Itô's Formula) *Let $X(t)$ be a continuous scalar Itô process given by (1.10) in Definition 1.6. Let $V \in C^{2,1}(\mathbb{R} \times \mathbb{R}^+; \mathbb{R})$. Then*

$$dV(X, t) = \left(V_t(X, t) + f(t)V_X(X) + \frac{1}{2}g(t)^2 V_{XX}(X) \right) dt + g(t)V_X(X)dW^{\mathbb{P}}(t),$$

for $t \in [0, T]$, where

$$V_X(X, t) = \frac{\partial V(X, t)}{\partial S}; \quad V_{XX}(X, t) = \frac{\partial^2 V(X, t)}{\partial X^2}; \quad V_X(X, t) = \frac{\partial V(X, t)}{\partial t}.$$

Itô's formula may be generalised to apply to d-dimensional systems of SDEs. Though we don't require the full generalisation here, we present a special case that provides a stochastic version of the product rule, to be used later in Chap. 2.

Proposition 1.8 (Stochastic Product Rule Formula) *Suppose that each component of the vector $W^{\mathbb{P}} = [W_1^{\mathbb{P}}, W_2^{\mathbb{P}}]^T$ is a scalar standard Brownian motion on a filtered probability space $(\Omega, \mathcal{F}, (\mathcal{F}_t)_{t \geq 0}, \mathbb{P})$, where $W^{\mathbb{P}}$ is adapted to $(\mathcal{F}_t)_{t \geq 0}$. Suppose also that $W_1^{\mathbb{P}}$ and $W_2^{\mathbb{P}}$ are mutually independent.*

Let X, Y be scalar Itô processes satisfying

$$\begin{aligned} dX(t) &= f_1(t)dt + g_1(t)dW_i^{\mathbb{P}}(t); \\ dY(t) &= f_2(t)dt + g_2(t)dW_j^{\mathbb{P}}(t), \end{aligned} \tag{1.12}$$

for some choice of $i, j \in \{1, 2\}$. Then

$$d[X(t)Y(t)] = Y(t)dX(t) + X(t)dY(t) + dX(t)dY(t). \tag{1.13}$$

Equation (1.13) can be restated by use of the formal multiplication table

$$\begin{aligned} dt\,dt &= 0; & dW^{\mathbb{P}}(t)dt &= 0; \\ dW_i^{\mathbb{P}}dW_i^{\mathbb{P}} &= dt; & dW_i^{\mathbb{P}}dW_j^{\mathbb{P}} &= 0, \ if\ i \neq j. \end{aligned}$$

Hence, if $i = j$,

$$d[X(t)Y(t)] = Y(t)dX(t) + X(t)dY(t) + g_1(t)g_2(t)dt. \tag{1.14}$$

Otherwise, if $i \neq j$,

$$d[X(t)Y(t)] = Y(t)dX(t) + X(t)dY(t). \tag{1.15}$$

We see from (1.14)–(1.15) that the stochastic product rule has an additional third term in the case where both SDEs in (1.12) are driven by the same scalar standard Brownian motion. This term is zero if both SDEs are instead driven by independent noises.

The multiplication table in the statement of Proposition 1.8 is referred to as formal in the sense that it describes a set of rules for computation involving stochastic differentials that leads to correct results, though it may be hard to assign a rigorous mathematical meaning to those rules as they are stated. We should continue to think of (1.13)–(1.15) as integral rather than differential equations.

Nonetheless, we can develop some intuition by considering dt as a small but positive step and then examining how each quantity behaves as $dt \to 0$. So $dt\, dt = (dt)^2$ vanishes at a higher rate than dt (quadratically rather than linearly) and may be considered in the limit as zero. Similarly, $dW^{\mathbb{P}}(t)dt$ is approximated over a step of length dt by the random variable $[W^{\mathbb{P}}(t + dt) - W^{\mathbb{P}}]dt \sim \mathcal{N}(0, dt^3)$. This has zero mean and standard deviation $dt^{3/2}$, and will tend to vanish at a higher rate than dt. So again we consider this in the limit to be zero. It follows that if dX is a stochastic differential that satisfies an SDE

$$dX(t) = f_1(t)dt + g_1(t)dW^{\mathbb{P}}(t)$$

then $dX(t)\, dt = 0$, since formally we can write the product of stochastic differentials as

$$dX(t)\, dt = f_1(t)\underbrace{dt\, dt}_{=0} + g_1(t)\underbrace{dW^{\mathbb{P}}(t)\, dt}_{=0} = 0. \tag{1.16}$$

Two related properties of the Itô integral are as follows:

Proposition 1.9 *Let $W^{\mathbb{P}}$ be a standard Brownian motion on a filtered probability space $(\Omega, \mathcal{F}, (\mathcal{F}_t)_{t\in[0,T]}, \mathbb{P})$ where $W^{\mathbb{P}}$ is adapted to $(\mathcal{F}_t)_{t\geq 0}$, and suppose that g is any $\mathcal{F}_t$-adapted process such that $\int_0^T g(s)^2 ds < \infty$ a.s, for all $T < \infty$. Then*

1. the Itô integral has zero expectation:

$$\mathbb{E}_{\mathbb{P}}\left[\int_0^t g(s)dW^{\mathbb{P}}(s)\right] = 0, \quad t \in [0, T].$$

2. Itô's isometry is satisfied:

$$\mathbb{E}_{\mathbb{P}}\left[\left(\int_0^t g(s)dW^{\mathbb{P}}(s)\right)^2\right] = \mathbb{E}_{\mathbb{P}}\left[\int_0^t g(s)^2 ds\right], \quad t \in [0, T].$$

Finally, we review the definition of an $\mathcal{F}_t$-martingale, which plays a key role in the pricing of financial derivatives via arbitrage theory and will be needed in Chap. 2, along with an important result on the representation of $\mathcal{F}_t$-martingales in terms of Itô integrals.

Definition 1.10 ($\mathcal{F}_t$-Martingale) Let M be a stochastic process defined on the filtered probability space $(\Omega, \mathcal{F}, (\mathcal{F}_t)_{t \in [0,T]}, \mathbb{P})$. Then M is an $\mathcal{F}_t$-martingale if and only all of the following hold:

1. M is adapted to the filtration $(\mathcal{F}_t)_{t \in [0,T]}$;
2. $\mathbb{E}_{\mathbb{P}}[|M(t)|] < \infty$ for all $t \in [0, T]$;
3. $\mathbb{E}_{\mathbb{P}}[M(t)|\mathcal{F}_s] = M(s)$ a.s, $0 \le s < t \le T$.

Theorem 1.11 (Martingale Representation Theorem) *Let $W^{\mathbb{P}}$ be a standard Brownian motion with respect to $\mathbb{P}$ on the filtered probability space $(\Omega, \mathcal{F}, (\mathcal{F}_t)_{t \ge 0}, \mathbb{P})$ satisfying Definition 1.3. Let M be an $\mathcal{F}_t$-martingale. Then there exists an $\mathcal{F}_t$-adapted process g such that*

$$M(t) = M_0 + \int_0^t g(u)dW^{\mathbb{P}}(u), \quad t \ge 0.$$

Proposition 1.12 *Let $W^{\mathbb{P}}$ be a standard Brownian motion on a filtered probability space $(\Omega, \mathcal{F}, (\mathcal{F}_t)_{t \in [0,T]}, \mathbb{P})$ where $W^{\mathbb{P}}$ is adapted to $(\mathcal{F}_t)_{t \ge 0}$, and suppose that g is any $\mathcal{F}_t$-adapted process such that $\int_0^T g(s)ds < \infty$ a.s, for all $T < \infty$. Then the process X defined at each time $t \in [0, T]$ by the pure Itô integral*

$$X(t) = \int_0^t g(s)dW^{\mathbb{P}}(s), \quad t \in [0, T],$$

is an $\mathcal{F}_t$-martingale.

1.2.7 Solving the Black-Scholes Asset Model SDE

To solve Eq. (1.11) we first appeal to the theory of linear SDEs of Itô type, which tells us that for any $0 < T < \infty$ there is a unique strong solution of (1.11) over the interval $[0, T]$ associated with each initial value $S_0 > 0$, and $S(t) > 0, t \in [0, T]$, a.s. Therefore the natural logarithm of the asset price process $\ln(S(t))$ is well defined a.s, for all $t \in [0, T]$.

Now apply Theorem 1.7 to determine the SDE governing $V(S, t) = \ln(S)$. The integral form of this SDE is

$$\ln(S(t)) = \ln(S_0) + \int_0^t (\mu - \sigma^2/2)dt$$

$$+ \int_0^t \sigma\, dW^{\mathbb{P}}(t)$$

$$= \ln(S_0) + (\mu - \sigma^2/2)t$$

$$+ \sigma W^{\mathbb{P}}(t),\ t \in [0, T],$$

since $W^{\mathbb{P}}(0) = 0$. Notice that both of the integrals on the RHS can be evaluated in closed form, and by taking the anti-log on both sides we arrive at the solution

$$S(t) = S_0 e^{(\mu - \sigma^2/2)t + \sigma W^{\mathbb{P}}(t)}, \quad t \in [0, T]. \tag{1.17}$$

In stochastic modelling this is referred to as a *geometric Brownian motion*, and in a financial setting it may be understood as the real-world form of the *Black-Scholes asset price model*. The log-return of the asset over an interval $[s, t]$ is defined as $\ln(S(t)/S(s))$, and can be expressed

$$\ln\left(\frac{S(t)}{S(s)}\right) = (\mu - \sigma^2/2)(t - s) + \sigma(W^{\mathbb{P}}(t) - W^{\mathbb{P}}(s)), \quad 0 \le s < t \le T.$$

By Definition 1.3 we can see that

$$\ln\left(\frac{S(t)}{S(s)}\right) \sim \mathcal{N}((\mu - \sigma^2/2)(t - s), \sigma^2(t - s)).$$

From this we see that the volatility σ can be interpreted as the standard deviation of the log-returns of the asset per unit time. Moreover the log-returns over non-overlapping intervals are mutually independent; this is consistent with the oft-repeated caveat *"Past performance is not indicative of future results"*.

1.3 Sampling and Simulation of the Black-Scholes Asset Model

Having constructed a continuous-time asset model with independent and normally distributed returns, we can ask how those returns may best be sampled in Python. This requires us to examine the nuts and bolts of pseudo-random number generation in that language, and to consider best practice with regard to the use of pseudo-random number generators (RNGs). RNGs are mathematical functions that use a deterministic process applied iteratively to an initial state (the *seed*) to generate a sequence of numbers with similar statistical properties to those of the random distribution from which we wish to sample.

1.3.1   Pseudo-Random Number Generation

In this section, we will introduce the plotting package PyPlot, part of the Matplotlib library, as a means of generating histograms and scatterplots.

```
In []: import matplotlib.pyplot as plt
```

API documentation for calls in this package may be found at https://matplotlib.org/stable/index.html.

1.3.1.1 The RNG and Direct Sampling

For pseudo-random number generation throughout the book, we will instantiate and use the following local random number generator:

```
In []: rng=np.random.default_rng()
```

We can easily generate samples from a large range of probability distributions by calling to rng. In this section we will focus on the uniform and normal distributions, since we will primarily require them for the first part of the book. Others will be introduced as necessary.

Produce a NumPy array containing 10^4 samples from a uniform distribution over $[0, 1]$:

```
In []: uniform=rng.uniform(0,1,10**4)
```

The first and second arguments of this call mark the start and end of the interval $[0, 1]$. The third argument specifies the number of samples returned, and we can pass multidimensional values to this argument in order to generate arrays of samples. While it is not so useful to examine the raw values that result from this call, we can check the shape of the density by plotting a histogram.

```
In []: plt.hist(uniform,bins=50,density=True)
```

Setting the `density` parameter to `True` rescales the histogram so that the area underneath it is one.

In order to generate samples from a standard normal distribution, we can invoke

```
In []: Z=rng.standard_normal(10**4)
```

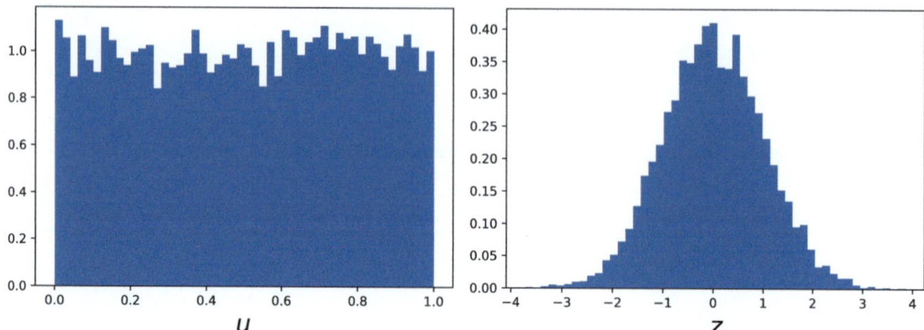

Fig. 1.3 Empirical density curves generated by the distribution of 10^4 samples from $U \sim$ unif[0, 1] (left) and on $Z \sim \mathcal{N}(0, 1)$ (right), sampled via calls to the rng random number generator

The only argument we use for this call is the sample size, since the mean and variance are fixed at 0 and 1 respectively. We can easily generate samples from a normal distribution with arbitrary mean mu and (positive) standard deviation sigma by transformation.

```
In []: mu=5; sigma=0.5
       norm1=mu+sigma*Z
```

Alternatively we could directly sample from the normal distribution:

```
In []: norm2=rng.normal(mu,sigma,10**4)
```

where we must also pass the mean mu and standard deviation sigma as parameters. Histograms generated as output from this section are displayed in Fig. 1.3.

1.3.1.2 Seeds, and Local Versus Global RNGs

If we wish to generate reproducible pseudo-random samples we would specify the initial state of rng by passing a seed as an argument. Let's illustrate this by instantiating a local RNG with a specific seed value:

```
In []: seed=10
       rng1=np.random.default_rng(seed)
```

along with another, different, local RNG for which the seed is not specified

```
In []: rng2=np.random.default_rng()
```

We can compare output from both RNGs by requesting (say) three samples from a uniform distribution on [0, 1]:

```
In []: print(rng1.uniform(0,1,3))
       print(rng2.uniform(0,1,3))
```

```
Out[]: [0.95600171, 0.20768181, 0.82844489]
        [0.41021121, 0.27696523, 0.69368594]
```

Both RNGs are producing distinct and independent output. Since both `rng1` and `rng2` are local RNGs, selecting the seed for `rng1` here will not affect the internal state of `rng2`, or indeed any other active RNGs in your code. This is an important consideration when working on large or collaborative projects: the use of calls to global RNGs such as (for example) `np.random.normal()` is to be avoided since a change in the internal state of the RNG in one part of a project can have unpredictable effects in another.

If we reset `rng2` with `seed` then we will see a repeat of the output from `rng1`, confirming that the sample has been reproduced:

```
In []: rng2=np.random.default_rng(seed)
       print(rng2.uniform(0,1,3))
```

```
Out[]: [0.95600171, 0.20768181, 0.82844489]
```

1.3.1.3 The Box-Muller Transformation

For many applications, the use of a local RNG and direct invocation of the calls above would be enough. However certain Monte Carlo techniques benefit from a deeper understanding of how samples from a normal distribution are produced using samples from a uniform distribution. We describe it now.

The following result tells us how we may transform independent pairs of samples from a uniform distribution into independent pairs of samples from a standard normal distribution.

Proposition 1.13 (The Box-Muller Transformation) *Let U_1 and U_2 be independent random variables distributed uniformly on $[0, 1]$. Define*

$$(\rho, \theta) = \left(\sqrt{-2 \log(U_1)}, 2\pi U_2 \right). \tag{1.18}$$

Then

$$(Z_1, Z_2) = (\rho \cos(\theta), \rho \sin(\theta)) \tag{1.19}$$

is a pair of independent standard normal random variables.

To implement this, let's start by passing the dimensions of a 500×2 array to the `uniform()` call.

```
In []: U=rng.uniform(0,1,(500,2))
```

This gives us an array U whose first and second columns correspond to vectors of 500 samples from each of U_1 and U_2 respectively. Now construct ρ and θ using the relation (1.18):

```
In []: rho=np.sqrt(-2*np.log(U[:,0]))
       theta=2*m.pi*U[:,1]
```

Finally, set up a 500×2 array to contain our normal sample pairs by instantiating an array of ones, then populate it via (1.19).

```
In []: normBM=np.ones((500,2))
       normBM[:,0]=rho*np.cos(theta)
       normBM[:,1]=rho*np.sin(theta)
```

Scatterplots of our samples of the pairs (U_1, U_2) and (Z_1, Z_2) may be generated as follows:

```
In []: plt.scatter(U[:,0],U[:,1])
       plt.scatter(normBM[:,0],normBM[:,1])
```

and are shown in Fig. 1.4.

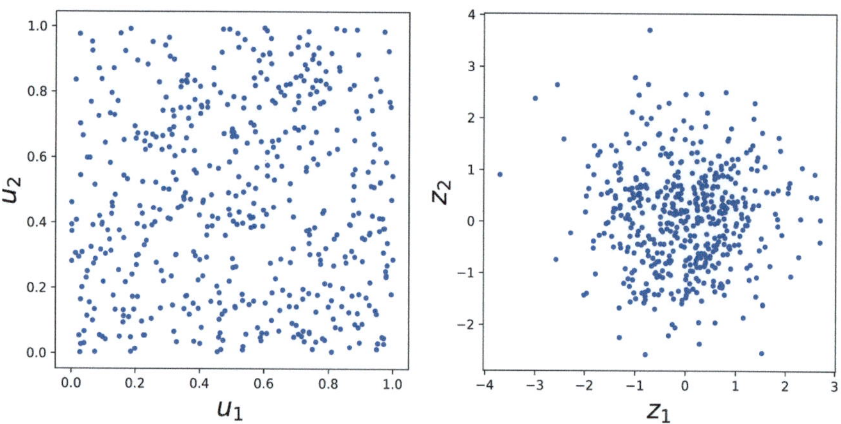

Fig. 1.4 The Box-Muller transformation takes 500 sample pairs from mutually independent uniform random variates, represented in the scatterplot on the left, and transforms them into 500 sample pairs from mutually independent standard normal random variates, represented in the scatterplot on the right

1.3.1.4 Marsaglia's Polar Variant, and Conditional Operations on NumPy Arrays

A modification of the Box-Muller transformation can reduce the compute time required to produce large samples from a normal distribution by avoiding the evaluation of trigonometric functions, which can be relatively slow.

Proposition 1.14 (Marsaglia's Polar Method)   *Let U_1 and U_2 be independent random variables distributed uniformly on* $[0, 1]$. *Define*

$$V_1 = 2U_1 - 1; \quad V_2 = 2U_2 - 1; \ R = V_1^2 + V_2^2.$$

If we only use values of V_1 and V_2 such that $0 < R < 1$ *then*

$$(Z_1, Z_2) = \left(V_1 \sqrt{-2\frac{\log(R)}{R}}, \ V_2 \sqrt{-2\frac{\log(R)}{R}} \right) \tag{1.20}$$

is a pair of independent standard normal random variables.

Notice that, although Marsaglia's method avoids trigonometric evaluations, observations on (V_1, V_2) are discarded unless $R < 1$. To implement this method, we will make use of NumPy's ability to broadcast conditional statements to array variables. First generate a vector of values for R:

```
In []: U=rng.uniform(0,1,(500,2))
       V=2*U-1
       R=np.sum(V**2,axis=1)
```

By setting `axis-1` we ensure that the `np.sum()` call is applied separately along each row of the array `V**2`. The array variable R contains some values that are greater than one, and these need to be removed. However we can see from (1.20) that corresponding values from V and R are combined and so we should keep track of the indices of the removed entries. We can use the `np.where()` call to extract the indices of R which fail to satisfy the condition and which therefore must be discarded for all of V_1, V_2, and R.

```
In []: indicesKeep=np.where(R<1)
       V=V[indicesKeep]
       R=R[indicesKeep]
```

If we inspect the value of `indicesKeep` we see that it is a one-dimensional array containing all values of i for which the boolean expression `R[i]` < 1 has value `true`. Passing this array as an index set to V and R has the effect of extracting only those terms which satisfy the necessary condition. Finally, apply (1.20) using

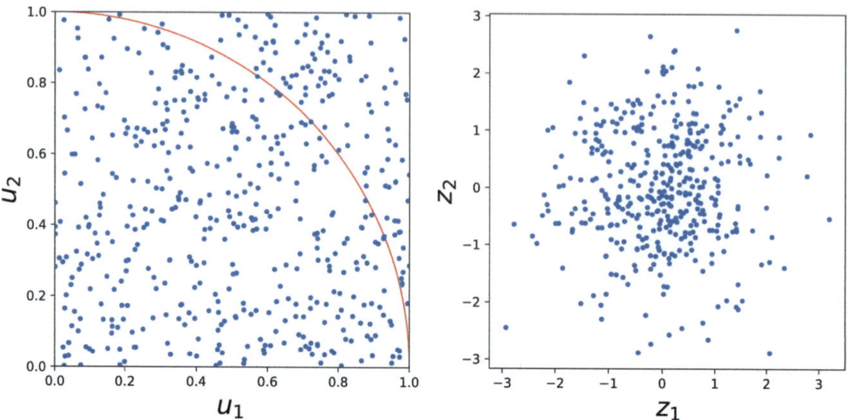

Fig. 1.5 On the left, the Marsaglia variant of the Box-Muller transformation discards all uniform sample pairs which do not fall inside the red circle. For this experiment, 103 sample pairs are discarded. The 397 pairs remaining are then transformed into standard normal sample pairs, represented in the scatterplot on the right

elementwise multiplication of the vector R with each of the columns of V to generate our Gaussian pairs

```
In []: normMars=np.ones((len(R),2))
       Rfac=np.sqrt(-2*np.log(R)/R)
       normMars[:,0]=V[:,0]*Rfac
       normMars[:,1]=V[:,1]*Rfac
```

Notice that the lengths of R and each column of V will remain equal to each other but less than 1000, and indeed this length will be random and therefore impossible to specify in advance. So we instantiate `normMars` as a $d \times 2$ dimensional array, where d is the length of R and is computed as `len(R)`. Scatterplots for (U_1, U_2) and (Z_1, Z_2) are displayed in Fig. 1.5.

1.3.1.5 Box-Muller in Practice: Reshaping and Resizing NumPy Arrays

The Box-Muller transformation presented in Sect. 1.3.1.3 takes an array of $M \times 2$ mutually independent samples, each uniformly distributed over the interval $[0, 1]$, and returns an array of $M \times 2$ independent samples from independent standard normal random variables. Grouping both inputs and outputs into pairs is a natural consequence of the transformation itself, which combines each sample pair of uniform variables to produce corresponding sample pairs from mutually independent standard normal random variables. In practice there may be no reason to have the output stored in pairs, but rather we may require (for example) a one-dimensional array of independent samples.

NumPy provides a convenient method to specify a new shape for data stored in an array. To illustrate this, lets consider the following array with dimension 3×2

```
In []: A=np.array([[1,2],
                   [3,4],
                   [5,6]])
```

If we want the same data to be stored in a one-dimensional array of length 6, try

```
In []: np.reshape(A,6)
```

```
Out[]: array([1, 2, 3, 4, 5, 6])
```

Notice that this has the effect of unraveling the array C, starting at the first entry in the first row and running along each row in turn. We could ask for it to be reshaped as a 2×3 array:

```
In []: np.reshape(A,(2,3))
```

```
Out[]: array([[1, 2, 3],
              [4, 5, 6]])
```

If we request an array dimension that is inconsistent with the number of entries in the array then Python will return an error.

This kind of array manipulation allows us to implement the Box-Muller transformation as a function that returns an $M \times 1$ array of standard normal samples when passed the sample size parameter M and a local RNG. Some considerations: if we are to return exactly M variates, then our function must handle both even and odd M. If M is even then we simply create $M/2$ pairs and reshape into an $M \times 1$ array before returning. If M is odd we create $(M+1)/2$ pairs to ensure we generate sufficiently many standard normal samples. However after reshaping this will result in an array of length $M+1$, so we use the `np.resize()` call to discard the surplus entry before returning.

What follows is a more complex chunk of code, relative to what has been seen so far in this chapter. So we will use comments in our code for the first time, each preceded by the # symbol instructing Python to ignore everything on the line that follows it.

```
In []: def boxMuller(M,rng):

           # First check if M is even or odd
           # before constructing enough uniform sample pairs.
           evenM=False
           if (M%2==0):
               evenM==True
               size=M/2
```

```
else:
    size=(M+1)/2
U=rng.uniform(0,1,(size,2))
normBM = np.ones((size, 2))

# Now apply the Box-Muller transform
rho=np.sqrt(-2 * np.log(U[:,0]))
theta=2*np.pi*U[:, 1]
normBM[:,0]=rho*np.cos(theta)
normBM[:,1]=rho*np.sin(theta)
normBM.reshape(2*size)

# If M is odd then there will be an extra sample
# to be discarded.
if (evenM==False):
    np.resize(normBM, len(normBM)-1)

# Return an array of length M
return(normBM)
```

Exercise 1.9 asks you to define a function with the same parameters that uses the Marsaglia variant defined in Proposition 1.14 to generate Gaussian variates.

1.3.2 Properties of the Distribution of the Asset Model

Let's now return to our examination of the asset model developed in Sect. 1.2, and identify some of its statistical properties.

1.3.2.1 Exact Expressions for the Density Function, Mean, and Variance

Having established our model of a risky asset, we see that it has a lognormal distribution. It is possible to identify an exact expression for the density function of $S(t)$ satisfying (1.17) for each $t \in [0, T]$. Since $W^{\mathbb{P}}(t) \sim \sqrt{t}Z$ where $Z \sim N(0, 1)$, express for any $a, b \in \mathbb{R}^+$

$$\mathbb{P}[a \le S(t) \le b] = \mathbb{P}\left[a \le S_0 e^{(\mu - \sigma^2/2)t + \sigma\sqrt{t}Z} \le b\right]$$

$$= \mathbb{P}\left[\frac{\ln(a/S_0) - (\mu - \sigma^2/2)t}{\sigma\sqrt{t}} \le Z \le \frac{\ln(b/S_0) - (\mu - \sigma^2/2)t}{\sigma\sqrt{t}}\right]$$

$$= \frac{1}{\sqrt{2\pi}} \int_{\frac{\ln(a/S_0) - (\mu - \sigma^2/2)t}{\sigma\sqrt{t}}}^{\frac{\ln(b/S_0) - (\mu - \sigma^2/2)t}{\sigma\sqrt{t}}} e^{-s^2/2} ds.$$

Making the substitution $x = S_0 e^{\sigma\sqrt{t}s + (\mu - \sigma^2/2)t}$, we arrive at

$$\mathbb{P}[a \leq S(t) \leq b] = \frac{1}{\sqrt{2\pi}} \int_a^b \frac{\exp\left(\frac{-\left(\log(x/S_0) - (\mu - \sigma^2/2)t\right)^2}{2\sigma^2 t}\right)}{x\sigma\sqrt{t}} dx, \qquad (1.21)$$

The form of the density function f may be read directly from (1.21), since

$$\mathbb{P}[a \leq S(t) \leq b] = \int_a^b f(x)dx,$$

and we can see that $S(t)$ has probability density function

$$f(x) = \frac{\exp\left(\frac{-\left(\log(x/S_0) - (\mu - \sigma^2/2)t\right)^2}{2\sigma^2 t}\right)}{x\sigma\sqrt{2\pi t}}, \qquad x > 0, \qquad (1.22)$$

with $f(x) = 0$ for $x \leq 0$. This density function can be plotted, and the effect of varying the volatility σ and time t on its shape is demonstrated in Fig. 1.6.

The mean, second moment, and variance of the Black-Scholes asset price model (1.17) are given by

$$\mathbb{E}_\mathbb{P}[S(t)] = S_0 e^{\mu t}; \qquad (1.23)$$

$$\mathbb{E}_\mathbb{P}[S(t)^2] = S_0^2 e^{(2\mu + \sigma^2)t};$$

$$\mathrm{Var}_\mathbb{P}[S(t)]] = S_0^2 e^{2\mu t}(e^{\sigma^2 t} - 1),$$

where $\mathbb{E}_\mathbb{P}$ denotes expectation under the probability measure $\mathbb{P}$. See Exercise 1.2, which asks you to confirm this.

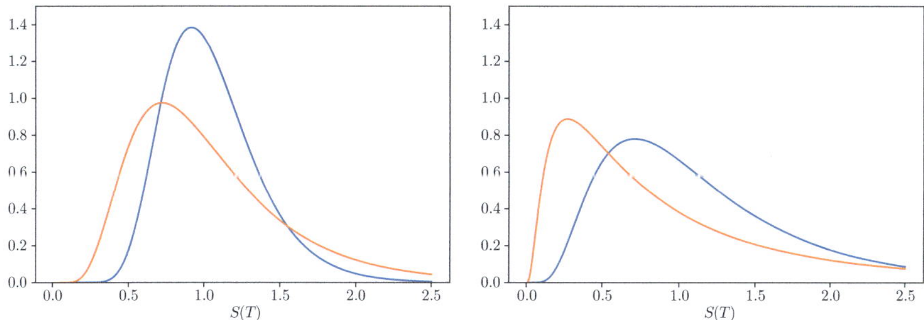

Fig. 1.6 Theoretical lognormal density curves (Eq. (1.22)) for $S_0 = 1$, $\mu = 0.05$, $\sigma = 0.3$ (blue lines), $\sigma = 0.5$ (orange lines), $t = 1$ (left plot), $t = 4$ (right plot)

1.3.2.2 Sampling from the Asset Model at Expiry

Let's try taking a statistical sample of M values of the asset model at expiry

$$S(T) = S_0 e^{(\mu - \sigma^2/2)T + \sigma W^{\mathbb{P}}(T)}, \tag{1.24}$$

and compare with the shape of the density function given by (1.22) and shown in Fig. 1.6. We make use of the fact that $W^{\mathbb{P}}(T) \sim \mathcal{N}(0, T)$ and use the random number generator rng that we created in Sect. 1.3.1.

```
In []: M=10**6; T=1
       WT=rng.normal(0,np.sqrt(T),M)
```

The variable WT now contains 10^6 independent samples from the distribution of $W^{\mathbb{P}}(T)$ where $T = 1$. We can convert this to a sample of $S(T)$ by applying exactly the transformation given by (1.24) as a function of $W^{\mathbb{P}}(T)$. Once we define some model parameters, the sample of asset values at time T can be generated as a single NumPy array:

```
In []: S0=1; mu=0.05; sig=0.3
       ST=S0*np.exp((mu-0.5*(sig**2))*T+sig*WT)
```

We can plot the empirical density of this sample and compare against the values presented in Fig. 1.6. In Fig. 1.7 we do this using the same parameter choices as for the theoretical density functions plotted in Fig. 1.6. You can see that the sampling approach is effective at building up an empirical distribution that closely matches the known theoretical distribution.

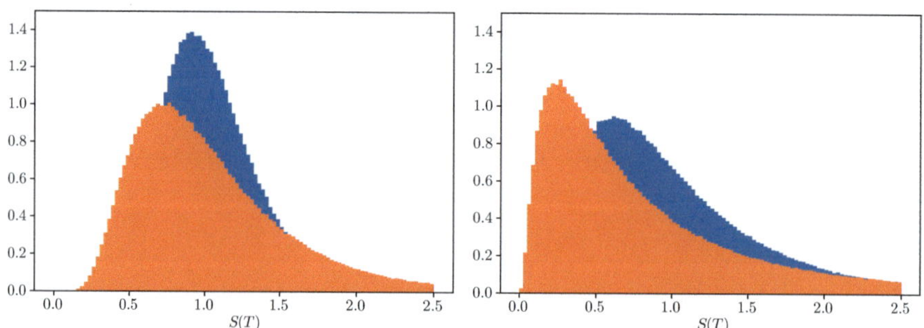

Fig. 1.7 Empirical density curves (Eq. (1.22)) generated by $M = 10^6$ samples of $S(T)$ for $S_0 = 1$, $\mu = 0.05$, $\sigma = 0.3$ (blue bars), $\sigma = 0.5$ (orange bars), $T = 1$ (left plot), $T = 4$ (right plot)

1.3.3 Simulating an Ensemble of Trajectories

In this section we will learn how to sample and visualise trajectories of the Black-Scholes asset price model $S(t)$ over the full range of values $t \in [0, T]$, rather than simply sampling directly from the asset prices at time T. This is useful if we are interested in valuing American or other path-dependent options by Monte Carlo, requiring us to generate a statistical sample (or *ensemble*) of M such trajectories, each computed on a discretised time-set with N steps. Since both M and N may be large, to do this efficiently in Python requires that we prioritise the use of vectorisation over loops.

For the purposes of demonstration, set $M = 10$, $N = 100$, and $T = 1$. Then, continuing to use the random number generator `rng` created in Sect. 1.3.1, generate an $M \times N$ array of independent normal samples with mean zero and variance $dt = T/N$:

```
In []: M=10; N=100; T=1; dt=T/N
        increments=rng.normal(0,np.sqrt(dt),size=(M,N))
```

Each value in the array `increments` represents a specific observation on

$$W^{\mathbb{P}}((n + 1)dt, \omega_i) - W^{\mathbb{P}}(ndt, \omega_i), \quad n = 0, \ldots, N - 1, \quad i = 0, \ldots, M - 1,$$

and each row in the array contains a single trajectory. Plot them overlaid (since `plt.plot()` will separate the observations by column rather than row we need to transpose `increments` here):

```
In []: plt.plot(increments.T)
```

Five trajectories generated in this way are shown in Fig. 1.8 (top). Note that the horizontal axis is indexed by the number of steps, rather than the value of t. Fix that by defining a time array, and plotting against that:

```
In []: t=np.linspace(dt,1,N)
        plt.plot(t,increments.T)
```

To generate M trajectories of $W^{\mathbb{P}}$ starting at time dt (not at zero), we now take the *cumulative sum* of the increment array along each row, and plot:

```
In []: W=np.cumsum(increments,axis=1)
        plt.plot(t,W.T)
```

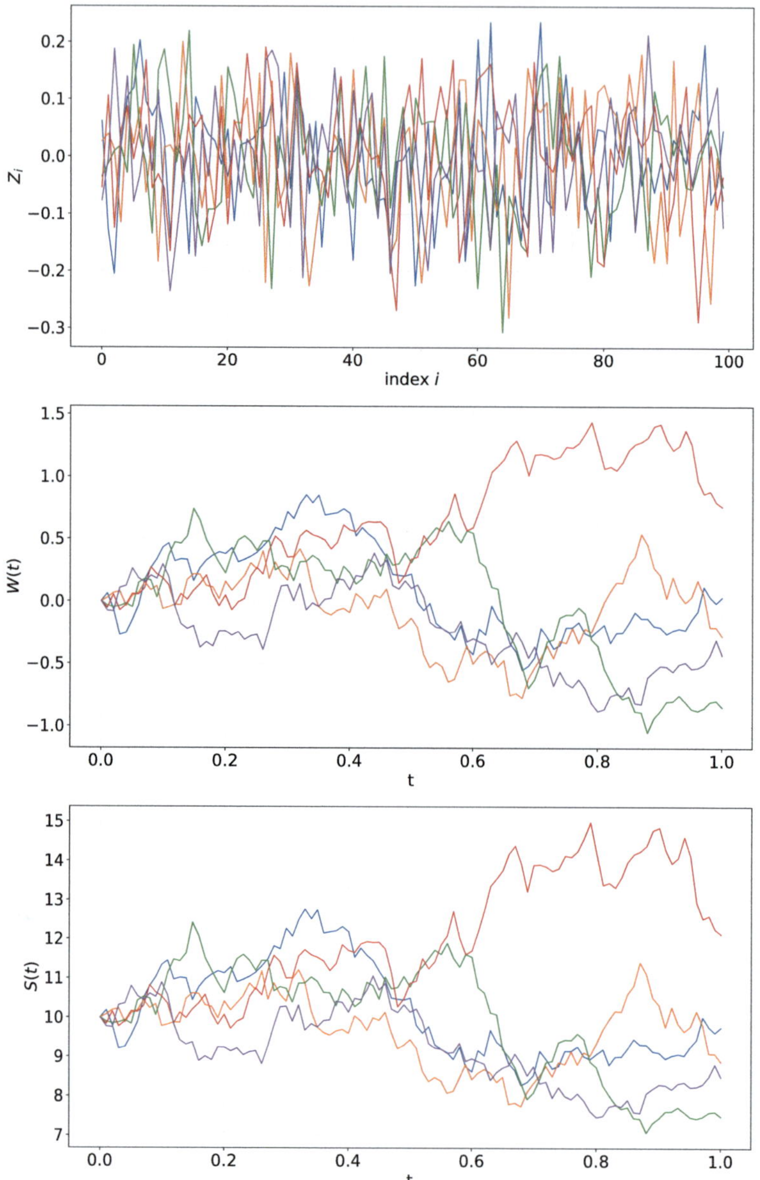

Fig. 1.8 Constructing an ensemble of $M = 5$ sampled trajectories from the Black-Scholes model over the interval $[0, 1]$ discretised into $N = 100$ timesteps, with $r = 0.01$ and $\sigma = 0.3$. At the top, we see 5 time plots, each of 100 independent observations on Z. These are then integrated over each trajectory to produce the 5 Brownian paths $W^{\mathbb{P}}(t)$ in the middle plot. Finally, we use the paths of $W^{\mathbb{P}}(t)$ to produce 5 trajectories on the asset model, displayed in the bottom plot

See Fig. 1.8 (middle). Finally, prepend $t = 0$ and $B(0) = 0$ to the arrays t and B respectively (the latter across all trajectories), and plot again

```
In []: t=np.insert(t,0,0)
       W=np.c_[np.zeros(M),W]
       plt.plot(t, W.T)
```

Since `np.exp()` can take an array as input and act on individual entries, we can convert W and t to the ensemble of Black-Scholes asset trajectories (for example, choose $S_0 = 10$ $\mu = 0.01$, and $\sigma = 0.3$), to get

```
In []: S0=10; mu=0.01; sig=0.3
       S=S0*np.exp((mu-0.5*sig**2)*t+sig*W)
```

See Fig. 1.8 (bottom), and compare to the real-world example of an asset price trajectory portrayed in Fig. 1.2. Asset price trajectories for the computation of expectations under the risk-neutral measure (see Chap. 2) can be generated by replacing the parameter mu with r. Problem 1.11 asks you to automate this process by constructing a function `bsEnsemble()` that allows you to generate an ensemble of asset prices with a single execution.

1.4   Further Reading

A mathematically rigorous, but nonetheless approachable, treatment of probability and stochastic processes without measure theory may be found in Grimmett and Stirzaker [26]. Following from this, a very good introduction to the theory and application of SDEs is provided by the book of Mao [53]. It contains a detailed treatment of the Itô calculus, including the construction of the Itô integral and proofs for the scalar and multidimensional forms of the Itô formula. There, you will find the existence and uniqueness theory for SDEs referenced in Sect. 1.2.7.

Finally, full derivations of the Box-Muller transform and Marsaglia variant for transforming pairs of indepedent uniform random variates to pairs of Gaussian random variates, as discussed in Sect. 1.3.1, may be found in Choe [15].

1.4.1   Exercises

1.1 Suppose that S is governed by the Black-Scholes asset model given by the SDE (1.11), and let $a, c \in \mathbb{R}$ and $n \in \mathbb{N}$. Use Theorem 1.7 (Itô's formula) to find the SDE satisfied by

1. $V(S) = aS$;
2. $V(S, t) = S^n + ct^2$.

1.2 Suppose that $X \sim \mathcal{N}(m, v^2)$, where $m \in \mathbb{R}$ and $v^2 > 0$.

1. Compute $\mathbb{E}[e^X]$ and $\mathbb{E}[e^{2X}]$.
2. Hence confirm that the mean and variance of the Black-Scholes asset price model (1.17) are given by
   (a) $\mathbb{E}_\mathbb{P}[S(t)] = S_0 e^{\mu t}$;
   (b) $\mathrm{Var}_\mathbb{P}[S(t)]] = S_0^2 e^{2\mu t}(e^{\sigma^2 t} - 1)$,
   where $\mathbb{E}_\mathbb{P}$ and $\mathrm{Var}_\mathbb{P}$ denote expectation and variance respectively under the probability measure $\mathbb{P}$.

1.3 Suppose that S is governed by the Black-Scholes asset model given by the SDE (1.11). Use Theorem 1.7 to write down the SDE for

$$Z(t) = \left[\frac{-\sigma^2}{2\mu + \sigma^2} \right] S(t)^{-2\mu/\sigma^2 + 1}.$$

Give an expression for $\mathbb{E}_\mathbb{P}[Z(t)]$ in terms of μ, σ^2, and S_0.

1.4 Review the explanation for the formal multiplication rules $dt\,dt = 0$ and $dW^\mathbb{P}dt = 0$ that appear in (1.16) in Sect. 1.2.6. Provide a similar intuitive explanation for $dW_i^\mathbb{P}dW_i^\mathbb{P} = dt$ and $dW_i^\mathbb{P}dW_j^\mathbb{P} = 0$ if $i \neq j$, from the table in Proposition 1.8.

1.5 Let $W^\mathbb{P}$ be a standard Brownian motion on a probability space $(\Omega, \mathcal{F}, \mathbb{P})$, and let $(\mathcal{F}_t)_{t \geq 0}$ be the natural filtration of $W^\mathbb{P}$. Show that each of the following processes is an $\mathcal{F}_t$-martingale under the measure $\mathbb{P}$:

1. $X(t) = \left(W^\mathbb{P}(t)\right)^2 - t$;
2. $X(t) = e^{t/2} \cos\left(W^\mathbb{P}(t)\right) - 1$;
3. $X(t) = e^{t/2} \sin\left(W^\mathbb{P}(t)\right)$;
4. $X(t) = \left(W^\mathbb{P}(t) + t\right) e^{-W^\mathbb{P}(t) - t/2}$.

1.6 Use Definition 1.10 to show that if M is an $\mathcal{F}_t$-martingale under the measure $\mathbb{P}$, then $\mathbb{E}_\mathbb{P}[M(t)]$ is a constant.

1.7 Define a 3×4 NumPy array consisting of the first 12 primes

```
In []: primes=np.array([[2,3,5,7],
                        [11,13,17,19],
                        [23,29,31,37]])
```

Apply the calls `np.reshape()` and `np.resize()` to `primes` to create a 2×5 array containing only the first 10 primes. Extract the third column of this new array.

1.8 Use the Python functions constructed in Sect. 1.2.4 to answer the following questions to 16 decimal places. Consider an asset following the Black-Scholes asset model (1.11) with $\mu = 0.03$, $\sigma = 0.4$, and $S_0 = 100$.

1. What is the probability that after 6 months have passed the *log-return*, given at time t by $\log(S(t)/S_0)$, is less than 0.05?
2. Suppose that after 3 months the asset is trading at a price of \$106. What is the present value of that price if the annualised rate of continuously compounded interest on a risk-free investment is 2%?

1.9 Define a Python function called `marsaglia()` that takes the sample size parameter M and a local RNG as parameters and returns exactly M standard normal variates using the Marsaglia variant of the Box-Muller transform described in Sect. 1.3.1.4.

1.10 Generate a sample of $M = 10^3$ observations on a standard normal distribution using each of the Python functions `boxMuller()` (constructed in Sect. 1.3.1) and `marsaglia()` (constructed in Exercise 1.9). Produce a histogram and Q-Q plot for each sample.

1.11 In Sect. 1.3.3, we walked through the steps required to generate an ensemble of trajectories of the asset model. Gather the relevant commands from this section into a Python function called `bsEnsemble()` taking arguments S_0, μ, σ, T, M, N, and a local RNG `rng` that returns a NumPy array with M rows and $N + 1$ columns representing an ensemble of M asset price trajectories observed over a uniform mesh with N uniform steps over the interval $[0, T]$.

Note: the parameter μ can take a numerical value representing either the real-world drift or the risk-free rate, depending on whether we are computing under the real-world or risk-neutral measure. See Chap. 2, next.

The Pricing of Financial Derivatives

2

A contingent claim, and particularly an option, may be viewed as a tool for transferring asset price risk between parties. So the value assigned to the option should correspond to the present value of the risk being transferred. This must be perceived as fair by both buyers and sellers in order for a market to exist.

There are two distinct but related ways to characterise option values mathematically. The first is sometimes known as the risk-neutral, or martingale characterisation and results in the representation of the option value at a time $t \in [0, T]$ as the expected payoff discounted back to t:

$$V(t) = e^{-r(T-t)} \mathbb{E}_{\mathbb{Q}}[V_T | \mathcal{F}_t] \tag{2.1}$$

where r is the annual rate of continuously compounded interest on the risk-free savings account, $\mathbb{E}_{\mathbb{Q}}$ represents an expectation taken under the risk-neutral measure, and V_T is the value of the option at expiry. We can compute values from this expression approximately using Binomial trees or Monte Carlo methods, the focus of Chaps. 3 and 4 of this book, respectively.

The second approach, valid in the case where the value of the option depends only on the current value of the underlying asset $s = S(t)$ and the time t, expresses option values as solutions of the Black-Scholes-Merton PDE:

$$V_t(s, t) + \frac{1}{2}\sigma^2 s^2 V_{ss}(s, t) + rs V_s(s, t) - rV(s, t) = 0, \quad t \in [0, T], \quad s \in \mathbb{R}^+, \tag{2.2}$$

with boundary conditions determined by the specific option to be valued. We can compute solutions of (2.2) and its variants approximately using finite difference methods and we explain how in Chap. 5 of this book.

Both of these approaches are linked mathematically. In this chapter we will review the construction of a self-financing portfolio in a complete market and appeal to the Fundamental Theorems of Asset Pricing, along with the no-arbitrage principle, to arrive at the martingale characterisation (2.1). We then show how PDE characterisation (2.2) emerges from (2.1) via an important result called the Feynman-Kac Theorem. In the case of European options, (2.2) can be solved explicitly, leading to the Black-Scholes pricing formulae for such options and their hedge parameters. We will review these and implement them in Python, show how the writer of a European option may dynamically hedge against the risk associated with the sale of the option, before finally examining how to approximate the market value of the volatility parameter σ; the only valuation parameter that must be estimated from market data.

Python Setup For the code in this chapter, we again need the NumPy package for numerical computation and PyPlot to produce visualisations of our output. Additionally we require the `scipy.stats` module. This module enables us to compute values from a variety of probability distributions, summary and frequency statistics, to carry out various statistical tests, and more. For our purposes here we will only use it to compute values from the CDF of the standard normal distribution.

In addition, we need a specific module from the `scipy.optimize` package, called `fsolve`, which is used to iteratively solve nonlinear algebraic equations, and which we will use to solve the root-finding problem that arises when we estimate σ from quoted option values.

```
In []: import numpy as np
       import matplotlib.pyplot as plt
       from scipy.optimize import fsolve
       from scipy.stats import norm
```

We will also make use of the user-defined functions from Chap. 1 listed in Table 2.1.

Table 2.1 User-defined function required in Chap. 2

Section	Function name	Function output	Parameter	Parameter meaning
1.2.4	dFac()	$e^{-r(T-t)}$	r	r
			dt	$T-t$
1.4.1	bsEnsemble()	$S(t_n, \omega_i)$	S0	S_0
		$n = 0, \dots N$	mu	μ or r
		$i = 0, \dots, M-1$	sig	σ
			T	T
			M	M
			N	N
			rng	Local RNG

2.1   Risk-Neutral Pricing of Financial Derivatives

2.1.1   A Simple Market Model Without Arbitrage

Consider a *simple market model* where prospective investors may choose to split their investment between two assets. One of these is a risk-free investment with price at time t given by $b(t)$, where $b(0) = 1$. The other is a share of non-dividend paying stock with price at time t denoted $S(t)$ and adapted to a filtration $(\mathcal{F}_t)_{t\geq0}$. The market model is characterised by the pair $\underline{S} = (b(t), S(t))_{t\geq0}$.

Suppose that at time $t \in [0, T]$, an investor has a portfolio of $\phi(t)$ shares in the stock and $\psi(t)$ risk-free bonds, where $\phi(t)$ and $\psi(t)$ are previsible stochastic processes. The portfolio is characterised by the pair $(\phi(t), \psi(t))_{t\geq0}$, and its value at time t is denoted $\Pi(t)$. We assume that the market admits no arbitrage opportunities by adopting the following principle:

Definition 2.1 (The No-Arbitrage Principle)   The greatest risk-free return that one can make on a portfolio of assets is the same as the return on an equivalent amount of cash that is placed in a bank earning a risk free rate of interest.

In this section, we review the provenance of the martingale pricing formula (2.1) using the example where b is a zero-coupon bond with unit issue price and fixed continuously compounded annual interest rate $r > 0$, and S is governed by a Black-Scholes asset model with drift μ and volatility σ. In this setting, we will refer to b as the *numeraire asset*. In a more general setting one need not require that r be constant or deterministic. Nor is it necessary for the numeraire asset to be a zero-coupon bond, or for there to be only a single risky asset.

The following is a roadmap for this review. In Sect. 2.1.2 we recall that the market $\underline{S}$ is complete if the payoff V_T of any derivative can be replicated by a self-financing portfolio. To avoid introducing opportunities for arbitrage, the price of that derivative $V(t)$ for any $t \in [0, T]$ must be the same as the price of the replicating portfolio at that time.

In Sects. 2.1.3 and 2.1.4, we see that if the market is complete and free of arbitrage opportunities we can find a unique measure, referred to as the risk-neutral measure, under which the discounted asset price process S/b is a martingale.

Finally, in Sect. 2.1.6, we demonstrate the completeness of the market by explicitly constructing the replicating portfolio. The price of this at time $t \in [0, T]$ is determined to be given by the RHS of (2.1), the discounted expected payoff under the unique risk-neutral measure. Hence, this must also be the price of the derivative.

2.1.2   Replicating Strategies for Derivatives

Consider the portfolio where at time $t \in [0, T]$ an investor holds $\phi(t)$ units of the risky asset S and $\psi(t)$ units of the risk-free asset b. Then the portfolio value is

$$\Pi(t) = \phi(t)S(t) + \psi(t)b(t), \quad t \in [0, T].$$

By an application of Proposition 1.8, the stochastic product rule, we can write

$$
\begin{aligned}
d\Pi(t) &= d(\phi(t)S(t)) + d(\psi(t)b(t)) \\
&= \big[\phi(t)\,dS(t) + \psi(t)\,db(t)\big] \\
&\quad + \big[S(t)\,d\phi(t) + d\phi(t)\,dS(t) + b(t)\,d\psi(t)\big], \quad t \in [0, T].
\end{aligned}
\tag{2.3}
$$

The portfolio is said to be *self-financing* if the overall change in its value over any time interval $[t, \delta t]$ comes only from the change in the prices of b and S: there are no cash inflows or outflows required as the value of the portfolio changes from $\Pi(t)$ to $\Pi(t + \delta t)$. This means that, mathematically, Π should satisfy the SDE

$$
d\Pi(t) = \phi(t)\,dS(t) + \psi(t)\,db(t), \quad t \in [0, T].
\tag{2.4}
$$

Comparing (2.4) and (2.3) we see that the condition for the portfolio $(\phi(t), \psi(t))_{t \geq 0}$ to be self-financing is

$$
S(t)\,d\phi(t) + d\phi(t)\,dS(t)) + b(t)\,d\psi(t) = 0, \quad t \in [0, T]
\tag{2.5}
$$

Definition 2.2 Let $V(t)$ be a the value of a derivative with exercise date T. A *replicating strategy* for V is a self-financing portfolio $(\phi(t), \psi(t))_{t \in [0,T]}$ satisfying the condition (2.5) whose value at time T is V_T:

$$
\Pi(T) = \phi(T)\,S(T) + \psi(T)\,b(T) = V_T.
$$

To avoid introducing an arbitrage opportunity, the value of the derivative at any previous time t has to equal the value of the replicating strategy, so that

$$
V(t) = \Pi(t), \quad t \in [0, T].
$$

We say that a market is *complete* if any derivative written on assets within it has such a replicating strategy. We will demonstrate market completeness for our simple market model example by explicitly constructing a replicating strategy in Sect. 2.1.6.

2.1.3 The Fundamental Theorems of Asset Pricing

Definition 2.3 Consider the market model given by $\underline{S}$. The probability measure $\mathbb{Q}$ is called an *equivalent martingale measure* for this model if it is equivalent to the probability measure $\mathbb{P}$, and the discounted asset price process

$$
X(t) := \frac{S(t)}{b(t)}, \quad t \in [0, T],
\tag{2.6}
$$

is an $\mathcal{F}_t$-martingale under the measure $\mathbb{Q}$.

Recall that two probability measures are equivalent if they agree on which events have probability zero. Strictly speaking, for $\mathbb{Q}$ to be an equivalent martingale measure in this context we need both of $b/b \equiv 1$ and S/b to be $\mathcal{F}_t$-martingales. However, since a deterministic constant is always an $\mathcal{F}_t$-martingale, the first of these is automatically true. Exercise 2.1 asks you to confirm this.

We can now link the existence of a unique equivalent martingale measure, referred to as the *risk-neutral measure*, to market completeness and the no-arbitrage principle.

The following summarises the fundamental theorems of asset pricing as they apply in our setting.

Theorem 2.4 (First and Second Fundamental Theorems of Asset Pricing)

1. The market model $\underline{S}$ is free of arbitrage if and only if there exists an equivalent martingale measure $\mathbb{Q}$ satisfying the conditions of Definition 2.3.
2. In the absence of arbitrage, the market model $\underline{S}$ is complete if and only if the martingale measure $\mathbb{Q}$ is unique.

We conclude from Theorem 2.4 that in a complete market without arbitrage opportunities the measure $\mathbb{Q}$, and therefore the price of any derivative, is uniquely determined. In the next subsection we examine how $\mathbb{Q}$ may be constructed for a specific example of our simple market model $\underline{S}$, and what this means in practice.

2.1.4 The Risk-Neutral Measure for Our Simple Market Model

Suppose that the assets in the simple market model $\underline{S}$ introduced in Sect. 2.1.2 satisfy

$$b(t) = e^{rt}; \quad S(t) = S_0 e^{(\mu - \sigma^2/2)t + \sigma W^{\mathbb{P}}(t)}, \quad t \in [0, T], \tag{2.7}$$

where $W^{\mathbb{P}}$ is a standard Brownian motion satisfying the conditions of Definition 1.3 with respect to the real-world probability measure $\mathbb{P}$.

2.1.4.1 $\mathbb{P}$ Is not the Risk-Neutral Measure

The expectation of the discounted asset price process is, by (1.23) in Chap. 1,

$$\mathbb{E}_{\mathbb{P}}[X(t)] = \mathbb{E}_{\mathbb{P}}\left[\frac{S(t)}{b(t)}\right] = \mathbb{E}_{\mathbb{P}}\left[e^{-rt} S(t)\right] = S_0 e^{(\mu - r)t}.$$

We can see from Exercise 1.6 in Chap. 1 that (unless $\mu = r$) the process X is not an $\mathcal{F}_t$-martingale under the real-world measure $\mathbb{P}$, since the expectation under $\mathbb{P}$ varies in time. We conclude that $\mathbb{P}$ does not satisfy Definition 2.3 and cannot be used to value an option without potentially introducing an arbitrage opportunity.

We must construct an alternative probability measure under which the conditions of Definition 2.3 hold.

2.1.4.2 The Market Price of Risk

The risky asset S satisfies the SDE

$$dS(t) = \mu S(t)dt + \sigma S(t)dW^{\mathbb{P}}(t), \quad t \in [0, T].$$

This can be rewritten as

$$dS(t) = rS(t)dt + \sigma S(t)\underbrace{\left[\left(\frac{\mu - r}{\sigma}\right)dt + dW^{\mathbb{P}}(t)\right]}_{=:d\widetilde{W}(t)}, \quad t \in [0, T].$$

We have not changed the model here, but nonetheless we have what looks like an SDE where the risk free rate r appears in the drift instead of μ. The process

$$\widetilde{W}(t) = \left(\frac{\mu - r}{\sigma}\right)t + W^{\mathbb{P}}(t), \quad t \in [0, T], \tag{2.8}$$

is adapted to the natural filtration generated by $W^{\mathbb{P}}$, but unfortunately it is not a standard Brownian motion under the real-world measure $\mathbb{P}$. It no longer satisfies the conditions of Definition 1.3: by rewriting the model we have introduced a drift into the driving noise, given by $(\mu - r)/\sigma$ and referred to as the *market price of risk*.

The idea now is to identify an equivalent probability measure $\mathbb{Q}$ under which $\widetilde{W}$ is once again a standard Brownian motion. If we can do this then it follows from the next theorem that $\mathbb{Q}$ is an equivalent martingale measure according to Definition 2.3.

Theorem 2.5 *Let $\mathbb{Q}$ be a probability measure under which the process $\widetilde{W}$ is a standard Brownian motion satisfying the conditions of Definition 1.3. The discounted asset price process $X(t)$, where X is as defined in (2.6) and b and S satisfy (2.7), is an $\mathcal{F}_t$-martingale under $\mathbb{Q}$.*

Proof The SDE (2.9) has solution

$$S(t) = S_0 e^{(r - \sigma^2/2)t + \sigma \widetilde{W}(t)},$$

and therefore we can write, for $0 \le s < t \le T$,

$$
\mathbb{E}_{\mathbb{Q}}\left[\frac{S(t)}{b(t)}\bigg|\mathcal{F}_s\right] = \mathbb{E}_{\mathbb{Q}}\left[e^{-rt}S(t)|\mathcal{F}_s\right]
$$

$$
= \mathbb{E}_{\mathbb{Q}}\left[S_0 e^{-\sigma^2 t/2 + \sigma \widetilde{W}(t)}|\mathcal{F}_s\right]
$$

$$
= S_0 e^{-\sigma^2 t/2}\mathbb{E}_{\mathbb{Q}}\left[e^{\sigma \widetilde{W}(t)}|\mathcal{F}_s\right]
$$

$$
= S_0 e^{-\sigma^2 t/2}\mathbb{E}_{\mathbb{Q}}\left[e^{\sigma((\widetilde{W}(t)-\widetilde{W}(s))+\widetilde{W}(s))}|\mathcal{F}_s\right]
$$

$$
= S_0 e^{-\sigma^2 t/2}e^{\sigma \widetilde{W}(s)}\mathbb{E}_{\mathbb{Q}}\left[e^{\sigma(\widetilde{W}(t)-\widetilde{W}(s))}|\mathcal{F}_s\right]
$$

$$
= S_0 e^{-\sigma^2 t/2}e^{\sigma \widetilde{W}(s)}\mathbb{E}_{\mathbb{Q}}\left[e^{\sigma(\widetilde{W}(t)-\widetilde{W}(s))}\right]
$$

where, taking out what is known, we move $\mathcal{F}_s$-measurable factors outside the conditional expectation and make use of the independence of $\widetilde{W}(t) - \widetilde{W}(s)$ from $\mathcal{F}_s$ at the last step.

By Definition 1.3, under the risk-neutral measure $\mathbb{Q}$ the random variable $\sigma(\widetilde{W}(t) - \widetilde{W}(s))$ has distribution $\mathcal{N}(0, \sigma^2(t - s))$ and therefore

$$
\mathbb{E}_{\mathbb{Q}}\left[e^{\sigma(\widetilde{W}(t)-\widetilde{W}(s))}\right] = e^{\frac{1}{2}\sigma^2(t-s)}.
$$

So we have

$$
\mathbb{E}_{\mathbb{Q}}\left[\frac{S(t)}{b(t)}\bigg|\mathcal{F}_s\right] = S_0 e^{-\frac{1}{2}\sigma^2 t}e^{\sigma \widetilde{W}(s)}e^{\frac{1}{2}\sigma^2(t-s)}
$$

$$
= S_0 e^{-\sigma^2 s/2 + \sigma \widetilde{W}(s)}
$$

$$
= e^{-rs}S(s)
$$

$$
= \frac{S(s)}{b(s)}, \quad 0 \le s < t \le T,
$$

as required. $\square$

Theorem 2.4 says that, as long as $\underline{S}$ is a complete market with no arbitrage opportunities, $\mathbb{Q}$ exists and is unique. Since $\mathbb{Q}$ has the effect of zeroing out the market price of risk, we refer to it as the *risk-neutral measure*.

For the specific market model $\underline{S}$ characterised by (2.7), we can explicitly construct $\mathbb{Q}$ via a result known as *Girsanov's Theorem*, and details are provided in Sect. 2.1.5 below.

In practice, when implementing any kind of valuation, theoretical or numerical, we can ensure that we calculate the expectation under $\mathbb{Q}$ simply by working with solutions of the SDE

$$dS(t) = r S(t)dt + \sigma S(t)d\widetilde{W}(t), \quad t \in [0, T], \tag{2.9}$$

as though the process $\widetilde{W}$ is a standard Brownian motion with independent, stationary increments and $\widetilde{W}(t) \sim \mathcal{N}(0, t)$. We normally denote $W^{\mathbb{Q}} := \widetilde{W}$ or, if the use of the risk-neutral measure is clear from context, we can omit mention of $\mathbb{Q}$ and simply call the process W.

2.1.5 Change-of-Measure and Girsanov's Theorem

We will look here in more detail at the construction of the equivalent martingale measure $\mathbb{Q}$. Fix $T > 0$, and let $W^{\mathbb{P}}$ be a standard Brownian motion on a probability space $(\Omega, \mathcal{F}, \mathbb{P})$ equipped with filtration $(\mathcal{F}_t)_{t \in [0,T]}$, and let θ be a process adapted to this filtration. Define the new process

$$\widetilde{W}(t) := W^{\mathbb{P}}(t) - \int_0^t \theta(u)du, \quad t \in [0, T]. \tag{2.10}$$

Comparing (2.10) with (2.8), we are interested in constructing the measure under which $\widetilde{W}$ will be a standard Brownian motion and particularly in the case where

$$\theta(t) \equiv \frac{r - \mu}{\sigma}. \tag{2.11}$$

2.1.5.1 The Radon-Nikodym Derivative and Novikov's Condition

For all events $A \in \mathcal{F}$, define a new probability measure $\mathbb{Q}$ as

$$\mathbb{Q}[A] := \int_A Z_T d\mathbb{P}, \tag{2.12}$$

where $Z_T := Z(T)$ and

$$Z(t) := \exp\left(\int_0^t \theta(u)dW^{\mathbb{P}}(u) - \frac{1}{2} \int_0^t \theta(u)^2 du \right), \quad t \in [0, T]. \tag{2.13}$$

In general, we require θ to satisfy the conditions of the following theorem:

Theorem 2.6 (Novikov's Theorem)   *Let $(Z(t))_{t\in[0,T]}$ be defined by (2.13). If for each $t \in [0, T]$,*

$$\mathbb{E}_{\mathbb{P}}\left[\exp\left(\frac{1}{2}\int_0^t \theta(u)^2 du\right)\right] < \infty, \tag{2.14}$$

then

$$\mathbb{E}_{\mathbb{P}}[Z(t)] = 1, \quad t \in [0, T], \tag{2.15}$$

and $(Z(t))_{t\in[0,T]}$ is a positive $\mathcal{F}_t$-martingale.

We don't present a general proof for the statement of Theorem 2.15, but Exercises 2.2 and 2.3 confirm that it holds when θ satisfies (2.11). Condition (2.14) is known as *Novikov's condition*.

The random variable Z_T is an example of a *Radon-Nikodym derivative*, and it allows us to switch between probability measures when computing expectations. In general, the Radon-Nikodym derivative that effects a change from one probability measure $\mathbb{P}_1$ to another $\mathbb{P}_2$, when it exists, is written $d\mathbb{P}_2/d\mathbb{P}_1$. In this specific case therefore, we denote

$$\frac{d\mathbb{Q}}{d\mathbb{P}} := Z_T.$$

So if X is a random variable on $(\Omega, \mathcal{F})$, we have $\mathbb{E}_{\mathbb{Q}}[X] = \mathbb{E}_{\mathbb{P}}[Z_T X]$.

The process $(Z(t))_{t\in[0,T]}$ is a *Radon-Nikodym derivative process* and, by Theorem 2.6, it satisfies the relation $\mathbb{E}_{\mathbb{P}}[Z_T | \mathcal{F}_t] = Z(t)$ for all $t \in [0, T]$. Then, we also denote

$$\left.\frac{d\mathbb{Q}}{d\mathbb{P}}\right|_{\mathcal{F}_t} := Z(t).$$

Suppose that $(X(t))_{t\in[0,T]}$ is a stochastic process adapted to the filtration $(\mathcal{F}_t)_{t\in[0,T]}$. Then by the tower property of conditional expectations,

$$\mathbb{E}_{\mathbb{Q}}[X(t)] = \mathbb{E}_{\mathbb{P}}[X(t)Z_T] = \mathbb{E}_{\mathbb{P}}\left[\mathbb{E}_{\mathbb{P}}[X(t)Z_T | \mathcal{F}_t]\right]$$

$$= \mathbb{E}_{\mathbb{P}}\left[X(t)\mathbb{E}_{\mathbb{P}}[Z_T | \mathcal{F}_t]\right]$$

$$= \mathbb{E}_{\mathbb{P}}[X(t)Z(t)], \quad t \in [0, T],$$

where we have taken out what is known at the penultimate step, and used the martingale property of the Radon-Nikodym derivative process $(Z(t))_{t\in[0,T]}$ from Theorem 2.6 at the final step.

2.1.5.2 Girsanov's Theorem

Girsanov's theorem simply confirms that $\widetilde{W}$ is a standard Brownian motion under the measure $\mathbb{Q}$ that was constructed in Sect. 2.1.5.1.

Theorem 2.7 (Girsanov's Theorem) *Let $(\widetilde{W}(t))_{t\in[0,T]}$ be a stochastic process constructed according to (2.10), where θ satisfies the conditions of Theorem 2.6. Let $\mathbb{Q}$ be a probability measure constructed according to (2.12) and (2.13). Then $\widetilde{W}$ is a standard Brownian motion under $\mathbb{Q}$.*

This theorem underpins all of the theory of option pricing in the no-arbitrage framework. It is sufficiently important to merit a proof, though we restrict ourselves to the special case where θ is deterministic and continuous (as is the case when θ satisfies (2.11)). We will show that the *joint moment generating function (MGF)* of the increments

$$\widetilde{W}(t_1),\, \widetilde{W}(t_2) - \widetilde{W}(t_1),\, \ldots,\, \widetilde{W}(t_n) - \widetilde{W}(t_{n-1}), \quad 0 = t_0 < t_1 < \cdots < t_n = T,$$

under the measure $\mathbb{Q}$ is the same as the joint MGF of the increments of $W^{\mathbb{P}}$ under $\mathbb{P}$, since it is known that $W^{\mathbb{P}}$ is a standard Brownian motion under that measure.

Proof We seek to prove that, for any $(\alpha_1, \ldots, \alpha_n) \in \mathbb{R}^n$,

$$\mathbb{E}_{\mathbb{Q}}\left[\exp\left(\sum_{k=1}^{n}\alpha_k(\widetilde{W}(t_k) - \widetilde{W}(t_{k-1}))\right)\right] = \mathbb{E}_{\mathbb{P}}\left[\exp\left(\sum_{k=1}^{n}\alpha_k(W^{\mathbb{P}}(t_k) - W^{\mathbb{P}}(t_{k-1}))\right)\right].$$

By (2.10), which relates $\widetilde{W}(t)$ to $W^{\mathbb{P}}(t)$ via θ, we get by substitution

$$\mathbb{E}_{\mathbb{Q}}\left[\exp\left(\sum_{k=1}^{n}\alpha_k(\widetilde{W}(t_k) - \widetilde{W}(t_{k-1}))\right)\right]$$

$$= \mathbb{E}_{\mathbb{Q}}\left[\exp\left(\sum_{k=1}^{n}\left\{\alpha_k\left(W^{\mathbb{P}}(t_k) - \int_0^{t_k}\theta(u)du\right)\right.\right.\right.$$

$$\left.\left.\left. -\alpha_k\left(W^{\mathbb{P}}(t_{k-1}) - \int_0^{t_{k-1}}\theta(u)du\right)\right\}\right)\right].$$

Rearranging yields

$$\mathbb{E}_{\mathbb{Q}}\left[\exp\left(\sum_{k=1}^{n}\alpha_k(\widetilde{W}(t_k) - \widetilde{W}(t_{k-1}))\right)\right]$$

$$= \mathbb{E}_{\mathbb{Q}}\left[\exp\left(\sum_{k=1}^{n}\alpha_k\left\{(W^{\mathbb{P}}(t_k) - W^{\mathbb{P}}(t_{k-1})) - \int_{t_{k-1}}^{t_k}\theta(u)du\right\}\right)\right].$$

Now we change the measure from $\mathbb{Q}$ to $\mathbb{P}$ using the Radon-Nikodym derivative Z_T and recalling that $T = t_n$:

$$\mathbb{E}_{\mathbb{Q}}\left[\exp\left(\sum_{k=1}^{n}\alpha_k(\widetilde{W}(t_k) - \widetilde{W}(t_{k-1}))\right)\right]$$

$$= \mathbb{E}_{\mathbb{P}}\left[Z(t_n)\exp\left(\sum_{k=1}^{n}\alpha_k(\widetilde{W}(t_k) - \widetilde{W}(t_{k-1}))\right)\right].$$

Substituting for (2.13) and simplifying, we get

$$\mathbb{E}_{\mathbb{Q}}\left[\exp\left(\sum_{k=1}^{n}\alpha_k(\widetilde{W}(t_k) - \widetilde{W}(t_{k-1}))\right)\right]$$

$$= \mathbb{E}_{\mathbb{P}}\left[\exp\left(\sum_{k=1}^{n}\left\{\int_{t_{k-1}}^{t_k}(\alpha_k + \theta(u))dW^{\mathbb{P}}(u) - \frac{1}{2}\int_{t_{k-1}}^{t_k}(2\alpha_k\theta(u) + \theta(u)^2)du\right\}\right)\right]$$

$$= \mathbb{E}_{\mathbb{P}}\left[\exp\left(\sum_{k=1}^{n}\left\{\int_{t_{k-1}}^{t_k}(\alpha_k + \theta(u))dW^{\mathbb{P}}(u) - \frac{1}{2}\int_{t_{k-1}}^{t_k}((\theta(u) + \alpha_k)^2 - \alpha_k^2)du\right\}\right)\right],$$

where we used the relation $2\alpha_k\theta(u) + \theta(u)^2 = (\theta(u) + \alpha_k)^2 - \alpha_k^2$ at the last step.

Combining the sums and integrals, and bringing deterministic factors outside the expectation, we can write

$$\mathbb{E}_{\mathbb{Q}}\left[\exp\left(\sum_{k=1}^{n}\alpha_k(\widetilde{W}(t_k) - \widetilde{W}(t_{k-1}))\right)\right]$$

$$= \exp\left(-\frac{1}{2}\int_{0}^{t_n}(\theta(u) + \bar{\alpha}(u))^2 du + \frac{1}{2}\int_{0}^{t_n}\bar{\alpha}(u)^2 du\right)$$

$$\times \mathbb{E}_{\mathbb{P}}\left[\exp\left(\int_{0}^{t_n}(\bar{\alpha}(u) + \theta(u))dW^{\mathbb{P}}(u)\right)\right], \qquad (2.16)$$

where $\bar{\alpha}(u) = \alpha_k$ for all $u \in [t_k, t_{k+1})$, $k = 0, \ldots, n - 1$. Since $W^{\mathbb{P}}$ is a standard Brownian motion under $\mathbb{P}$, by Eq. (1.9) in Chap. 1,

$$\int_{0}^{t_n}(\bar{\alpha}(u) + \theta(u))dW^{\mathbb{P}}(u) \sim \mathcal{N}\left(0, \int_{0}^{t_n}(\bar{\alpha}(u) + \theta(u))^2 du\right),$$

and therefore

$$\mathbb{E}_{\mathbb{P}}\left[\exp\left(\int_{0}^{t_n}(\bar{\alpha}(u) + \theta(u))dW^{\mathbb{P}}(u)\right)\right] = \exp\left(\frac{1}{2}\int_{0}^{t_n}(\bar{\alpha}(u) + \theta(u))^2 du\right)$$

Substituting back into (2.16) and cancelling terms gives

$$
\mathbb{E}_{\mathbb{Q}}\left[\exp\left(\sum_{k=1}^{n}\alpha_k(\widetilde{W}(t_k)-\widetilde{W}(t_{k-1}))\right)\right]=\exp\left(\sum_{k=1}^{n}\frac{\alpha_k^2}{2}(t_k-t_{k-1})\right)
$$

$$
=\prod_{k=1}^{n}\exp\left(\frac{\alpha_k^2}{2}(t_k-t_{k-1})\right)
$$

$$
=\mathbb{E}_{\mathbb{P}}\left[\exp\left(\sum_{k=1}^{n}\alpha_k(W^{\mathbb{P}}(t_k)-W^{\mathbb{P}}(t_{k-1}))\right)\right],
$$

as required. Since the joint MGF of the increments of $\widetilde{W}$ under $\mathbb{Q}$ is the same as that of the increments of $W^{\mathbb{P}}$ under $\mathbb{P}$, we may conclude that $\widetilde{W}$ is a standard Brownian motion under $\mathbb{Q}$. $\qquad\square$

2.1.6 Market Completeness and the Risk-Neutral Pricing Formula

It remains to confirm that our simple market model $\underline{S}$, where b and S satisfy (2.7), is a complete market. We will do this by explicitly constructing the replicating strategy for any derivative with payoff V_T.

2.1.6.1 The Discounted Value of the Expected Payoff

Suppose we discount the final value of the derivative back to time $t \in (0, T)$ and take the conditional expectation under the risk neutral measure $\mathbb{Q}$:

$$
\widetilde{Z}(t) = e^{-r(T-t)}\mathbb{E}_{\mathbb{Q}}\left[V_T|\mathcal{F}_t\right], \quad a.s, \quad t \in (0, T).
$$

This could be interpreted as our best guess at time $t \in (0, T)$ for the present value of the final payout V_T given the information currently available. Now discount $\widetilde{Z}(t)$ again back to time 0, to get

$$
D(t) = e^{-rt}\widetilde{Z}(t) = e^{-rT}\mathbb{E}_{\mathbb{Q}}[V_T|\mathcal{F}_t], \quad a.s, \quad t \in [0, T), \tag{2.17}
$$

and set $D(T) = e^{-rT}V_T$.

Proposition 2.8 *The process D is an $\mathcal{F}_t$-martingale on the interval $[0, T]$.*

Proof First, since $\mathbb{E}_{\mathbb{Q}}[V_T|\mathcal{F}_t]$ is an $\mathcal{F}_t$-measurable random variable, and the discount factor e^{-rT} is deterministic, by (2.17) $D(t)$ is also an $\mathcal{F}_t$-measurable random variable for each $t \in [0, T]$.

Second, for $0 \le s < t$, and bringing the deterministic discount factor out of the conditional expectation

$$\mathbb{E}_{\mathbb{Q}}[D(t)|\mathcal{F}_s] = e^{-rT}\mathbb{E}_{\mathbb{Q}}[\mathbb{E}_{\mathbb{Q}}[V_T|\mathcal{F}_t]|\mathcal{F}_s]$$

$$= e^{-rT}\mathbb{E}_{\mathbb{Q}}[V_T|\mathcal{F}_s]$$

$$= D(s), \quad a.s, \quad 0 \le s < t \le T,$$

where we have used the tower property of conditional expectation at the second step, and (2.17) at the last step. Therefore D satisfies the conditions of Definition 1.10.

$\square$

Next, define the discounted asset price process under the risk-neutral measure $\mathbb{Q}$ to be $X(t) = S(t)/b(t) = e^{-rt}S(t)$, where S satisfies

$$S(t) = S_0 e^{(r-\sigma^2/2)t + \sigma W^{\mathbb{Q}}(t)}, \quad t \in [0, T].$$

We see from Theorem 2.5 in Sect. 2.1.4.2 that $X(t)$ is an $\mathcal{F}_t$-martingale.

2.1.6.2 A Replicating Strategy for a Financial Derivative

Now we will demonstrate completeness of the market model $\underline{S}$ by explicitly constructing a self-financing portfolio that replicates the payoff of the derivative. Since both D and X are $\mathcal{F}_t$-martingales, it is a consequence of Theorem 1.11, the Martingale Representation Theorem, that there exists a previsible process $\phi(t)$ such that

$$dD(t) = \phi(t)dX(t), \quad t \in [0, T].$$

Set

$$\psi(t) = D(t) - \phi(t)X(t), \quad t \in [0, T]. \tag{2.18}$$

Consider again the portfolio where at time $t \in [0, T]$ an investor holds $\phi(t)$ units of the risky asset S and $\psi(t)$ units of the zero-coupon bond b. The value of this portfolio is

$$\Pi(t) = \phi(t)S(t) + \psi(t)b(t) = \phi(t)S(t) + \psi(t)e^{rt}$$

$$= \phi(t)S(t) + (D(t) - \phi(t)X(t))e^{rt}$$

$$= e^{rt}D(t) + \phi(t)(S(t) - e^{rt}X(t))$$

$$= e^{rt}D(t) = \tilde{Z}(t),$$

where we have used the relations $X(t) = e^{-rt}S(t)$ and $D(t) = e^{-rt}\widetilde{Z}(t)$. At expiry the value of the portfolio is

$$\phi(T)S(T) + \psi(T)e^{rT} = \widetilde{Z}(T) = \mathbb{E}_{\mathbb{Q}}[V_T|\mathcal{F}_T]$$
$$= V_T, \quad a.s,$$

since V_T is $\mathcal{F}_T$-measurable. We have confirmed that this portfolio replicates the value of the derivative at expiry. In order to confirm that it is self-financing, we need to check that Condition (2.5) holds. By (2.18),

$$d\psi(t) = dD(t) - d(\phi(t)X(t))$$
$$= \phi(t)dX(t) - [\phi(t)dX(t) + X(t)d\phi(t) + (d\phi(t))(dX(t))]$$
$$= -X(t)d\phi(t) - (d\phi(t))[e^{-rt}dS(t) - re^{-rt}S(t)dt]$$
$$= -e^{-rt}[S(t)d\phi(t) + (d\phi(t))(dS(t))],$$

since $d\phi(t)dt = 0$, by (1.16) in Chap. 1. Thus,

$$S(t)(d\phi(t)) + e^{rt}(d\psi(t)) + (dS(t))$$
$$= S(t)(d\phi(t)) - S(t)(d\phi(t)) - (d\phi(t))(dS(t)) + (d\phi(t))(dS(t)) = 0,$$

as required.

2.1.6.3 The Risk-Neutral Pricing Formula for Financial Derivatives

The portfolio of $\phi(t)$ stocks and $\psi(t)$ bonds constructed in Sect. 2.1.6.2 is both self-financing and replicates the derivative payoff at expiry. By the no-arbitrage principle, stated in Definition 2.1, the value of the derivative at any time t before expiry has to equal the value $V(t)$ of this replicating portfolio. We have established the following:

Proposition 2.9 (Risk-Neutral Pricing Formula for Derivatives) *The value $V(t)$ of a derivative written on the underlying asset S whose payoff at expiry is V_T is given by*

$$V(t) = e^{-r(T-t)}\mathbb{E}_{\mathbb{Q}}[V_T|\mathcal{F}_t], \quad t \in [0, T]. \tag{2.19}$$

2.2   The Black-Scholes-Merton PDE

In the case where the value of the derivative $V(S, t)$ depends only on the current time $t \in [0, T]$ and the price of the underlying asset S, we can express the value of the option as the solution of a partial differential equation (PDE) known as the Black-Scholes-Merton PDE.

2.2.1 The Theorem of Feynman-Kac

While there are several ways to derive this PDE, we choose to present here an important result from stochastic analysis that builds a direct path between the conditional expectation on the RHS of the risk-neutral pricing formula (2.19) in Proposition 2.9 and the deterministic solution of a PDE.

Theorem 2.10 (Feynman-Kac) *Let $V, f, g : [0, T] \times \mathbb{R} \to \mathbb{R}$ be functions of t and s. Consider a PDE*

$$F_t(t, s) + f(t, s)F_s(t, s) + \frac{1}{2}g^2(t, s)F_{ss}(t, s) = 0, \quad t \in [0, T], \tag{2.20}$$

with final time condition $F(T, s) = F_T(s)$. Let $X(t)$ be a stochastic process defined on a probability space $(\Omega, \mathcal{F}, \mathbb{Q})$, adapted to a filtration $(\mathcal{F}_t)_{t \in [0,T]}$ and satisfying the SDE

$$dX(t) = f(t, X(t))dt + g(t, X(t))dB(t), \quad t \in [0, T]. \tag{2.21}$$

Then we have

$$F(t, s) = \mathbb{E}_{\mathbb{Q}}\left[F_T(X(T))|\mathcal{F}_t\right]\Big|_{X(t)=s} = \mathbb{E}_{\mathbb{Q}}\left[F_T(X(T))|X(t) = s\right].$$

For a function $F(t, s)$ satisfying the PDE (2.20) define the function

$$V(t, s) = e^{-r(T-t)}F(t, s).$$

Since

$$V_t(t, s) = re^{-r(T-t)}F(t, s) + e^{-r(T-t)}F_t(t, s),$$

the function V satisfies the PDE

$$V_t(t, s) + f(t, s)V_s(t, s) + \frac{1}{2}g^2(t, s)V_{ss}(t, s) = rV(t, s),$$

with a final time condition

$$\Lambda(s) := V(T, s) = e^{-r(T-t)}F(T, s).$$

2.2.2 Application to Our Simple Market Model

Consider again the simple market model $\underline{S} = (b, S)$ where $b(t) = e^{rt}$ for some $r > 0$ and S satisfies the SDE (2.21) with $f(t, x) = rx$ and $g(t, x) = \sigma x$. Recall from Sect. 2.1.4.2 that by this choice of drift f we ensure that expectations are being taken under the risk-neutral measure $\mathbb{Q}$.

By Theorem 2.10

$$V(t, s) = e^{-r(T-t)} F(t, s) = e^{-r(T-t)} \mathbb{E}_{\mathbb{Q}}[\Lambda(S(T)) | S(t) = s].$$

We recognise this from the statement of Proposition 2.9 as the value of a derivative written on S with payoff $V_T = \Lambda(S(T))$. Therefore, if V represents the value of a derivative written on S with payoff Λ, then V satisfies the PDE

$$V_t(t, s) + rs V_s(t, s) + \frac{1}{2}\sigma^2 s^2 V_{ss}(t, s) = rV(t, s), \tag{2.22}$$

which is known as the Black-Scholes-Merton PDE. This is true for any derivative whose payoff depends only on the final value of the asset S, and the precise derivative being valued may be specified by the boundary and final time conditions for (2.22).

2.2.3 The Black-Scholes Pricing Formulae for European Options

The PDE (2.22) can be explicitly solved to give a pricing formula for many options that fall into the category of digital and binary options. Among these are the European call and put. In this section we review the pricing formulae for these options and implement them in Python in order to carry out option price computations.

2.2.3.1 Boundary Conditions and the Price of a European Call

The boundary and final-time conditions associated with a European call option are

$$\Lambda(s) = V(s, T) = \max(s - E, 0); \tag{2.23}$$

$$\lim_{s \to 0^+} V(s, t) = 0; \tag{2.24}$$

$$V(s, t) \sim s \quad \text{as } s \to \infty. \tag{2.25}$$

The motivation for (2.23) is clear: the final-time condition must correspond to the value of the option at expiry and hence to the payoff. Condition (2.24) holds since as the value of the underlying asset approaches zero, the possibility that it will recover to a value above the strike price E in the time remaining to expiry, so that the holder can exercise and make a profit, vanishes. Condition (2.25) holds since as the value of the underlying asset grows large relative to the strike, the possibility that the asset will fall below the strike price vanishes, and additionally the value of $s - E$ is indistinguishable from s in the limit. Informally, consider that, if $E = 1$ and $S(T) = 10^6$ (in dollars), the holder of the call will receive $S(T) - E = 999,999$ upon exercising. For all practical purposes they made a million dollars.

Solving (2.22) with the boundary and final time conditions (2.23)–(2.25) yields the Black-Scholes formula for a the value of a European call option,

$$C(S, t) = SN(d_1) - Ee^{-r(T-t)}N(d_2), \tag{2.26}$$

where

$$N(x) = \frac{1}{\sqrt{2\pi}} \int_{-\infty}^{x} e^{-\frac{1}{2}y^2} dy, \tag{2.27}$$

and

$$d_1 = \frac{\log(S/E) + (r + \sigma^2/2)(T - t)}{\sigma\sqrt{T - t}}; \tag{2.28}$$

$$d_2 = \frac{\log(S/E) + (r - \sigma^2/2)(T - t)}{\sigma\sqrt{T - t}}. \tag{2.29}$$

We can implement this in Python, starting with the functions d_1 and d_2:

```
In []: def d1(E,S0,r,sig,T):
           logSE=np.log(S0/E)
           return((logSE+(r+0.5*sig**2)*T)/(sig*np.sqrt(T)))

       def d2(E,S0,r,sig,T):
           return(d1(E,S0,r,sig,T)-sig*np.sqrt(T))
```

Notice that we made use of the relation $d_2 = d_1 - \sigma\sqrt{T - t}$ in the construction of d2(), and both functions have been defined so that they could take an array-valued input from any one of the key parameters S_0, E, or T, and produce a corresponding array-valued output.

Now define a function corresponding to the call valuation formula (2.26). For this, we will use norm.cdf(), which has been imported from the stats package and which is a vectorised implementation of the standard normal CDF given by N in (2.27).

```
In []: def euroCallBenchmark(E,S0,r,sigma,T):
           enDee1=norm.cdf(d1(E,S0,r,sigma,T))
           enDee2=norm.cdf(d2(E,S0,r,sigma,T))
           return(S0*enDee1-E*dFac(r,T)*enDee2)
```

2.2.3.2 Put-Call Parity and the Price of a European Put

Consider a portfolio consisting of one unit of the underlying, one long position in a European Put and one short position in a European call, both with the same strike price E and expiry date T. The value of the portfolio at expiry is given by

$$\Pi(T) = S(T) + P(S(T), T) - C(S(T), T)$$
$$= S(T) + (E - S(T))^+ - (S(T) - E)^+$$
$$= E.$$

Since the final value of the portfolio is E, we can discount at the risk free rate r in order to get prior values: $\Pi(t) = e^{-r(T-t)} E$ for all $t \in [0, T]$. This leads to the *put-call-parity relation* for European options:

$$e^{-r(T-t)} E = S + P(S, t) - C(S, t), \quad t \in [0, T]. \tag{2.30}$$

The put-call-parity relation (2.30) can be applied directly to the Black-Scholes formula for a European call option to confirm the pricing formula for the European put is given by

$$P(S, t) = E e^{-r(T-t)} N(-d_2) - S N(-d_1). \tag{2.31}$$

We make use of put-call-parity to define another Python function for the put:

```
In []: def euroPutBenchmark(E,S0,r,sig,T):
           BSC=euroCallBenchmark(E,S0,r,sig,T)
           return(E*dFac(r,T)-S0-BSC)
```

Alternatively, we could use (2.31) directly. Figure 2.1 illustrates the pricing formulae given by (2.26) and (2.31) .

2.2.4   Option Price Sensitivities: The Greeks

In this section we introduce a collection of quantities, known as the Greeks, that characterise the sensitivity of an option price to changes in one of its parameters. These are also referred to as hedge parameters and we will show how one of them, denoted Δ (*delta*), can be used to minimise or eliminate the asset-price risk associated with writing (selling) an option.

2.2.4.1 Delta for European Options

Let $V(S, t)$ be the value of an option written on an underlying asset S and consider a portfolio with a value

$$\Pi(S, t) = V(S, t) - \Delta S,$$

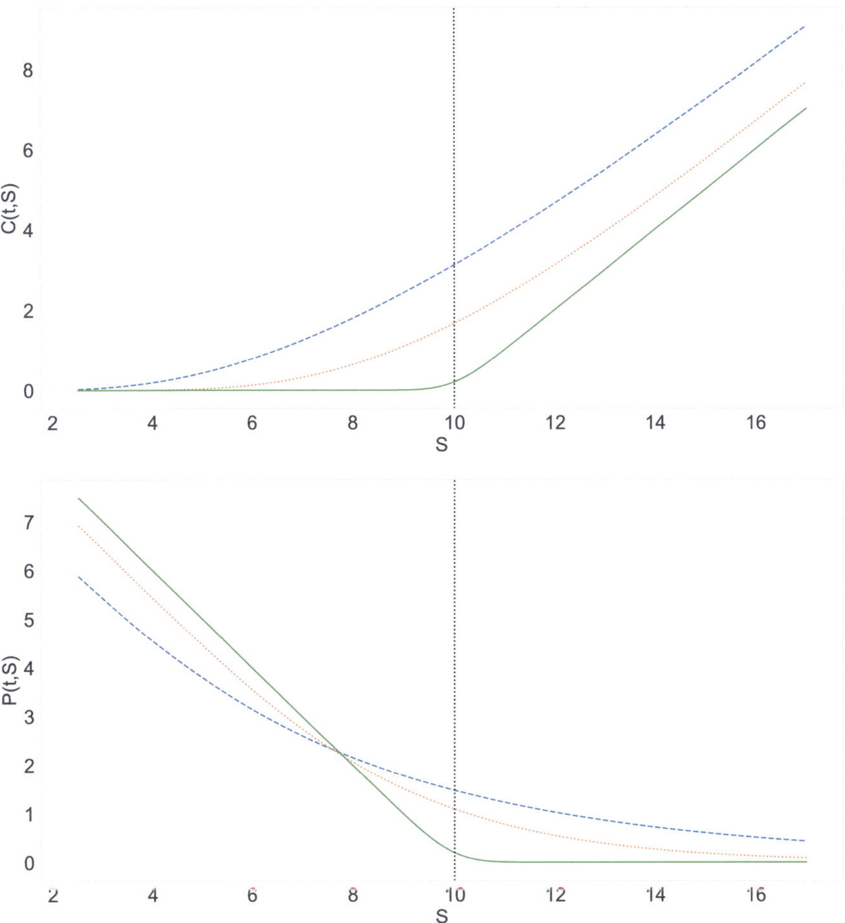

Fig. 2.1 Black-Scholes values of European call option (top) and a European put option (bottom), as a function of S, with $E = 10$, $r = 0.12$, $\sigma = 0.5$ and for reducing values of the time to expiry $T - t$. In each case, the blue dashed line represents option values when $T - t = 1.5$, the orange dotted line when $T - t = 0.5$, and the green line when $T = 0.01$. The strike is represented by a vertical black dotted line

so that the size of an uncovered position in S is given by $-\Delta$. It is clear that if $\Delta < 0$, the portfolio has a long position in S, whereas if $\Delta > 0$ that position is short.

We can make the value of this portfolio insensitive to changes in the price of the underlying asset by choosing Δ such that

$$\Pi_s(S, t) = V_s(S, t) - \Delta = 0.$$

This will be the case if, at any particular time $t \in [0, T]$ and asset price S,

$$\Delta = V_s(t, S).$$

Clearly, it will not be possible to achieve our goal if we maintain the same position in the underlying asset for all $t \in [0, T]$. $\Delta = \Delta(t)$ must be updated from moment to moment, and should be thought of as time-dependent. In the case of a European call option, we can directly differentiate the Black-Scholes formula (2.26) to show that

$$\Delta = N(d_1). \tag{2.32}$$

Given that we already have a Python function for $\mathtt{d1()}$, it is straightforward to define one for Δ as well:

```
In []: def deltaCall(E,S0,r,sig,T):
           return(norm.cdf(d1(E,S0,r,sig,T)))
```

For a European put, it is a consequence of the put-call parity relation (2.30) that

$$\Delta = N(d_1) - 1. \tag{2.33}$$

Δ is referred to as a hedge parameter, and is one of the Greeks: a collection of partial derivatives representing the sensitivity of option prices to parameter changes. Figure 2.2 illustrates Δ for European call and put options. Exercise 2.7 asks you to verify (2.32) and (2.33).

Other important portfolio sensitivities are as follows:

Definition 2.11 (The Greeks)

1. The sensitivity of a portfolio to changes in the underlying asset value is called the *delta of the portfolio*, and is given by

$$\Delta = \frac{\partial \Pi(S, t)}{\partial S}.$$

2. The sensitivity of a portfolio to second-order changes in the underlying asset value is called the *gamma of a portfolio*, and is given by

$$\Gamma = \frac{\partial^2 \Pi(S, t)}{\partial S^2}.$$

3. The rate of decay in time of the value of a portfolio is called the *theta of a portfolio*, and is given by

$$\Theta = -\frac{\partial \Pi(S, t)}{\partial t}.$$

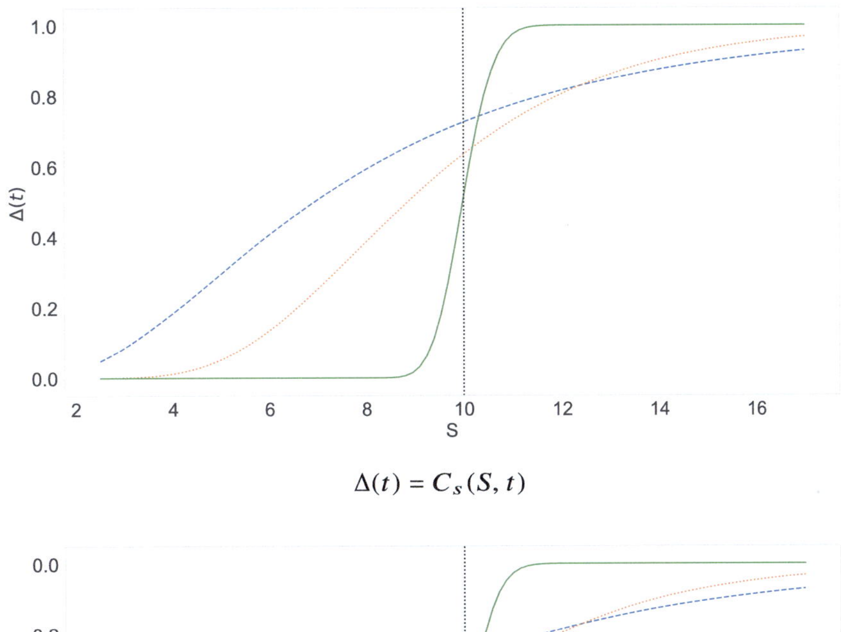

Fig. 2.2 Graphs of Δ for a European call option (top) and a European put option (bottom), with $E = 10$, $r = 0.12$, $\sigma = 0.5$. The blue dashed line corresponds to $T - t = 0.5$, the orange dotted line to $T - t = 0.1$ and the green line to $T - t = 0.01$. The strike is represented by a vertical black dotted line. Notice the shift on the vertical axis between graphs

4. The sensitivity of a portfolio to volatility is called the *vega of a portfolio*, and is given by

$$\frac{\partial \Pi(S, t)}{\partial \sigma}.$$

5. The sensitivity of a portfolio to interest rate is called the *rho of a portfolio*, and is given by

$$\rho = \frac{\partial \Pi(S, t)}{\partial r}.$$

2.2.4.2 A Dynamic Hedging Strategy

Consider an investor who, at time $t = t_0 = 0$, writes one European option (put or call) on an underlying asset, and takes an uncovered position with $-\Delta$ units of the underlying. Then the value of a portfolio consisting of the proceeds from the sale of the European option along with the uncovered position in the underlying asset at time t_0 is given by

$$\Pi(t_0) = V(S_0, t_0) - \Delta S_0. \tag{2.34}$$

As time passes, the investor seeks to adjust the size of the uncovered position in S through the hedging parameter Δ in order to offset the risk associated with asset price movements.

Suppose that they update their position at each time t_i for $i = 0, \ldots, N$, where $t_0 = 0$, $t_N = T$, and each $t_{i+1} - t_i = \delta t = T/N$. At time $t_1 = \delta t$, they receive interest on the previous value of the portfolio, close their previous position in the underlying asset, now valued at $-\Delta(t_0)S(t_1)$, and open a new position valued at $-\Delta(t_1)S(t_1)$. The portfolio value becomes

$$\Pi(t_1) = \Pi(t_0)e^{r\delta t} - (\Delta(t_1) - \Delta(t_0))S(t_1).$$

where $\Pi(t_0)$ is given by (2.34). Repeating this procedure at each t_i gives the recursion

$$\Pi(t_0) = V(S_0, t_0) - \Delta(t_0)S(t_0)$$

$$\Pi(t_i) = \Pi(t_{i-1})e^{r\delta t} - (\Delta(t_i) - \Delta(t_{i-1}))S(t_i), \quad i = 1, \ldots, N-1. \tag{2.35}$$

At time $t_N = T$, when the European put option expires, the investor, who wrote it, becomes liable for the payoff. The value of their portfolio at $t_N = T$ is then given by

$$\Pi(t_N) = \Pi(t_{N-1})e^{r\delta t} - (E - S(t_N))^+ + \Delta(t_{N-1})S(t_N). \tag{2.36}$$

Ideally, the final value of this portfolio should be close to zero on average, i.e. $\mathbb{E}_\mathbb{P}[\Pi(t_N)] \approx 0$, where the expectation is taken with respect to the real-world measure $\mathbb{P}$ since the values of $S(t_i)$ used are observations on the market price of the asset. In this way, the writer of the option will make a consistent profit if they charge a premium, known as the *spread*, over the Black-Scholes price of the option at the outset.

2.2.4.3 A Python Demonstration of Dynamic Hedging

Suppose we have written a European call option on a non-dividend-paying asset currently priced at $S_0 = 10$, with strike $E = 10$, expiry $T = 1$, $r = 0.01$, and $\sigma = 0.5$. We wish to hedge this position using the dynamic hedging strategy described in Sect. 2.2.4.2.

First, we define our variables, set up a random number generator, and use the `bsEnsemble()` function from Exercise 1.11 in Chap. 1 to generate a single trajectory of the underlying asset.

```
In []: # Option parameters
       S0=10; E=10; r=0.01; mu=0.03; sig=0.5; T=1;

       # Simulation parameters and instantiate local RNG
       M=1; N=2**8
       dt=T/N; t=np.linspace(0,1,N+1)
       rng=np.random.default_rng()

       # Generate real-world form of asset price trajectory
       ens=bsEnsemble(S0,mu,sig,T,M,N,rng); ens=ens[0,:]
```

Notice that we call `bsEnsemble` with $M = 1$ to generate a single trajectory; the second command of the last line serves to convert the output to a 1-dimensional array if necessary. Moreover the values contained in `ens` are simulated observations of the market price of the asset and should therefore be modelled using the real-world form of the asset model with drift μ as given in (2.7).

Now, set up the initial value of the hedging portfolio according to (2.34), and instantiate an array called `pnl` (named for the profit-and-loss account used to manage a hedging portfolio in practice) to hold the values of this portfolio as it evolves in time. We will use the `deltaCall()` function defined in Sect. 2.2.4.1 here, as well as `euroCallBenchmark()` from Sect. 2.2.3.1.

```
In []: # Cash acquired from sale of call option
       C0=euroCallBenchmark(E,S0,r,sig,T)

       # Initial value of delta and hedging portfolio
       delta0=deltaCall(E,S0,r,sig,T)
       pnl0=C0-delta0*S0

       # Instantiate array to hold portfolio values over time
       pnl=pnl0*np.ones(len(t))
```

Next execute the recursion given by (2.35) up to time t_{N-1} using a `for` loop. Indexing the loop is carried out using the `range()` call, which should start at zero and finish so that `pnl` has been updated up to the $(N-1)$th entry.

```
In []: for i in range(0,len(t)-2):

           # Update current and previous times-to-expiry
           nwTExp=T-t[i+1]; odTExp=T-t[i]

           # Compute Delta at current and previous times
           nwAsset=ens[i+1]; odAsset=ens[i]
           nwDel=deltaCall(E,nwAsset,r,sig,nwTExp)
           odDel=deltaCall(E,odAsset,r,sig,odTExp)

           # Update hedging portfolio
           pnl[i+1]=pnl[i]*dFac(-r,dt)-(nwDel-odDel)*nwAsset
```

Notice that we pass `-r` rather than `r` to `dFac()` since we are applying compound interest to the previous value of the portfolio rather than discounting it. Compute the final value of the portfolio by closing the uncovered position in the underlying asset and incorporating the potential liability associated with the exercise of the option, according to (2.36).

```
In []: finalHedge=pnl[N-1]*dFac(-r,dt)+nwDel*ens[N]
       pnl[N]=finalHedge-np.maximum(ens[-1]-E,0)
```

In Figs. 2.3, 2.4, and 2.5, we illustrate this process for three asset price trajectories, Fig. 2.3 illustrates the case where the European call option finishes in-the-money (ITM). Notice that over the life of the option $\Pi(t)$ closely tracks the changes in the value of the option so that it finishes at a value that is effectively cancelled by the exercise of the option at expiry. Fig. 2.4 illustrates the case where the European call finishes out-of-the-money (OTM). Here, again notice how the value of $\Pi(t)$ tracks changes in the value of $C(t, S)$, but as it becomes apparent that the option is unlikely to be exercised, $\Pi(t)$ becomes small and finishes close to zero. By contrast, we observe in Fig. 2.5 an example where the underlying asset crosses the strike price E several times in quick succession just before the option expires, and is close to finishing at-the-money (ATM). Here, Δ is far more variable around $\Delta = 0.5$ as t approaches T, and a lot of rebalancing activity is required to maintain the hedge.

In each case, the actual final value of $\Pi(t)$ is random, and in Fig. 2.6 we illustrate its distribution numerically, along with its relationship to the frequency at which the hedging portfolio is rebalanced. We can see that, although all sample ensembles result in values for $\Pi(T)$ that are approximately zero, as N grows larger and hence the more frequently the portfolio is rebalanced, the smaller the sample variance around zero that results. In practice, a compromise must be struck between

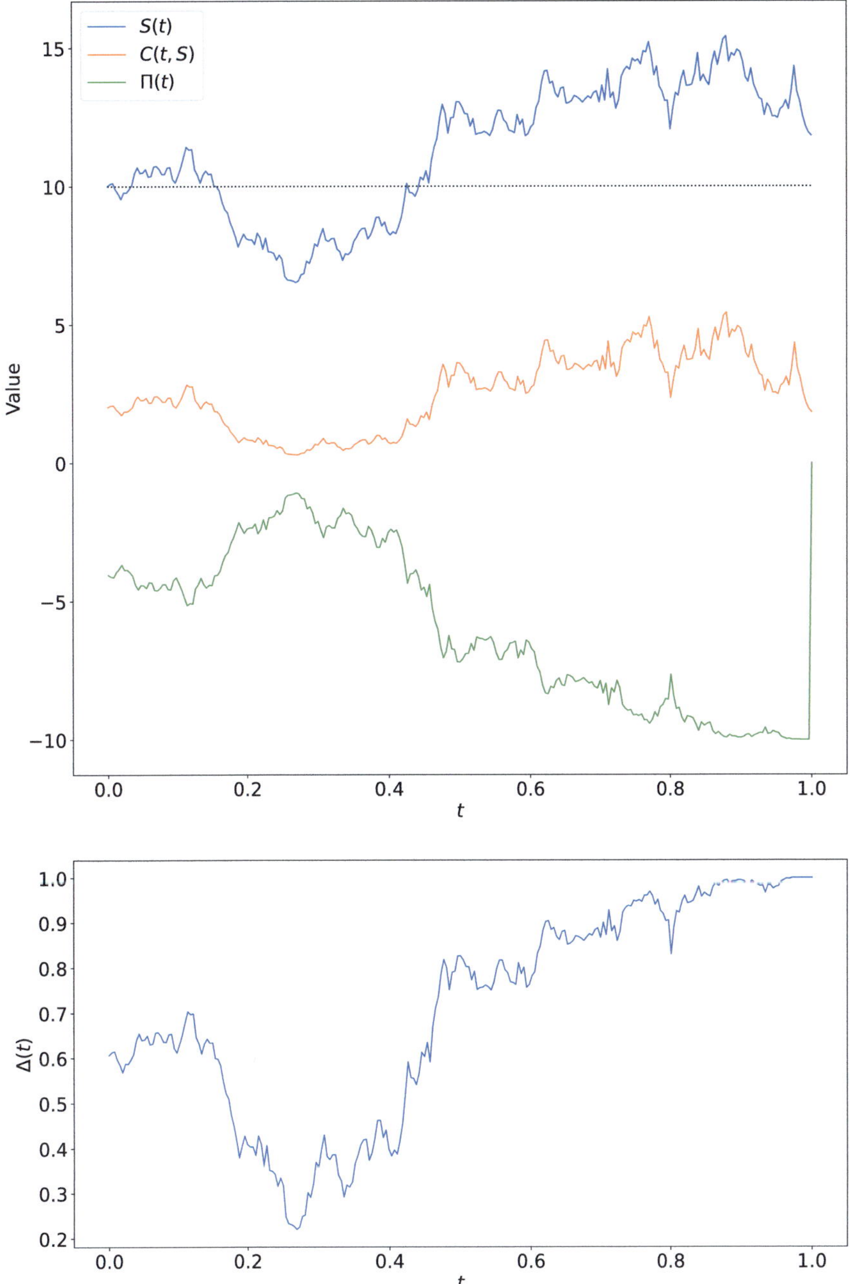

Fig. 2.3 The dynamic hedging process for the writer of a European call option with $S_0 = E = 10$, $T = 1, r = 0.01, \mu = 0.02, \sigma = 0.5$, where the hedging portfolio is rebalanced $N = 2^8 = 256$ times over the course of a year. In the top pane, the strike is indicated by the black dotted line, and we see the asset price trajectory (blue) finishing in-the-money (ITM), the corresponding price of the European call (orange), and the value of the hedging portfolio as it changes over the life of the option (green). In the bottom pane, we see the evolution of $\Delta(t)$ (blue)

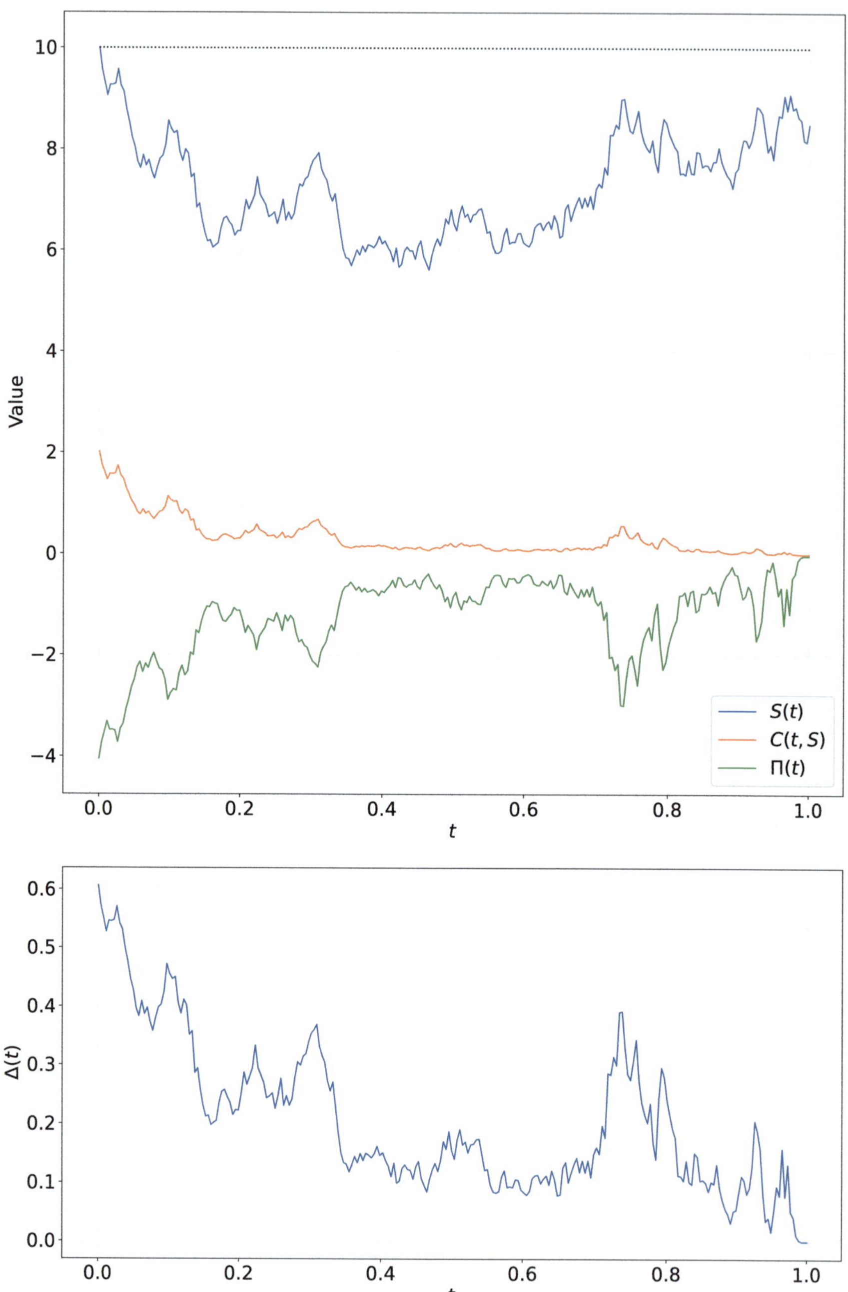

Fig. 2.4 The dynamic hedging process for the writer of a European call option with $S_0 = E = 10$, $T = 1$, $r = 0.01$, $\mu = 0.02$, $\sigma = 0.5$, where the hedging portfolio is rebalanced $N = 256$ times over the course of a year. In the top pane, the strike is indicated by the black dotted line, and we see the asset price trajectory (blue) finishing out-of-the-money (OTM), the corresponding price of the European call (orange), and the value of the hedging portfolio as it changes over the life of the option (green). In the bottom pane, we see the evolution of $\Delta(t)$ (blue)

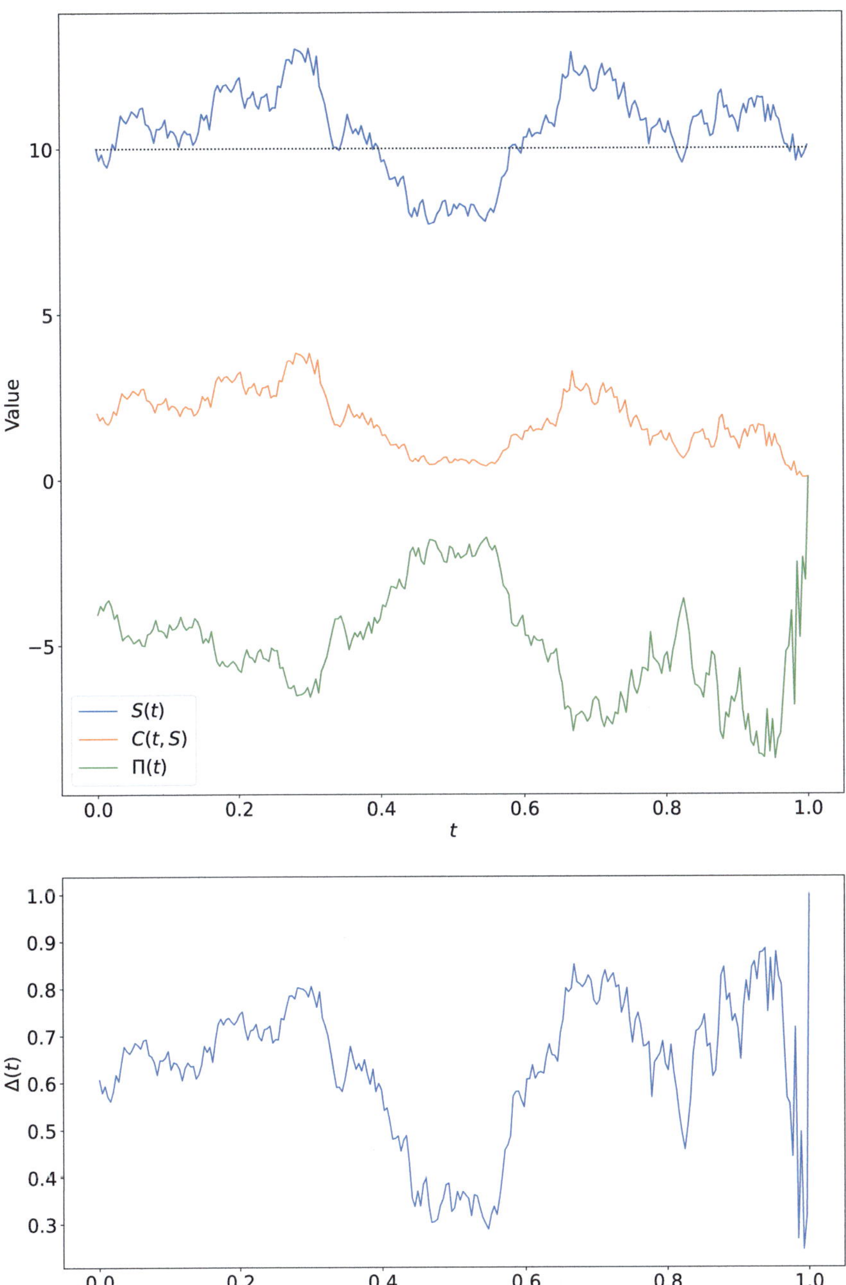

Fig. 2.5 An illustration of the dynamic hedging process for the writer of a European call option with $S_0 = E = 10$, $T = 1$, $r = 0.01$, $\mu = 0.02$, $\sigma = 0.5$, where the hedging portfolio is rebalanced $N = 256$ times over the course of a year. In the top pane, the strike is indicated by the black dotted line, and we see the asset price trajectory (blue) finishing very close to being ATM, the corresponding price of the European call (orange), and the value of the hedging portfolio as it changes over the life of the option (green). In the bottom pane, we see the evolution of $\Delta(t)$ (blue)

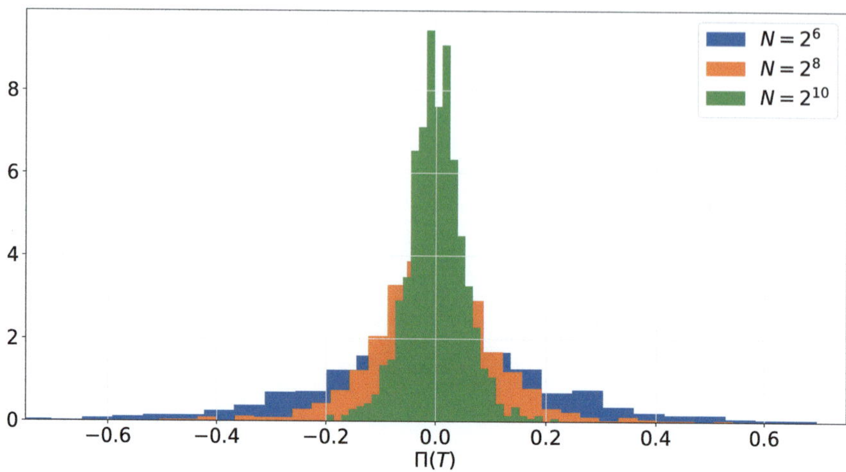

Fig. 2.6 Overlaid density histograms of the final value of the portfolio $\Pi(T)$ resulting from the dynamic hedging process sampled over $M = 1000$ trajectories of the asset price model. For $N = 2^6$ we have a sample mean of -0.0074 and sample standard deviation of 0.2123. For $N = 2^8$ we have a sample mean of -0.0034 and a sample standard deviation of 0.1102. For $N = 2^{10}$ we have a sample mean of 0.0004 and a sample standard deviation of 0.051348

frequency of rebalancing and the transaction costs associated with maintaining the hedge. Exercise 2.9 asks you to vary the real-world drift parameter μ and show that its value does not affect the distribution of the final value of the portfolio.

We see also that on the ITM trajectory, $\lim_{t \to T^-} \Delta(t) = 1$, and on the OTM trajectory $\lim_{t \to T^-} \Delta(t) = 0$. If the trajectory in Fig. 2.5 were to finish exactly ATM (an event with probability zero), then we would have $\lim_{t \to T^-} \Delta(t) = 0.5$. Part 1 of Exercise 2.8 asks you to confirm theoretically that the limiting behaviour of Δ observed in these figures is characteristic in each case for European call options. Part 2 asks you to show a corresponding result for European put option.

2.3 Estimation of the Black-Scholes Implied Volatility

To execute a dynamic hedging strategy of the kind described in Sect. 2.2.4, which involves the valuation of an option as well as values for its Δ, we need numerical values for the current value of the asset, $S(t)$, the strike E, the expiry time T, the risk free rate r, and the volatility σ.

2.3.1 Characterising σ as the Solution of a Nonlinear Equation

The values of $S(t)$, r, E, and T can all be looked up. However σ must be estimated from available data. Consider the value of a European option as a function of σ only.

Suppose that we look up the current market value of an option trading with the same parameters, denoted C^*. Then the value of σ chosen such that

$$C(\sigma) = C^* \tag{2.37}$$

where $C(\sigma)$ defined by (2.26)–(2.29), is referred to as the Black-Scholes implied volatility, or simply *implied volatility*. C is a nonlinear function of σ, and there is no way to compute the value of σ that solves (2.37) explicitly. We must instead rewrite (2.37) in the form

$$C(\sigma) - C^* = 0$$

and apply an iterative root-finding method to the function $g(\sigma) = C(\sigma) - C^*$.

2.3.2   Root-Finding with a Built-in Solver

2.3.2.1 Using `fsolve()`: A 1-Dimensional Example

Suppose that $C^* = 14.44$ is the published price of a European call with strike $E = 100$, time to expiry $T = 0.5$, where the underlying asset is $S_0 = 102$, and the risk free rate is $r = 0.02$. Fixing these as global variables in the context of this computation, we can then define a function for g with the sole argument σ:

```
In []: Cstar=14.44; E=100; T=0.5; S0=102; r=0.02

       def g(sig):
           return(euroCallBenchmark(E,S0,r,sig,T)-Cstar)
```

Now apply the `fsolve()` call from the `scipy.optimization` subpackage to find the root of the function g with initial guess g_0. Since σ typically takes values over the interval $[0.1, 0.5]$, we could choose $g_0 = 0.35$.

```
In []: g0=0.35
       sigImplied=fsolve(g,g0)
       print(sigImplied[0])

Out[]: 0.4565617766852346
```

Notice that for this example `fsolve` returns a NumPy array of length 1, since `g()` is passed (and returns) a single value.

2.3.2.2 The Volatility Smile

We can extend this technique to investigate the realism of the assumption implicit in the use of the Black-Scholes asset model (2.9) that asset price volatility σ is constant. Consider an asset with current price $S_0 = 138.88$, and suppose that we check the value of a call on this underlying asset with $T = 0.3$, $r = 0.04$ for each of 10 strike prices ranging in increments of 5 from 115 to 160:

```
In []:  S0=138.88; T=0.3; r=0.04
        E=np.linspace(115,160,10)
        Cstar=[29.20,24.37,21,10,16.95,
               13.80,10.95,8.62,6.48,4.70]
```

The value of a call struck at $E = 115$ is $C^* = 29.20$, the value at $E = 120$ is $C^* = 24.37$, and so on. By vectorisation, we can use `fsolve()` almost exactly as before to compute an array of implied volatilities:

```
In []:  g0=0.35*np.ones(len(E))
        sigImplied=fsolve(g,g0)
```

The only difference in syntax is that the initial value passed to `fsolve()` as the second argument must be of the same dimension as the data returned from `g()`. `sigImplied` is now a NumPy array with 10 entries, each a value of σ^* with a corresponding entry in E.

Figure 2.7 displays a plot of these against E, with the current value of the asset S_0 marked as a vertical line. We can see that volatilities are not constant as the difference between S_0 and E changes, and the relationship with E shows an asymmetric *volatility smile* (often called a *smirk*), with implied volatilities appearing to grow larger for options which are further OTM.

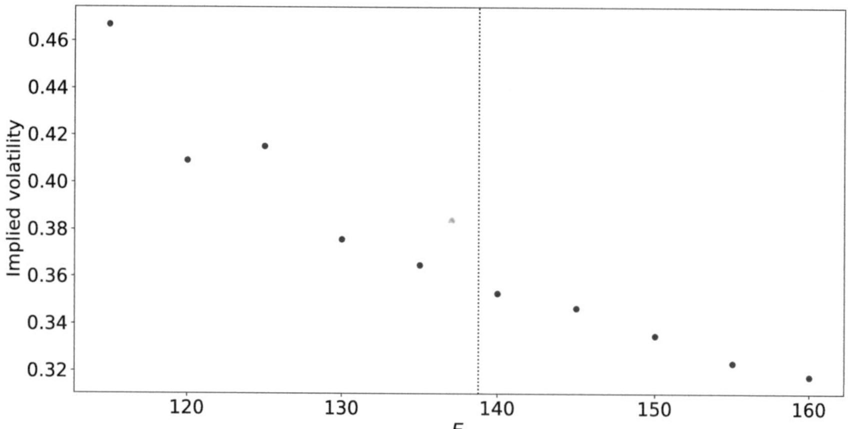

Fig. 2.7 A graph of implied volatilities against strike for call options over the range [115, 160] in increments of 5. The black dotted line shows the current asset price of $S_0 = 138.88$

2.3.2.3 Local and Stochastic Volatility Models

One way to accommodate an apparent dependence of the implied volatility on the relative position of E and S_0 is to consider σ as a function of asset price and time (since implied volatilities may also change as the expiry date approaches). This leads to so-called local-volatility asset models of the form:

$$dS(t) = rS(t) + \sigma(t, S(t))dW^{\mathbb{Q}}(t), \quad t \in [0, T].$$

As simple example, we could choose $\sigma(t, S(t)) = \bar{\sigma}S(t)^{\gamma}$ for some $\bar{\sigma}, \gamma \geq 0$, leading to a *constant elasticity of variance (CEV) model*. Another approach is that of stochastic volatility models, where σ is described as a stochastic process with its own driving source of noise. The Heston and SABR volatility models fall into this category. They can be analytically difficult to work with and often must be solved numerically, a topic we will address in Chap. 8.

2.4 Further Reading

We hope we have succeeded in striking a balance between mathematical rigour and readability in this chapter. Nonetheless this treatment is primarily intended to set out a context for the computational techniques presented in the rest of the book; for readers interested in learning more about both the financial and mathematical aspects of this topic there are several excellent texts available at varying levels of technical sophistication.

Those without a specifically mathematical background approaching the subject for the first time may consult Hull [36]. The textbook of Higham [32] provides an excellent and concise mathematical introduction to the subject and in its later chapters introduces some of the numerical techniques discussed in the early parts of this book, showing how they may be implemented in Matlab.

A by-now classical mathematical introduction to the world of option valuation and the Black-Scholes framework may be found in Wilmott et al. [69], with more advanced treatments available in the monograph by Bjork [7] and the recent textbook of Oosterlee and Grzelak [58]. We also point readers interested in stochastic analysis to the textbook of Øksendal [56, Chapter 8] for a more detailed treatment of Radon-Nikodym derivative processes and Girsanov's theorem in a general setting.

2.4.1 Exercises

2.1 Let $(\Omega, \mathcal{F}, (\mathcal{F}_t)_{t\geq 0}, \mathbb{P})$ be a filtered probability space. Show that any finite deterministic constant is an $\mathcal{F}_t$-martingale.

2.2 Let $(\Omega, \mathcal{F}, (\mathcal{F}_t)_{t\in[0,T]}, \mathbb{P})$ be a filtered probability space and suppose that $W^{\mathbb{P}}$ is an $\mathcal{F}_t$-adapted standard Brownian motion. Suppose also that θ satisfies (2.11). Confirm that

1. the process Z defined in (2.13) is given by

$$Z(t) = \exp\left\{ -\left(\frac{\mu - r}{\sigma}\right) W^{\mathbb{P}}(t) - \frac{1}{2}t\left(\frac{\mu - r}{\sigma}\right)^2 \right\}, \quad t \in [0, T]. \qquad (2.38)$$

2. Condition (2.14) (Novikov's condition) in the statement of Theorem 2.6 holds for this Z.

2.3 Use Itô's formula to derive an SDE for Z in (2.38) and hence show that

1. Condition (2.15) holds.
2. Z is a positive $\mathcal{F}_t$-martingale.

2.4 Following the style of the discussion of boundary conditions for a European call option in Sect. 2.2, write out a motivation for the final and boundary conditions associated with a European put option:

$$\Lambda(s) = V(s, T) = \max(E - s, 0);$$

$$\lim_{s\to 0^+} V(s, t) = Ee^{-r(T-t)};$$

$$\lim_{s\to\infty} V(s, t) = 0.$$

2.5 Let S be the value of a tradeable asset paying a continuous dividend yield of D_0. The value of a European call option written on S with strike E, expiry T, risk-free rate r and volatility σ is given at time $t \in [0, T]$ by the formula

$$C(S, t) = e^{-D_0(T-t)} S N(d_{10}) - E e^{-r(T-t)} N(d_{20});$$

$$d_{10} = \frac{\log(S/E) + (r - D_0 + \frac{\sigma^2}{2})(T - t)}{\sigma\sqrt{T - t}};$$

$$d_{20} = d_{10} - \sigma\sqrt{T - t},$$

where N is defined in (2.27). Produce a variant of the user-defined Python function `euroCallBenchmark()` from Sect. 2.2.3 that accepts a parameter D specifying a dividend yield and produces a valuation according to this formula.

2.6 For European options written on the same underling asset S paying a constant dividend yield of D_0, both with the same strike price, the put-call parity relation becomes

$$P(S, t) = C(S, t) + Ee^{-r(T-t)} - Se^{-D_0(T-t)}, \quad t \in [0, T]. \tag{2.39}$$

1. Use (2.39) and the formula for $C(S, t)$ given in Problem 2.5 to derive the corresponding formula for $P(S, t)$.
2. Produce a variant of the function `euroPutBenchmark()` constructed in Sect. 2.2.3 that accepts a parameter D specifying a dividend yield and produces a valuation according to this formula.

2.7 1. Prove that for a European call option written on a non-dividend-paying asset $\Delta = N(d_1)$.
2. Use the put-call-parity relation to show that for a European put option on the same underlying asset, $\Delta = N(d_1) - 1$.

2.8 Show that, if S is a non-dividend-paying asset and E is the strike price,

1. for a European call option,

$$\lim_{t \to T^-} \Delta(t) = \begin{cases} 0, & S < E; \\ 1/2, & S = E; \\ 1, & S > E. \end{cases}$$

2. for a European put option,

$$\lim_{t \to T^-} \Delta(t) = \begin{cases} 1, & S < E; \\ 1/2, & S = E; \\ 0, & S > E. \end{cases}$$

2.9 Reproduce the dynamic hedging simulation from Sect. 2.2.4.3 over an ensemble of $M = 1000$ trajectories, and hence produce an empirical density curve for the values of $\Pi(T)$. Repeat the experiment with the number of rebalancing events set at each of $N = 2^6, 2^8, 2^{10}$ and change the value of μ over the range $\mu \in [0.1, 0.5]$. Does changing the real-world drift μ have an affect on the resulting empirical density?

2.10 Consider the value of a European call option as a function of σ only with all other parameters fixed.

1. Use (2.26)–(2.29) to show that

$$\lim_{\sigma \to \infty} d_{1,2}(\sigma) = \pm\infty.$$

2. Hence show that

$$\lim_{\sigma \to \infty} C(\sigma) = S.$$

3. In a similar way, show that

$$\lim_{\sigma \to 0^+} C(\sigma) = \max(S - Ee^{-r(T-t)}, 0).$$

2.11 Suppose that historical asset price data $S(t_i)$ is available at equally spaced times $t_i = i\delta t$, $0 \le i \le n$, and that $t_n = n\delta t$ is the present time. Define the log return of an asset on the ith timestep as

$$U_i = \ln\left(\frac{S(t_i)}{S(t_{i-1})}\right), \quad 1 \le i \le n.$$

The maximum likelihood estimator for the annualised historical volatility is

$$\sigma^* = \sqrt{\frac{1}{\delta t} \frac{1}{M} \sum_{k=1}^{M} U_{n-k+1}^2}$$

Write a Python function that takes an array of n values of S, a timestep δt, the moving window size M, and returns σ^*.

Part II

Computational Pricing Methods in the Black-Scholes Framework

Binomial Tree Methods

3

In this chapter, we introduce the binomial algorithm for pricing European options, based upon the martingale characterisation of the value of the option at time $t \in [0, T]$ as

$$V(t) = e^{-r(T-t)}\mathbb{E}\left[V_T | \mathcal{F}_t\right],$$

where the conditional expectation on the RHS is understood to be computed with respect to a risk neutral measure.

Binomial methods are relatively straightforward to implement, and flexible enough to accommodate path-dependent options such as American options and barrier options. For options written on a single underlying asset they can be efficient, but suffer from what is known as the *curse of dimensionality*, becoming increasingly computationally expensive for multi-asset options as the number of underlying assets grows large.

Using this approach, the underlying asset price process is modelled, not as a geometric Brownian motion of the form (1.11), but rather as a geometric random walk on a recombining binomial tree. We will investigate how to calibrate the parameters of this discrete model to market data. Included will be a discussion of how we construct a risk-neutral measure in this setting, and an investigation of how deviating from that measure in the calibration of our model affects the precision of the computational method.

We look at how to adjust the tree approach for underlying assets that pay dividends, and conduct an error analysis in the case of European options by comparing valuations on binomial trees with an increasing number of steps against benchmark Black-Scholes valuations. We show how to implement a modified

© The Author(s), under exclusive license to Springer Nature Switzerland AG 2024

C. Kelly, *Computation and Simulation for Finance*,
Springer Undergraduate Texts in Mathematics and Technology,
https://doi.org/10.1007/978-3-031-60575-8_3

Table 3.1 User-defined functions required in Chap. 3

Section	Function name	Function output	Parameter	Parameter meaning
1.2.4	dFac()	$e^{-r(T-t)}$	r	r
			dt	$T-t$
2.2.3	d1()	d_1	S0	$S(t)$
	d2()	d_2	E	E
	euroCallBenchmark()	$C(t, S)$	r	r
	euroPutBenchmark()	$P(t, S)$	sig	σ
			T	$T-t$

framework for the treatment of options with an early exercise facility, such as American or Bermudan options, and adapt these techniques to efficiently value both knock-out- and knock-in- style barrier options.

Python Setup To follow the Python code in this chapter, you will need to import the NumPy and PyPlot packages, as well as the `scipy.stats.norm` module.

```
In []: import numpy as np
       from scipy.stats import norm
       import matplotlib.pyplot as plt
```

We will also make use of the user-defined functions from Chaps. 1 and 2 listed in Table 3.1.

3.1 An Asset Model Evolving in Discrete-Time

First, discretise the interval $[0, T]$ into a uniform mesh of M steps, each separated from its neighbours by a time step of length $\delta t = T/M$. Since we include the first point $t = 0$ and the last point $t = T$, there are $M+1$ points in the mesh. Let $t_i = i\delta t$, for $i = 0, \ldots, M$, represent the ith meshpoint. We will only consider asset values at these times.

3.1.1 The Recombining Random Walk

Let S_0 be the initial asset value and suppose that, over each step of length δt, the asset will either move up by a factor of u or down by a factor of d. These movements will occur with probability p and $1 - p$ respectively, where $p \in (0, 1)$, and they are mutually independent over disjoint steps. This leads to the recombining tree structure illustrated in Fig. 3.1. Notationally, we define the asset value at time t_i on a trajectory where the asset has moved up n times and down $i - n$ times to be

$$S_n^i = d^{i-n} u^n S_0, \quad n = 0, \ldots, i, \quad i = 0, \ldots, M,$$

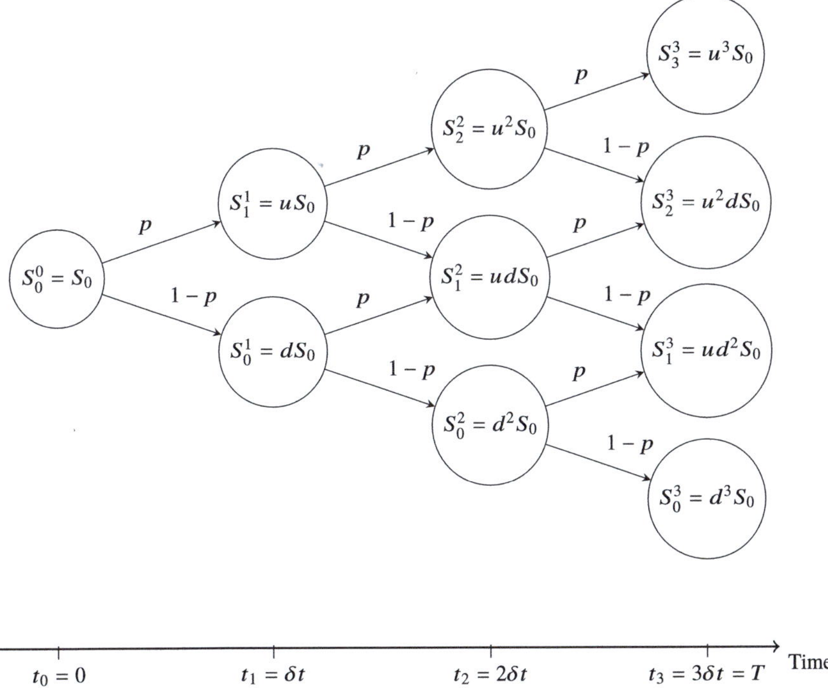

Fig. 3.1 Recombining tree structure of the binomial asset price model with $M = 3$

so that $S_0 := S_0^0$. At each point t_i there are $i + 1$ possible values of S_n^i, each corresponding to a different number of upward steps taken:

$$\left\{ S_0^i, S_1^i, \ldots S_{i-1}^i, S_i^i \right\}.$$

3.1.2   Calibrating the Model Parameters

The next question is how to select the model parameters (u, d, p). Tracking the framework described in Chap. 2 we are guided by the following considerations:

1. the values of u and d should be consistent with the no-arbitrage principle,
2. p should be chosen so that expectations are computed with respect to the risk-neutral measure,
3. the binomial tree asset model aligns to some degree with the risk-neutral form of the Black-Scholes asset model where $r > 0$ is the risk-free rate, and $\sigma > 0$ is the volatility.

3.1.2.1 Constraints on u and d Imposed by the No-Arbitrage Principle

The asset price process is characterised by the triple (u, d, p), and it is natural to take $d < u$, but in order not to introduce arbitrage opportunities we also require that $u > e^{r\delta t}$ and $d < e^{r\delta t}$, where r is the risk-free rate, and $p \in (0, 1)$.

To see this, consider that if $d \geq e^{r\delta t}$ then $dS_0 \geq e^{r\delta t}S_0$, and so the asset price after a decline is no less than would be achieved by investment at the risk-free rate. In the case where $d > e^{r\delta t}$, one could finance the purchase of the asset using funds by borrowing S_0 at rate r and sell after one timestep of length δt to lock in a risk-free profit of at least $(d - e^{r\delta t})S_0 > 0$. In the worst case, where $d = e^{r\delta t}$, one has probability $p > 0$ of making a profit $(u - e^{r\delta t})S_0 > 0$ and still no possibility of loss.

Similarly, if $u \leq e^{r\delta t}$ then $uS_0 \leq e^{r\delta t}S_0$ and we could take a short position to the value of S_0 in the asset and lend the funds thus raised at the risk-free rate, again leading to a risk-free profit of at least $(e^{r\delta t} - u)S_0 > 0$ in the case where $u < e^{r\delta t}$. If $u = e^{r\delta t}$ then there is probability $1 - p > 0$ of making a profit $(e^{d\delta} - d)S_0 > 0$, with again no possibility of loss.

3.1.2.2 A Risk-Neutral Choice of p

Suppose that $S^{(n)}$ represents the (random) value of the underlying asset at time $t = n\delta t$, regardless of the model used to describe it. The discounted price process should satisfy the martingale property

$$\mathbb{E}\left[e^{-r\delta t}S^{(n+1)}|S^{(n)}\right] = S^{(n)},$$

or equivalently, since r is a deterministic constant,

$$\mathbb{E}\left[S^{(n+1)}|S^{(n)}\right] = e^{r\delta t}S^{(n)}. \tag{3.1}$$

If $S^{(n)}$ follows the binomial model described in Sect. 3.1 with parameters (u, d, p) then the first conditional moment over a single step takes the form

$$\mathbb{E}\left[S^{(n+1)}|S^{(n)}\right] = (pu + (1 - p)d)S^{(n)}. \tag{3.2}$$

How can we can select p so that expectations are computed under the risk-neutral measure, ensuring the martingale property in the discounted price process for this model? Comparing (3.1) and (3.2) we see that this can be achieved if $pu + (1 - p)d = e^{r\delta t}$, so that

$$p = \frac{e^{r\delta t} - d}{u - d}. \tag{3.3}$$

The pair of probability weights $\{p, 1 - p\}$ characterise the risk-neutral probability measure in this discrete-time setting. It is important to emphasise that this is not the same measure as the risk-neutral probability $\mathbb{Q}$ constructed in Chap. 2 for the Black-Scholes asset model.

It's tempting to move directly on to the question of how to choose u and d but let's take a little time first to explore the relationship between this risk-neutral probability p, and the empirical real-world probability of an upward movement p^* and downward movement $1 - p^*$.

3.1.2.3 Change-of-Measure for the Discrete-Time Model

In principle, the pair $\{p^*, 1 - p^*\}$ could be estimated from historical prices by counting the frequency with which our asset makes a net upward movement over the intervals between meshpoints. As we saw in Chap. 2, we cannot use it to price the derivative, since the resulting value would deviate from that of the replicating portfolio (which must be equal to the value of the option, to avoid an arbitrage opportunity). Nonetheless, just as we did in Sect. 2.1.5 for the Black-Scholes model, we can explicitly link the real-world probability measure to the risk-neutral probability measure, characterised by the pair $\{p, 1 - p\}$, and construct a method of moving between them.

Define a new random variable Z which takes the possible values

$$(z_1, z_2) = \left(\frac{p}{p^*}, \frac{1 - p}{1 - p^*} \right).$$

If we denote the expectation taken under the real-world probability weights $\{p^*, 1 - p^*\}$ as $\mathbb{E}_*$, then

$$\mathbb{E}_*[Z] = z_1 p^* + z_2(1 - p^*) = \frac{p}{p^*} \cdot p^* + \frac{1 - p}{1 - p^*} \cdot (1 - p^*) = p + 1 - p = 1.$$

The conditional expectation of the product of Z and the discounted asset price over one step is

$$\mathbb{E}_* \left[Z e^{-r\delta t} S^{(n+1)} | S^{(n)} \right] = e^{-r\delta t} z_1 u S^{(n)} p^* + e^{-r\delta t} z_2 d S^{(n)}(1 - p^*)$$

$$= \left(\frac{p}{p^*} u p^* + \frac{(1 - p)}{1 - p^*} d(1 - p^*) \right) e^{-r\delta t} S^{(n)}$$

$$= (pu + (1 - p)d)e^{-r\delta t} S^{(n)}$$

$$= \mathbb{E} \left[e^{-r\delta t} S^{(n+1)} | S^{(n)} \right].$$

The final expectation on the RHS is taken under the risk-neutral weights $\{p, 1 - p\}$ and is denoted simply $\mathbb{E}$. Constructed in this way, Z is very useful as it allows us to change from an expectation under the real-world measure to one under the risk-neutral measure by a simple multiplication of our discounted asset price process. In a more general setting it corresponds to the Radon-Nikodym derivative introduced in Sect. 2.1.5. We will return to this idea in the next chapter when we apply the technique of importance sampling for Monte Carlo approximation, using a similar change-of-measure.

3.1.2.4 Moment Consistency with the Black-Scholes Model

There are various ways to construct our binomial model for consistency with Black-Scholes and we only present a few here. The first approach seeks to choose u and d so that the first two moments of the binomial model coincide with those of the Black-Scholes model.

Recall that the Black-Scholes asset price model is given by

$$S(t) = S_0 e^{(r-\sigma^2/2)t + \sigma W^{\mathbb{P}}(t)}, \quad t \in [0, T], \tag{3.4}$$

where $W^{\mathbb{P}}$ is a standard Brownian motion under the risk neutral measure, r is the risk-free rate and σ is the volatility. Suppose that $S^{(n)}$ is modelled via (3.4) with $S^{(n)} = S(n\delta t)$. A similar calculation to that required by Exercise 1.2 in Chap. 1, tells us that the first two moments of $S^{(n+1)}$ conditional upon $S^{(n)}$ are given by

$$\mathbb{E}_{\mathbb{Q}}\left[S^{(n+1)}\big|S^{(n)}\right] = e^{r\delta t}S^{(n)}; \quad \mathbb{E}_{\mathbb{Q}}\left[\left(S^{(n+1)}\right)^2\big|S^{(n)}\right] = e^{(2r+\sigma^2)\delta t}\left(S^{(n)}\right)^2.$$

$$\tag{3.5}$$

Under the binomial model, the first two moments of $S^{(n+1)}$ conditional upon $S^{(n)}$ are given by

$$\mathbb{E}\left[S^{(n+1)}\big|S^{(n)}\right] = (pu + (1-p)d)S^{(n)},$$

$$\mathbb{E}\left[\left(S^{(n+1)}\right)^2\big|S^{(n)}\right] = (pu^2 + (1-p)d^2)\left(S^{(n)}\right)^2.$$

Equating these moments to those given in (3.5) leads, after some cancellation, to

$$pu + (1-p)d = e^{r\delta t};$$

$$pu^2 + (1-p)d^2 = e^{(2r+\sigma^2)\delta t}.$$

Since p has already been identified in terms of u and d in (3.3), this is a system of two equations with two unknowns, and admits the solution

$$u = A + \sqrt{A^2 - 1}, \quad d = A - \sqrt{A^2 - 1}, \quad A = \frac{1}{2}\left[e^{(r+\sigma^2)\delta t} + e^{-r\delta t}\right]. \tag{3.6}$$

It is easy to check that $ud = 1$, and so the possible asset values at step i have the convenient shortcut representations

$$S_n^i = u^{2n-i}S_0 = d^{i-2n}S_0. \tag{3.7}$$

3.1.2.5 Cox-Ross-Rubenstein Parameters

A slightly different approach to choosing (u, d, p) is that of the original Cox-Ross-Rubenstein method [17], given by

$$u = e^{\sigma\sqrt{\delta t}}; \quad d = e^{-\sigma\sqrt{\delta t}}; \quad p = \frac{e^{r\delta t} - d}{u - d}. \tag{3.8}$$

You can see that the risk-neutral value of p is maintained here, but u and d are chosen in a way that mimics the variance of log-asset values under the Black-Scholes model over each step, which is $\sigma^2 \delta t$, rather than trying to match the first two moments precisely. It is still the case that $ud = 1$, so that (3.7) holds, and in fact the distribution of the binomial model with these values converges to that of the Black-Scholes model as the value of M grows large.

3.1.3 Deviation from the Risk-Neutral Choice of p

Even if p does not preserve risk-neutrality over each step of the tree, it is nonetheless possible to choose (u, d, p) so that the distribution of asset price values at expiry converges in distribution to that of the Black-Scholes model as M grows large. Select

$$u = e^{\sigma\sqrt{\delta t}+(r-\sigma^2/2)\delta t}; \quad d = e^{-\sigma\sqrt{\delta t}+(r-\sigma^2/2)\delta t}; \quad p = \frac{1}{2}. \tag{3.9}$$

We will refer to (3.9) as the *lognormal* parameter choice. Exercise 3.5 asks you to show that this version of the asset model does not lead to computation of expectation under the risk neutral measure, and so in Sect. 3.4 we explore the nature of the error thus introduced. Exercise 3.6 asks you to determine an upper bound on the size of the time step δt so that the no-arbitrage condition $d < e^{r\delta t} < u$ is satisfied.

We can write the final value of the asset price process as

$$S_T = S^{(M)} = S^{(M-1)} e^{(r-\sigma^2/2)\delta t + \sigma\sqrt{\delta t}Y_{M-1}},$$

where Y_{M-1} is a random variable taking values $\{-1, 1\}$ with equal probability, so that $\mathbb{E}[Y_{M-1}] = 0$ and $\text{Var}[Y_{M-1}] = 1$, for all M. Here we have matched the expected returns from the binomial asset model to those of the Black-Scholes model.

Iterating back to the initial price we can write

$$S^{(M)} = S_0 \prod_{i=0}^{M-1} \left[e^{(r-\sigma^2/2)\delta t + \sigma\sqrt{\delta t}Y_i} \right].$$

Taking the natural logarithm of both sides,

$$\ln\left(\frac{S^{(M)}}{S_0}\right) = \sum_{i=0}^{M-1} \left[(r - \sigma^2/2)\delta t + \sigma\sqrt{\delta t}Y_i \right]. \tag{3.10}$$

The right-hand-side (RHS) of (3.10) is a sum of independent and identically distributed (i.i.d.) random variables, each with expectation $(r - \sigma^2/2)\delta t$ and variance $\sigma^2 \delta t$. So for small δt and hence large M, we apply the central limit theorem to get in the large-M limit

$$\ln\left(\frac{S^{(M)}}{S_0}\right) \sim \mathcal{N}\left(M(r - \sigma^2/2)\delta t, \, M\sigma^2 \delta t\right),$$

But $T = M\delta t$ and therefore the sequence of random variables $\ln(S^{(M)}/S_0)$ converges in distribution to a normally distributed random variable with mean $(r - \sigma^2/2)T$ and variance $\sigma^2 T$. Since $S_T = S^{(M)}$ for all M, we can write in the large-M limit

$$S_T \sim S_0 e^{(r - \sigma^2/2)T + \sigma\sqrt{T}Z}, \quad Z \sim \mathcal{N}(0, 1).$$

In distribution, this is the same as the Black-Scholes asset model (3.4) under the risk neutral measure, since a standard Brownian motion satisfies $W^{\mathbb{Q}}(T) \sim \mathcal{N}(0, T)$. Note that here,

$$ud = e^{(2r - \sigma^2)\delta t},$$

and therefore $ud \leq 1$ iff $2r - \sigma^2 \leq 0$ with equality when $2r = \sigma^2$, and $ud > 1$ iff $2r - \sigma^2 > 0$.

3.1.4 Dynamical Consistency Between Discrete- and Continuous-Time Models

Using the choice of (u, d, p) from Sect. 3.1.3, it is instructive to look at the change in log-asset prices over two steps of the tree, particularly in the scenario where exactly one upward step and one downward step have been taken. Consider

$$\ln\left(S_1^2\right) = \ln(udS_0) = \ln(ud) + \ln(S_0).$$

So $\ln\left(S_1^2\right) > \ln(S_0)$ if $ud > 1$ (equivalently, if $r - \sigma^2/2 > 0$) and $\ln\left(S_1^2\right) < \ln(S_0)$ if $ud < 1$ (equivalently, if $r - \sigma^2/2 < 0$). The drift of the log-asset value defined by the Black-Scholes model, given over a single step as

$$\ln(S(\delta t)) = \ln(S_0) + (r - \sigma^2/2)\delta t + \sigma B(\delta t), \tag{3.11}$$

is now reflected as a skew in the structure of the tree itself. See Fig. 3.2 for an illustration. Notice that it is possible for $r > 0$ and still have $r - \sigma^2/2 < 0$, so our tree will skew downwards even though under the risk-neutral measure for the Black-Scholes model $\mathbb{E}[S(\delta t)] = S_0 e^{r\delta t}$ (notice the positive exponent in the form

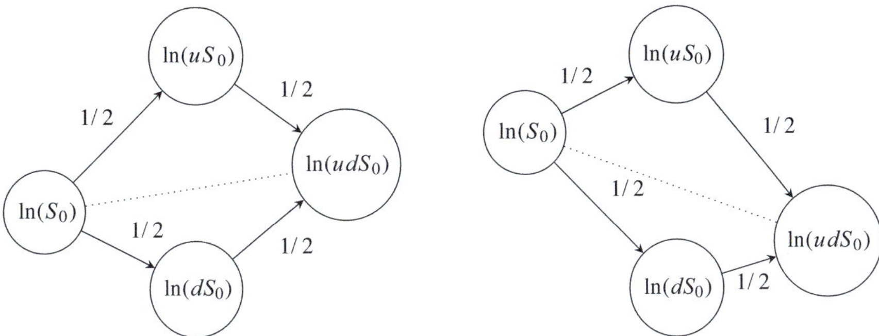

Fig. 3.2 Detail of the binomial model showing structural drift in the log-asset values when $p = 1/2$. On the left, $2r - \sigma^2 > 0$ so that $ud > 1$. On the right, $2r - \sigma^2 < 0$ so that $ud < 1$

of the expected value). The directional skew built into the structure of the tree is instead reflective of the asymptotic behaviour of trajectories of the Black-Scholes asset model, since we are equally likely to take an upward step as a downward step. To see this, suppose that our model is defined without any expiry date, so that there is no upper limit on the time domain:

$$S(t) = S_0 e^{(r - \sigma^2/2)t + \sigma B(t)}, \quad t \in [0, \infty)$$

Then

$$\frac{1}{t} \ln(S(t)) = \frac{\ln(S_0)}{t} + (r - \sigma^2/2) + \sigma \frac{B(t)}{t} \quad t \in [0, \infty),$$

and since the Strong Law of Large Numbers implies that $\lim_{t \to \infty} B(t)/t = 0$ a.s. (see, for example, Mao [53, Theorem 3.4, Chapter 1]), we have

$$\lim_{t \to \infty} \frac{1}{t} \ln(S(t)) = r - \frac{1}{2}\sigma^2, \quad a.s.$$

The value on the RHS captures the common asymptotic rate of exponential growth or decay of the model across almost all trajectories. If we consider that

$$\lim_{t \to \infty} \frac{1}{t} \ln(e^{\lambda t}) = \lambda,$$

then we can see that trajectories of S will have a tendency to decay exponentially over long time periods if $2r - \sigma^2 < 0$ (when the binomial model has a tree with a downward skew), and to grow exponentially if $2r - \sigma^2 > 0$ (when the binomial model has a tree with an upward skew). In the case where $2r = \sigma^2$, S is a recurrent process on $\mathbb{R}^+$ in the sense that

$$\liminf_{t \to \infty} S(t) = 0; \quad \limsup_{t \to \infty} S(t) = \infty.$$

Fig. 3.3 Detail of the binomial model showing structural symmetry in the log-asset values when $p = (e^{r\delta t} - d)/(u - d)$

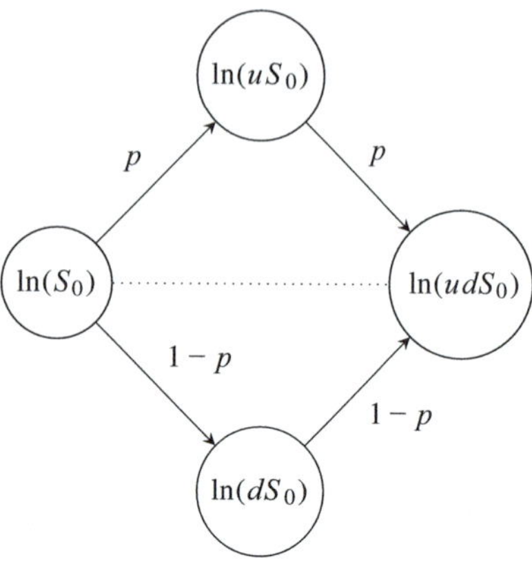

By contrast, consider the case where $p = (e^{r\delta t} - d)/(u - d)$, so that the discounted asset process is a martingale. Then $ud = 1$ and (since $\ln(1) = 0$)

$$\ln\left(S_1^2\right) = \ln(ud) + \ln(S_0) = \ln(1) + \ln(S_0) = \ln(S_0).$$

The log-asset values form a tree that is symmetric about $\ln(S_0)$: this is illustrated in Fig. 3.3. The drift of the Black-Scholes log-asset values given by (3.11) is not apparent here, but instead we rely on the effect of p on the weighting of trajectories through the tree to ensure consistency with the Black-Scholes model.

3.2   The Binomial Algorithm

Now that we have established a set of discrete models for the underlying asset, we can describe the *binomial algorithm for option valuation*. We start by presenting it for a general option with payoff Λ that depends only on the value of the underlying asset at expiry and for which there is no early exercise facility. Later in the chapter we extend the algorithm for the valuation of more complex options.

3.2.1   Mathematical Characterisation

Suppose we wish to value an option written on an underlying asset S with known initial value $S_0^0 = S_0$, expiry T, and a payoff Λ that depends only on the final value of S and for which there are no early exercise opportunities. Suppose the asset model evolves over a uniform mesh with

- M independent steps, each of length δt,
- at each step, upward movements occurring with probability p by a factor of u,

- at each step, downward movements occurring with probability $1 - p$ by a factor of d.

Then the binomial algorithm is characterised by the following relations:

1. $S_n^i = d^{i-n} u^n S_0^0$, for $0 \le n \le i$;
2. $V_n^M = \Lambda(S_n^M)$, for $0 \le n \le M$;
3. $V_n^i = e^{-r\delta t}[p V_{n+1}^{i+1} + (1-p) V_n^{i+1}]$, for $0 \le n \le i$ and $0 \le i \le M - 1$.

So the structure of this algorithm is naturally divided into three stages. The first stage describes the evolution of the asset price process, and computes the set of all possible values that the asset value at expiry $S^{(M)}$ could take. The second stage applies the option payoff to the vector of final asset values. The third stage moves backwards through the tree, recursively computing the discounted expected value of the option at each node, until the initial price V_0^0 is reached.

In this setting, it is also possible to compute the initial value V_0^0 directly without working backwards through the tree one step at a time via the binomial sum

$$V_0^0 = e^{-rT} \sum_{i=0}^{M} \binom{M}{i} p^i (1-p)^{M-i} \Lambda \left(S_0^0 u^i d^{M-i} \right).$$

However we will show the general implementation of the algorithm stage by stage here, so that it is easily adapted to more complex problems such as the valuation of path-dependent options, an example of which is covered later in the chapter.

3.2.2 Python Implementation for a European Option

Let's implement the binomial algorithm to value a European put option written on a non-dividend-paying asset with current price $S_0 = 10$, strike price $E = 10$, expiry $T = 1$, $r = 0.04$, and $\sigma = 0.5$. Before constructing the binomial model of the asset price process, we choose how we will compute (u, d, p) from the possibilities introduced in Sect. 3.1.2, and we must also decide on how many steps M should be in our mesh.

We will demonstrate the implementation of each stage of the binomial algorithm in turn, using the Cox-Ross-Rubenstein values for the model parameters (u, d, p), and initially taking $M = 4$.

3.2.2.1 Stage 1: Evolution of the Asset Price Process
Set up the model and discretisation parameters as follows:

```
In []: # Option parameters
        S0=10; E=10; T=1; r=0.04; sig=0.5
```

```
# Mesh parameters
M=4; dt=T/M

# Cox-Ross-Rubenstein values for binomial model
u=np.exp(sig*np.sqrt(dt)); d=np.exp(-sig*np.sqrt(dt))
p=(np.exp(r*dt)-d)/(u-d)
```

We can take advantage of vectorisation in Python to compute the asset values at any time t_i. Suppose for example that we wish to look at asset values at time t_2, then

$$
\begin{aligned}
[S_0^2, S_1^2, S_2^2] &= [d^2 S_0, ud S_0, u^2 S_0] \\
&= [d^2 S_0, d S_0, S_0] * [1, u, u^2] \\
&= (S_0 * [d^2, d, 1]) * [1, u, u^2] \\
&= (S_0 * d * *[2, 1, 0]) * (u * *[0, 1, 2])
\end{aligned}
$$

In Python, this looks like

```
In []: S0*d**np.arange(2,-1,-1)*(u**np.arange(0,3,1))
```

Generalising to an M-step tree, it will be convenient for the remainder of the chapter to set up two vectors corresponding to all powers of d and u applied to S_0:

$$
\begin{aligned}
\text{upwr} &= [1, u, u^2, \ldots, u^{M-1}, u^M] \\
\text{dpwr} &= [d^M, d^{M-1}, \ldots, d^2, d, 1]
\end{aligned}
$$

We can then compute the final asset values (at step M) in a single line of code as follows:

```
In []: # Stage 1 implementation: asset values at step M
       upwr=u**np.arange(0, M+1)
       dpwr=d**np.arange(M,-1,-1)

       assetM=S0*dpwr*upwr
```

and this will be sufficient for European-style options.

3.2.2.2 Stage 2: Option Value at Expiry

Here, we compute the payoff $V_T = \Lambda(s)$ for each possible asset price s at expiry, and store in an array variable. For example, for a European put option with strike E, the payoff is $\Lambda(s) = \max\{E - s, 0\}$, and computed at each possible asset value as

```
In []: # Stage 2 implementation: apply payoff for European put
       putVals=np.maximum(E-assetM,0)
```

Following the notational convention used to define asset values S_n^i, the value of the option at expiry is denoted $V_n^M = \Lambda(S_n^M)$, and our overarching goal is to compute the value of the option at the outset, denoted V_0^0.

3.2.2.3 Stage 3: Backwards Iteration of Option Values to Initial Value

Work back through the binomial tree, at each step computing the discounted expected value associated with each possible asset price, to determine the initial value of the option. Recall that for any observed asset value s and any time $t \in [0, T]$ we have

$$V(s, t) = \mathbb{E}\left[e^{-r(T-t)} \max\{E - S(T), 0\} \,\middle|\, S(t) = s \right].$$

At each possible asset value S_n^i in the tree, the next step is determined by the outcome of a Bernoulli trial with success probability p and therefore we may compute V_n^i recursively as

$$V_n^i = e^{-r\delta t}[p V_{n+1}^{i+1} + (1 - p) V_n^{i+1}]. \tag{3.12}$$

To see how this works in practice, suppose that $M = 2$. After Stage 2 the variable putValues will contain the values at expiry represented in the array $[V_0^2, V_1^2, V_2^2]$. To compute $[V_0^1, V_1^1]$ we would overwrite the variable putValues with

```
In []: valsUp=putVals[range(1,3)]
       valsDown=putVals[range(0,2)]
       putVals=dFac(r,dt)*(p*valsUp+(1-p)*valsDown)
```

To compute V_0^0 we would overwrite again with

```
In []: valsUp=putVals[range(1,2)]
       valsDown=putVals[range(0,1)]
       putVals=dFac(r,dt)*(p*valsUp+(1-p)*valsDown)
```

Now construct a **for** loop to handle the recursion for general M and, after it has completed, output the initial value of the option:

```
In []: # Stage 3 implementation: backwards iteration
       for i in range(int(M)-1,-1,-1):
           valsUp=putVals[range(1,i+2)]
           valsDown=putVals[range(0,i+1)]
           putVals=dFac(r,dt)*(p*valsUp+(1-p)*valsDown)

       print(putValues[0])
```

Exercise 3.2 asks you to consolidate the code from this section into a function called `euroPutTree()` that returns the value of a European put option written on an underlying asset with current value S_0, strike E, risk-free rate r, volatility σ, time to expiry T, using a binomial tree method with M steps.

3.3 Modifications for Options Written on Dividend-Paying Assets

In theory, the value of a share in the company is the value of all future dividend payments discounted to the present. If the asset distributes a continuous dividend yield then this must be factored into the SDE asset price model as a reduction in the intrinsic underlying growth rate described by the drift coefficient. Alternatively, if S is the price in cents of a share of stock immediately before payment of a one-off dividend d of cents per share, the asset price upon payment should undergo a jump to $S - d$ cents per share. In either case, to incorporate dividends into our pricing algorithm, we must modify the asset price model.

3.3.1 Assets Paying a Continuous Dividend Yield

If we assume that there is a continuous dividend yield of D_0 paid on the underlying asset, how can we adjust the binomial tree model? In the Black-Scholes model the effective drift coefficient becomes $r - D_0$, leading to the corresponding asset model

$$S(t) = S_0 e^{(r-D_0-\sigma^2/2)t+\sigma B(t)}, \quad t \in [0, T].$$

Since we are motivated to set up our binomial asset model to be consistent with this, our selections of (u, d, p) can be similarly modified. For example in Stage 1 using the Cox-Ross-Rubenstein parameters we choose

$$u = e^{\sigma\sqrt{\delta t}}; \quad d = e^{-\sigma\sqrt{\delta t}}; \quad p = \frac{e^{(r-D_0)\delta t} - d}{u - d}.$$

Here, the drag of the dividend yield on the drift is reflected in the choice of probability p, and Exercise 3.4 asks you to show that the discounted asset price model is a martingale with respect to its natural filtration under this choice of p.

For the lognormal parameters given in Sect. 3.1.3, the tree structure itself captures this drag and so we would use

$$u = e^{\sigma\sqrt{\delta t}+(r-D_0-\sigma^2/2)\delta t}; \quad d = e^{-\sigma\sqrt{\delta t}+(r-D_0-\sigma^2/2)\delta t}; \quad p = \frac{1}{2}.$$

In either case, payment of the dividend affects only the asset price model, and not the discounting in Stage 3, which is still computed using the discount factor $e^{-r\delta t}$.

3.3.2 Assets Paying a Discrete Dividend

What if the asset pays a dividend on a specific date? This is the most common scenario for options written on publicly traded shares. Consider the case of a European option due to expire in one year on an underlying asset that pays a single dividend at the 6 month mark with a yield of $D_1 \in (0, 1)$. We need to make a once-off adjustment to the asset prices at the dividend date, and this requires a change to Stage 1 of the pricing algorithm. First, to ensure there is a meshpoint at the dividend date we would choose an even value for M, so that the dividend is paid at time $t_{M/2+1}$ precisely. Suppose $M = 4$, then at time t_2 if there were no dividend the asset would take the values $[d^2 S_0, ud S_0, u^2 S_0]$. However since the dividend is paid on that date the asset values will shift downwards to $[d^2 S_0(1 - D_1), ud S_0(1 - D_1), u^2 S_0(1 - D_1)]$. After that we continue to build the tree as before and the asset values at the final time $T = t_4$ would be

$$[d^4 S_0(1 - D_1), d^3 u S_0(1 - D_1), d^2 u^2 S_0(1 - D_1), du^3 S_0(1 - D_1), u^4 S_0(1 - D_1)].$$

See Fig. 3.4 for a schematic illustration of the effect of a single dividend payment on the structure of a binomial tree.

Here we see a nice consequence of modelling with a geometric random walk. Since all factors commute, it doesn't matter to the final asset values exactly when the dividend was paid, as long as it was paid at a meshpoint. So the only change to Stage 1 is to compute the final time asset values as

```
In []: divAssetM=S0*(1-D1)*d**dpwr)*(u**upwr)
```

Note that for a discrete dividend this modification is enough: D_1 does not affect our choice of (u, d, p) and these are unchanged. Also, if there are two dividends with a yields of D_1 and D_2 at two different dividend dates (as might be the case for dividends paid semi-annually), we would replace the factor $(1 - D_1)$ in the above with $(1 - D_1)(1 - D_2)$. Nonetheless, we will see in Sect. 3.5 that in the case of American options, and indeed other path-dependent options, Stage 3 of the valuation depends on the asset values prior to expiry. For such options the dates at which dividends are paid do need to be tracked.

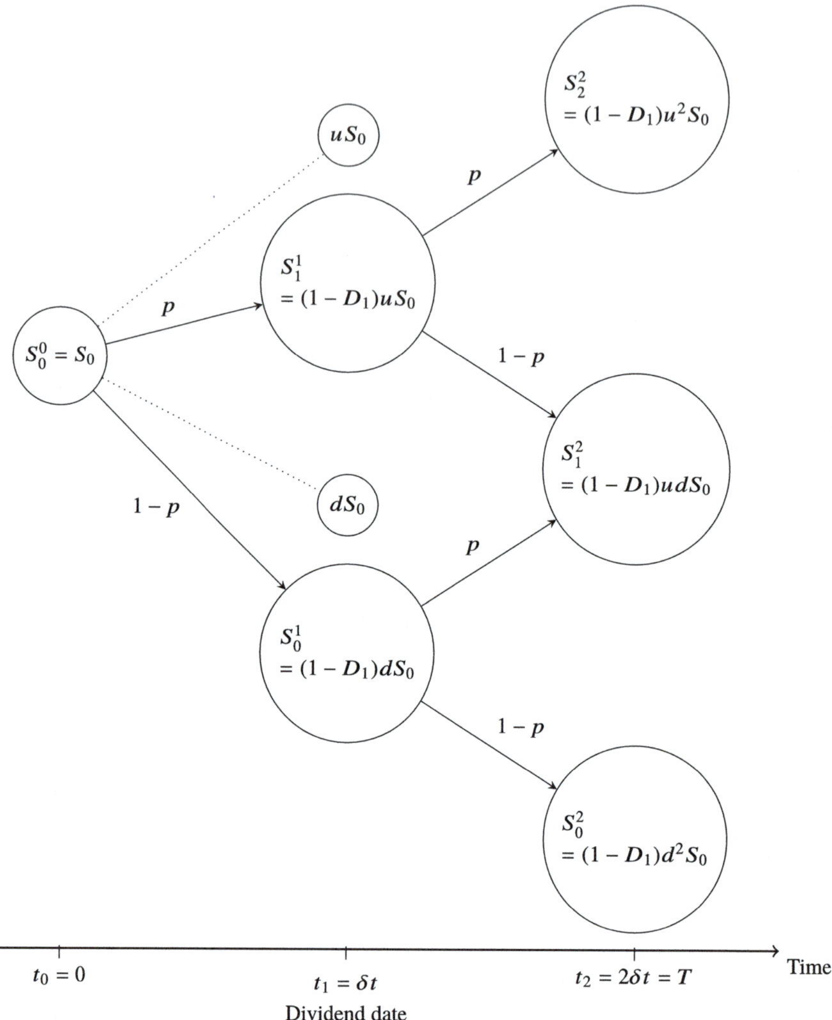

Fig. 3.4 Effect of a discrete dividend payment at 6 months ($t = \delta t$) on a binomial tree model with $T = 1$ and $M = 2$. The nodes labelled uS_0 and dS_0, which represent the values the asset would take in the absence of a dividend payment, are disregarded

3.4   Error Analysis

If we view the binomial model as a discrete approximation to a continuous market, then we may prefer to use the choices of (u, d, p) introduced in Sect. 3.1.2.2, as they maintain the martingale property of the discounted asset price process over each step of the tree. We might object to the lognormal values introduced in Sect. 3.1.3, since the corresponding probability $p = 1/2$ does not preserve the martingale property, and so in Stage 3 of the algorithm we are not calculating expectations with respect to a risk-neutral probability measure.

However if we are prepared to take the viewpoint of the numerical analyst, deviations from the risk-neutral framework may be seen as a bias, and part of the overall option price error induced by the discretisation. In this situation it makes sense to ask how that error behaves as the number of steps in the tree M grows large. If we observe *convergence*, so that the error tends to zero as we increase M, then in some sense risk-neutrality is recovered in the limit and in practice we can suppress deviations from a mathematically acceptable discrete market model by increasing the resolution of our mesh. Let's explore this further.

3.4.1 Defining the Error for European Options

Let $P_{BS}(S_0, 0)$ be the benchmark value of a European put option with strike price E produced by the Black-Scholes formula, and let $P_{BT}^M(S_0, 0)$ be the approximate value of the same option produced using a binomial model with M steps. The *absolute error of the binomial tree method* for a given number of steps M is

$$e_M = \left| P_{BS}(S_0, 0) - P_{BT}^M(S_0, 0) \right|.$$

To compute this error, we make use of the functions euroPutBenchmark() constructed in Sect. 2.2.3 of Chap. 2, and euroPutTree() constructed in Exercise 3.2.

For any given set of parameter values we can now compute the error as

```
In []: bMark=euroPutBenchmark(E,S0,r,sigma,T)
       treeVal=euroPutTree(E,S0,r,sigma,T,M)
       np.abs(bMark-treeVal)
```

3.4.2 Convergence Plots

To investigate convergence, we generate a plot of these errors against M for $M = 100, \ldots, 400$:

```
In []: # Compute benchmark value from Black-Scholes formula
       bMark = euroPutBenchmark(E,S0,r,sig,T)

       # Instantiate array to contain error values
       M=np.arange(100,401)
       error=np.ones(len(M))

       # Compute error for M=100,...,400
       for i in range(M[0],M[-1]):
```

```
treeVal=euroPutTree(E,S0,r,sigma,T,i)
error[i-100]=np.abs(bMark-treeVal)

# Plot array of errors against M
plt.plot(M,error)
```

In Fig. 3.5, we show values of e_M for a European put option with small and large values of the volatility parameter σ, comparing in each case a tree model that preserves risk neutrality (where (u, d, p) satisfy (3.6) and (3.3)) against one that does not (where (u, d, p) satisfy (3.9)).

Notice that, in all cases as M increases, the error tends to decrease towards zero, and so convergence appears to be taking place. This is the case for both parameter sets, even if $p = 1/2$ does not lead to the computation of expectations under the risk-

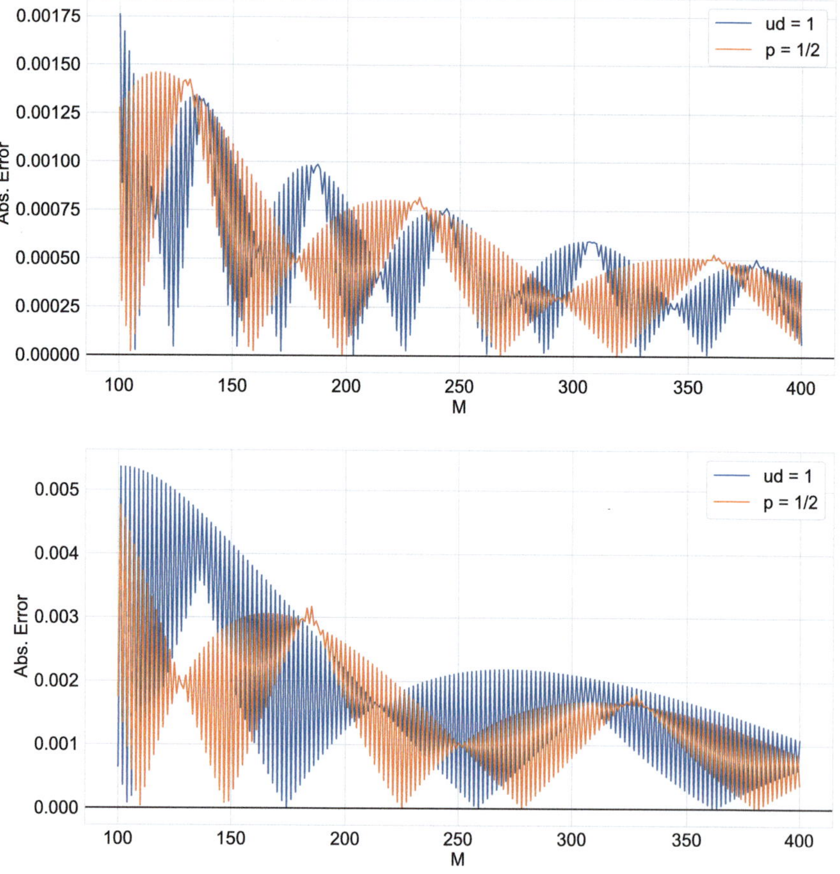

Fig. 3.5 Plots of absolute error against M for a European put option with $S_0 = 19$, $E = 20$, $r = 0.03$, $T = 1$, and no dividend yield. On the top $\sigma = 0.1$, on the bottom $\sigma = 0.3$

neutral probability measure. This convergence is non-monotone, and characterised by sawtooth oscillations between small and large error values within a wavelike bounding envelope.

We can see that it is not automatically the case that by choosing a larger value of M we can reduce the deviation from the benchmark Black-Scholes value. The error might go up before it goes down. Moreover as σ increases, the amplitude of the outer bounding envelope tends to be larger, and the length of each oscillation longer. It is clear that, from the point of view of convergence to the benchmark value, both models appear to eventually converge as M grows large, and neither of the two models is consistently preferable to the other. In fact, it can be shown that there exists a constant K such that $e_M \leq K/M$. This unusual oscillatory behaviour can be motivated by recasting the binomial asset model as an explicit finite difference approximation of solutions of the Black-Scholes PDE that is on the cusp of instability, and we return to this characterisation at the end of Chap. 5.

3.5 The Binomial Algorithm for Options with an Early Exercise Facility

3.5.1 American and Bermudan Options

American and Bermudan options differ from European options in that the holder has the right to exercise prior to the expiry date. The holder of an American option can exercise at any time $t \in [0, T]$, whereas the holder of a Bermudan option can exercise only at a finite number of predetermined times $\{0 = t_0, t_1, \ldots, t_N = T\}$. Extra rights may enhance the value of the option to the holder and so the binomial algorithm requires modification.

The exception to this is that of an American call option written on a non-dividend paying asset, which can be valued in the same way as its European counterpart. To see this consider the viewpoint of the holder of such an option at time $t \in (0, T)$ if $S(t) > E$. If they choose to exercise now, they will receive the value of $S(t) - E$. However, they could alternatively take a short position in the asset receiving payment of $S(t) > E$ at time t, then repurchase at time T. Since they held on to the option, they can do this either by exercising the option, buying at the strike E, or if $S(T) < E$ buying at the market price. The net gain from this strategy will be $S(t) - \min(E, S(T)) \geq S(t) - E$. Exercising the American call early will never be a better strategy and so the option can be valued as though the facility to do so doesn't exist.

3.5.2 The Dynamic Programming Representation of Option Value

Since the binomial model is discrete we will approximate the American case if we only allow early exercise at a finite number of times with $t_{j+1} - t_j = \delta t \ll 1$. The same method applies for Bermudan options that may be exercised only at the meshpoints. An effective valuation technique for either must approximate the optimal exercise strategy. In order to understand what this is, we introduce the following recursive representation of the value of an American option.

Suppose that we make no assumptions about our asset price model S, and let $V_i(s)$ be the value of an American option at time t_i, given $S(t_i) = s$, for any $s \in \mathbb{R}^+$, and assuming the option has not previously been exercised. The *dynamic programming representation* for the initial value $V_0(S_0)$ is defined as

$$V_N(s) = \Lambda(s);$$

$$V_{i-1}(s) = \max\left\{\Lambda(s), e^{-r\delta t}\mathbb{E}\left[V_i(S(t_i))|S(t_{i-1}) = s\right]\right\}, \quad i = 1, \ldots, N.$$

$$(3.13)$$

In this representation, the holder of the option must decide whether or not to exercise at each time t_i for $i = 1, \ldots, N-1$, and each of these decisions is made by comparing the value gained by exercising immediately to the value of all subsequent decisions if the option is not exercised at that moment.

Equation (3.13) is called the *Bellman equation* for the problem, and the second term in the maximum on the RHS is called the *continuation value*, denoted

$$C_i(s) := e^{-r\delta t}\mathbb{E}\left[V_i(S(t_i))|S(t_{i-1}) = s\right].$$

The *optimal exercise time* can be shown to be

$$\tau = \min\left\{i \ : \ \Lambda(S(t_i)) > C_i(S(t_i))\right\}.$$

Note that since τ is path-dependent, it is a random variable, and indeed is a stopping time with respect to the filtration generated by S: see Definition 1.4 and Fig. 3.6. The *optimal exercise strategy* is to exercise at time $t = \tau$. Since $\Lambda(S(t_i))$ is computable at time t_i, we can approximate the optimal exercise strategy by estimating the continuation value $C_i(s)$ in order to recognise τ. Broadly speaking, this reflects the decision that must be made at each time t_i by the option holder on whether they should exercise or continue to hold, and says that they will exercise if the intrinsic (early exercise) value of the option is greater than the discounted expected value, conditional on the current value of the asset, achieved by holding the option for one more step.

3.5.2.1 Computing Continuation Values in the Binomial Tree

When the binomial model is used for the asset price, the continuation value is computed for each value of $S(t_{i-1})$ as the discounted expectation of the only two possible option values associated with asset values that can be reached in a single step. Therefore, using the binomial method, valuing American options proceeds in the same way as for European options for Stages 1 and 2. However, due to the early exercise facility, Stage 3 is different and, motivated by (3.13) in the dynamic programming representation with the binomial model of the asset price,

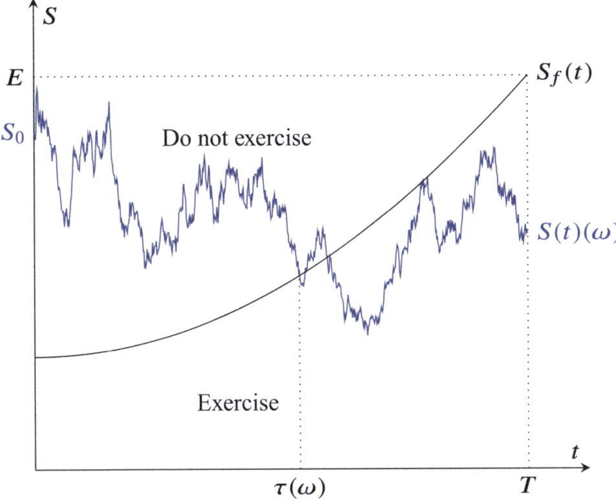

Fig. 3.6 Schematic presentation of the optimal exercise boundary $S_f(t)$ for an American put option with strike E and expiry T. A single trajectory of a Black-Scholes asset price process $S(t)(\omega)$ is overlaid, with the optimal exercise time $\tau(\omega)$ for this particular trajectory marked

we modify (3.12) to read

$$V_n^i = \max\left\{\Lambda(S_n^i),\, e^{-r\delta t}\left[pV_{n+1}^{i+1} + (1-p)V_n^{i+1}\right]\right\},$$

$$n = 0, \ldots, i,\ i = 0, \ldots, M-1. \qquad (3.14)$$

3.5.2.2 Modifications to the Binomial Algorithm

Equation (3.14) tells us to compare the intrinsic value of the option given by $\Lambda(S_n^i)$ to the discounted expected value associated with holding for at least one more step at each backward iteration in Stage 3 and so clearly it is insufficient to generate the asset values only at expiry.

We can then compute the asset values at any time step t_i for all $i = 0, \ldots, M$, by extracting the appropriate subsequence from each of the arrays `dpwr` and `upwr` defined in Sect. 3.2.2.1.

If `upwr` and `dpwr` have been set up as global variables, this operation can be coded into a short function that will compute intrinsic values for us when passed the parameters `S0, E, M, i`:

```
In []: def intrinsicPut(S0, E, M, i):
           siValues=S0*upwr[range(0,i+1)]*dpwr[range(M-i,M+1)]
           return np.maximum(E-siValues, 0)
```

The Stage 3 loop from Sect. 3.2 then becomes

```
In []: # Stage 3 variation for American options
       for i in range(int(M)-1,-1,-1):
           valsUp=putValues[range(1,i+2)]
           valsDown=putValues[range(0,i+1)]
           putVals=np.maximum(intrinsicPut(S0,E,M,i),
                       dFac(r,dt)*(p*valsUp+(1-p)*valsDown))
```

Exercise 3.7 asks you to construct a function `amerPutTree()` that implements this variant, which also values Bermudan options when the nodes are selected to coincide with the exercise times. Exercise 3.8 asks you to design a variant that values Bermudan options when exercise is not allowed at every node.

Finally, since the dynamic programming representation is model-agnostic, we can use it to value American-style options under other models (including Black-Scholes) as long as efficient methods can be found to approximate the conditional expectation in the continuation value $C_i(s)$. Our next chapter, on Monte Carlo methods, provides a set of tools to do this.

3.6 The Binomial Algorithm for Barrier Options

Barrier options have a payoff that is accessible or not depending on whether $S(t)$ breaches a pre-agreed value (the barrier) at some point $t \in [0, T]$. For example a *down-and-out call (resp. put) option* has a payoff that is zero if the asset price falls below some pre-defined barrier $B < S_0$ at some time $t \in [0, T]$. If the barrier is not crossed the payoff is that of a European call (resp. put). These are sometimes referred to as knock-out options, since the option is knocked out if the asset price reaches the barrier. Since these options have payoff that depends on the history of the underlying asset, they fall into the broad category of exotic option, and we consider valuation techniques for other such options in Chap. 4.

In this section we examine the use of binomial tree methods to value a knock-out barrier call option. A down-and-out call option on an underlying asset S with strike price E, barrier $B < E$, and expiry time T, has value at time $t \in [0, T]$ given by

$$V_{\text{DO}}(S, t) = C(S, t) - \left(\frac{S}{B}\right)^{-(k-1)} C\left(\frac{B^2}{S}, t\right), \tag{3.15}$$

where $C(S, t)$ is the value of a European call option on the same underlying asset with the same strike and expiry time, and $k = 2r/\sigma^2$.

3.6.1 Python Implementation Using Conditional Indexing

We adapt the technique used in Sect. 3.5 for quickly accessing asset values in the interior of the tree, combined with the use of conditional indexing as introduced in

Sect. 1.3.1.4 in Chap. 1. We will use the Cox-Ross-Rubenstein parameters on a tree with 400 steps. Everything proceeds as for the European call option in Stage 1.

```
In []: # Set up the problem and method parameters
       S0=40; E=40; B=35; T=1; r=0.04; sig=0.4
       M=400; dt=T/M
       u=np.exp(sig*np.sqrt(dt));
       d=np.exp(-sig*np.sqrt(dt));
       p=(np.exp(r*dt)-d)/(u-d);

       # Possible values of the underlying asset at expiration:
       upwr=u**np.arange(0,M+1)
       dpwr=d**np.arange(M,-1,-1)
       assetAtExpiry=(S0*dpwr)*(upwr)
```

For Stages 2 and 3, after we compute the value of the option associated with all nodes at a given timestep, we use conditional indexing via `np.where()` to identify those nodes where the asset price has reached the barrier or dropped below it, and reset those values to zero. So at expiry we would compute

```
In []: # Stage 2 variation for down-and-out call option
       callVals=np.maximum(assetAtExpiry-E,0)
       for j in np.where(assetAtExpiry<B):
           putVals[j]=0
```

In Fig. 3.7 we can see that the result of this computation is that the option value at the node corresponding to S_0^2 has been set to zero, and this node is coloured grey. The procedure for zeroing out values where the asset price has breached the barrier can also be applied at each timestep during the Stage 3 iteration:

```
# Stage 3 variation for down-and-out call option
for i in range(int(M)-1,-1,-1):
    valsUp=putVals[range(1,i+2)]
    valsDown=putVals[range(0,i+1)]
    callVals=dFac(r,dt)*(p*valsUp+(1-p)*valsDown)

    siValues=S0*upwr[range(0,i+1)]*dpwr[range(M-i,M+1)]
    for j in np.where(siValues<B):
        callVals[j]=0

print(callVals[0])
```

```
Out[]: 4.591008328835646
```

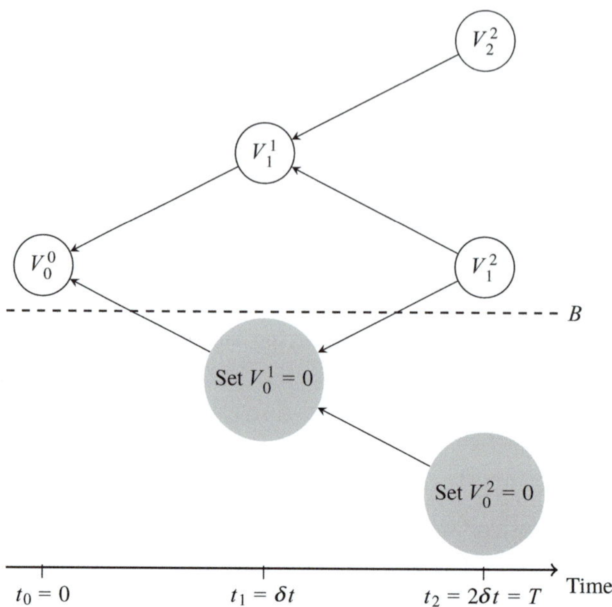

Fig. 3.7 Incorporating a knock-out barrier B in stages 2 and 3 of the valuation of a down-and-out option using a binomial tree with $M = 2$

In Fig. 3.7 we can see that the result of this computation is that the option value at the node corresponding to S_0^1 has been set to zero, and this node is coloured grey. All other nodes in the tree correspond to asset values above the barrier B.

3.6.2   A New Source of Error: The Specified Versus the Effective Barrier

The continuous-time value of this option using formula (3.15) is 4.451568760722852, and so there is an error of approximately 0.13944 for a tree with 400 steps.

Using (3.15) we can carry out a numerical convergence analysis for increasing M along the lines conducted for European options in Sect. 3.4, and the results of this for $M = 100, \ldots, 1000$ are shown in Fig. 3.8. We see that as M increases, the error displays a sawtooth shape, with deviations that are quite large and don't fall off appreciably by $M = 1000$. In fact this is characteristic of such options, particularly where the barrier is close to the strike.

Why might this be the case? In general the *specified barrier* value B will fall between asset values in the tree. We can see this in Fig. 3.7 where we are clearly valuing an option with an *effective barrier* of S_0^1 (the asset value corresponding to the grey node highest in the tree) rather than B. The deviation of the effective barrier

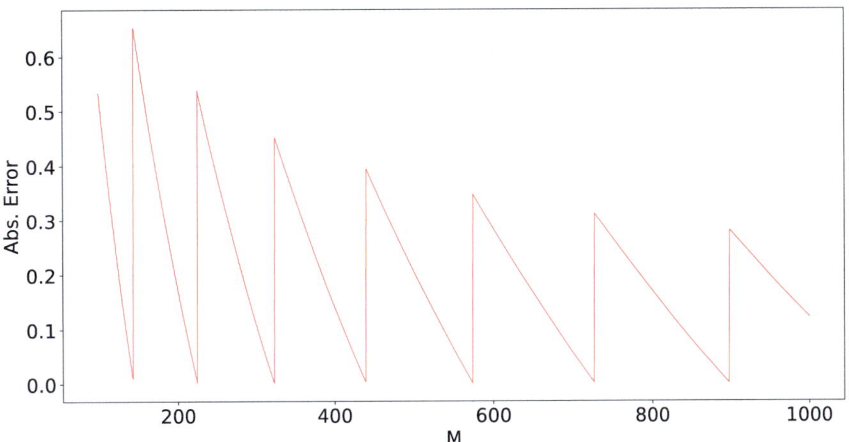

Fig. 3.8 A plot of absolute error against M for a down-and-out barrier call option with $S_0 = E = 40$, $B = 35$, $r = 0.04$, $\sigma = 0.4$, $T = 1$

from the specified barrier leads to what is known as *specification error*, slowing down convergence and introducing bias in the computed option price even for large M.

We can minimise specification error by choosing M so that the specified barrier falls as close as possible to a horizontal layer of nodes. In the Cox-Ross-Rubenstein tree, where the locations of the nodes are independent of the risk-free rate r, this can be achieved when δt is such that

$$S_0 e^{-m\sigma\sqrt{\delta t}} = B$$

for some $m = 1, 2, \ldots$ corresponding to the number of downward steps between S_0 and B. Using the fact that $\delta t = T/M$ we would then choose M as

$$M = \left\lfloor \frac{Tm^2\sigma^2}{\log^2(B/S_0)} \right\rfloor. \tag{3.16}$$

By increasing m over $\mathbb{N}$ on the RHS of (3.16), we construct an increasing set of values of M which minimizes specification error. We can do this for our example for $m = 0, \ldots, 11$ by

```
In []: m=np.arange(11)
        niceM=np.floor((T*(m**2)*sig**2)/(np.log(B/S0))**2)
        print(niceM)
```

```
Out[]: [ 0.    8.   35.   80.  143.  224.  323.  439.  574.  726.  897.]
```

Indeed the values of M in this array that lie on the interval $[100, 1000]$ correspond to the values of M in Fig. 3.8 where the absolute error is seen to be closest to zero. Computing the value of the option with (for example) $M = 439$ steps leads to a value of 4.45511379843372 and an absolute error of 0.00354504.

3.6.3 The In-Out Parity Relation for Single-Barrier Options

Finally, recall that a *down-and-in* call option has a payoff that is zero unless the asset price falls below the barrier $B < S_0$ at some time $t \in [0, T]$. Since the time $t = 0$ value of European call option $C(S_0, 0)$ can be replicated by holding one down-and-out-call with value $V_{DO}(S_0, 0)$ and one down-and-in call with value $V_{DI}(S_0, 0)$ written on the same underlying asset with the same strike and barrier, we can immediately use the code in this section to value the latter by the relation

$$V_{DI}(S_0, 0) = C(S_0, 0) - V_{DO}(S_0, 0).$$

We will revisit barrier options in Chap. 4, where we will use Monte Carlo sampling to value variants where the barrier is monitored only at a discrete set of times.

3.7   Further Reading

Sections 3.1 and 3.2 in this chapter owe a particular debt to the treatment given in Higham [32] for binomial trees for European options, and he also treats the convergence of these methods for European options to the benchmark Black-Scholes solution. Going deeper into the literature for the binomial method itself, we refer to the original article by Cox et al. [17], as well as the later articles by Rogers and Stapleton [59] and Leisen and Reimer [45], who analysed the converge of these methods along the lines discussed in Sect. 3.4. The dynamic programming representation by which we value American and Bermudan options, here and in the next chapter, has its roots in the book of Bellman [5], and a demonstration of its use in a Monte Carlo setting for the pricing of American and Bermudan options is given in Glasserman [25].

The method for choosing M to control specification error in the valuation of single barrier options described in Sect. 3.6 was introduced in Boyle and Lau [10], and later extended to the case of double barrier knock-out options in Costabile [16].

3.7.1 Exercises

3.1 Consider a trinomial share price model with zero risk-free rate in which S_n is the value of the stock at market close on day n. Then there are three possible price outcomes at close on day $n + 1$, $\{uS_n, S_n, dS_n\}$, associated with the probabilities $\{p_1, p_2, p_3\}$ respectively. Suppose that $S_n = 100$, $u = 1.05$, $d = 0.9$, $p_1 = 5/12$, $p_2 = 1/3$, and $p_3 = 1/4$. We will denote expectations taken under these probability weights as $\mathbb{E}_*$. Show that, to 2 decimal places,

1. $\mathbb{E}_*[S_{n+1}|S_n] = 99.58$;
2. $\mathrm{Var}_*[S_{n+1}|S_n] = 35.91$.
3. Follow the steps shown in Sect. 3.1.2.3 to identify a new probability measure characterised by three weights $\{q_1, q_2, q_3\}$ corresponding to each of the three outcomes $\{uS_n, S_n, dS_n\}$, such that if we take expectations under this new measure

$$\mathbb{E}[S_{n+1}|S_n] = 100, \quad \text{and} \quad \mathrm{Var}[S_{n+1}|S_n] = 35.91.$$

3.2

1. Write a Python function called `euroPutTree()` that will value a European put when passed the parameters E, S_0, r, σ, T, M using the Cox-Ross-Rubenstein method.
2. Produce another function `euroCallTree()` that values a European call when passed the same parameters.
3. Incorporate the lognormal parameter selection given by (3.6) and (3.9) as possible choices of model parameters in your binomial tree code for valuing European options.

3.3

1. Produce variants of the functions `euroPutTree()` and `euroCallTree()` that accept a parameter D specifying a dividend yield and use it to incorporate the payment of either a continuous dividend yield from the underlying asset or discrete dividend payments to be made either annually or semi-annually.
2. Introduce an additional user defined parameter `freq` that specifies how often the dividend is paid. So (for example), the user would specify `freq = 0` for a continuous dividend yield, `freq=1` for annual payment, `freq=2` for semi-annual payment.

3.4 The Cox-Ross-Rubenstein method described in Sect. 3.1.2 is characterised by the choice of

$$u, d = e^{\pm \sigma \sqrt{\delta t}},$$

with

$$p = \frac{e^{r\delta t} - d}{u - d}$$

for options written on non-dividend paying assets, and

$$p = \frac{e^{(r-D_0)\delta t} - d}{u - d}$$

for options written on assets paying a continuous dividend yield of D_0. In both cases, show that the expectation is being computed with respect to the risk-neutral measure for the asset price model.

3.5 Show that, for the lognormal parameter values given by (3.9),

$$\mathbb{E}\left[S^{(n+1)}|S^{(n)}\right] \neq e^{r\delta t} S^{(n)}, \quad n = 0, \ldots, M - 1,$$

and therefore $p = 1/2$ does not lead to computation of expectation under risk-neutral measure for this version of the asset price model.

3.6 Show that, for the lognormal parameter values given by (3.9), the no-arbitrage condition

$$d < e^{r\delta t} < u$$

holds if and only if $\delta t < 4/\sigma^2$.

3.7 Produce a Python function with name `amerPutTree()` that values an American put option when passed the parameters E, S_0, r, σ, T, M.

3.8 Produce a variant of `amerPutTree()` that values Bermudan options where exercise is not allowed at every node.

3.9 In the case of an American put option, there is no equivalent of the Black-Scholes formula against which to compare our binomial tree valuation. Is there anything that can be demonstrated about the convergence or otherwise of this method in the absence of such an explicit benchmark?

3.10

1. Reproduce the plots in Fig. 3.5, and compare to a plot of errors generated by the Cox-Ross-Rubenstein parameter selection (3.8).

2. Now investigate the effect of
   (a) varying $|S_0 - E|$;
   (b) varying r and/or introducing a continuous dividend yield D_0.

3.11 Write a piece of code that uses the formula (3.15) to compute the closed-form Black-Scholes value for the down-and-out call option. Then use it to

1. Produce a numerical convergence plot of absolute error against M and examine the effect of the absolute difference $|E - B|$ on the mode of convergence.
2. Repeat, using only values of M computed according to (3.16).

Simulation I: Monte Carlo Methods

4

Monte Carlo methods work by statistically estimating the conditional expectation on the RHS of the risk-neutral characterisation of option price from Proposition 2.9 in Chap. 2:

$$V(t) = e^{-r(T-t)}\mathbb{E}_{\mathbb{Q}}[V_T|\mathcal{F}_t], \quad t \in [0, T]. \tag{4.1}$$

Since V_T is a random variable with a conditional distribution that is unknown to us, we can use simulation to generate a sufficiently large statistical sample, then estimate the expectation by taking an arithmetic average of sample values. This approach generalises easily to the valuation of exotic options, options written on multiple underlying assets, and the estimation of option price sensitivities (*the Greeks*) required for the dynamic hedging of a portfolio that includes long or short positions in one or more derivatives. It is also highly flexible, allowing us to select from any asset model we like, as long as we know how to sample from the universe of possible asset price trajectories. Finally, unlike binomial tree methods, it does not suffer from the curse of dimensionality, and is better suited to the valuation of multi-asset options.

Monte Carlo methods can also be applied to the pricing of American-style options, and in this chapter we will revisit the question of how to approximate the continuation value

$$C_i(s) := e^{-r\delta t}\mathbb{E}_{\mathbb{Q}}[V_i(S(t_i))|S(t_{i-1}) = s], \tag{4.2}$$

allowing S to be represented by models other than the binomial.

We start by introducing the basic idea of Monte Carlo (MC) for a digital option with value given by (4.1). Then we show how to apply these techniques

© The Author(s), under exclusive license to Springer Nature Switzerland AG 2024

C. Kelly, *Computation and Simulation for Finance*,
Springer Undergraduate Texts in Mathematics and Technology,
https://doi.org/10.1007/978-3-031-60575-8_4

to the estimation of (4.2) above, and implement the Longstaff-Schwarz method for the valuation of an American put. Next we look at the application of Monte Carlo techniques to path-dependent (exotic) options, and strategies for reducing the sample variance of Monte Carlo estimators, improving the computational efficiency of the approach. Finally we look at how these methods combine with finite difference approximation to estimate the Greeks, and point out some practical issues.

Python Setup To follow the Python code in this chapter, you will need to import the NumPy and PyPlot packages, as well as the the `scipy.stats.norm` module. We also introduce the `scipy.linalg` package, which allows us to manipulate vectors and matrices according to the rules of linear algebra. Finally, in our implementation of the Longstaff-Schwarz algorithm we will use the `numpy.polynomial` package to fit regression polynomials of Legendre type.

```
In []:  import numpy as np
        import scipy.linalg as linalg
        from scipy.stats import norm
        import matplotlib.pyplot as plt
        from numpy.polynomial import Legendre
```

For pseudo-random number generation throughout this chapter, we will instantiate and use the following local random number generator (RNG)

```
In []:  rng=np.random.default_rng()
```

making use of the `numpy.random` module. We will also need the user-defined functions from Chaps. 1–3 listed in Table 4.1.

Table 4.1 User-defined functions required in Chap. 4

Section	Function name	Function output	Parameter	Parameter meaning
1.2.4	dFac()	$e^{-r(T-t)}$	r	r
			dt	$T-t$
1.4.1	bsEnsemble()	$S(t_n, \omega_i)$	S0	S_0
		$n = 0, \ldots N$	mu	μ or r
		$i = 0, \ldots, M-1$	sig	σ
			T	T
			M	M
			N	N
			rng	local RNG
2.2.3	d1()	d_1	S0	$S(t)$
	d2()	d_2	E	E
	euroCallBenchmark()	$C(t, S)$	r	r
			sig	σ
			T	$T-t$

4.1 Brute-Force Monte Carlo

We introduce the basic idea of Monte Carlo estimation and illustrate its implementation in Python.

4.1.1 Estimation by Independent Statistical Sampling

Let X be a random variable with $\mathbb{E}[X] = a$ and $\mathrm{Var}[X] = b^2$, where a and b are unknown. To generate a confidence interval for the unknown expectation a using sampled values of X, use the following estimators:

$$a_M = \frac{1}{M} \sum_{i=1}^{M} X_i; \quad b_M = \frac{1}{M-1} \sum_{i=1}^{M} (X_i - a_M)^2.$$

It is a consequence of the central limit theorem that $a_M - a \sim \mathcal{N}(0, b^2/M)$, and therefore we can write an approximate 95% confidence interval for $\mathbb{E}[X]$ as

$$\left[a_M \pm \frac{1.96 b_M}{\sqrt{M}} \right]$$

The length of this confidence interval is proportional to $1/\sqrt{M}$, and so if we wish to shorten it by a factor of 10 (i.e. to gain one extra decimal place of precision) we will need to increase our statistical sample size by a factor of 100. This is computationally expensive and even on modern computers can lead to slow execution times. We will come back to this issue later, in Sect. 4.4, when we discuss ways to mitigate the problem by variance reduction techniques. For now, let's look at how to apply Monte Carlo techniques in their unmodified (*brute-force*) form to a simple derivative pricing problem, where we estimate the expectation on the RHS of the martingale representation of an option price (4.1).

4.1.2 Python Implementation for a European Option

Consider a European option with $E = 52$, $T = 1$, $r = 0.06$, $\sigma = 0.1$, $S_0 = 50$ and $\Lambda(S) = (S - E)^+$. We can use the function `euroCallBenchmark()` from Sect. 2.2.3 in Chap. 2 to compare our numerical output against the benchmark Black-Scholes valuation. We will walk through the steps required to code the MC algorithm in python for these parameter values, using $M = 10$ samples. The idea is to sample from the distribution of the random variable

$$S(T) = S_0 e^{(r - \sigma^2/2)T + \sigma \sqrt{T} Z}, \quad Z \sim \mathcal{N}(0, 1),$$

which is identical to that of the Black-Scholes asset price model at time $t = T$, given by (3.4) in Sect. 3.1.2.4. We will use `rng.normal()` to generate a sample of M independent standard normal random variables.

```
In []: # Step 1. Define parameters
       S=50.0; E=52.0; T=1.0; r=0.06; sigma=0.1
       M=10


       # Step 2. Compute call payoffs for each sample
       Z=rng.normal(0,1,M)
       Sfinal=S*np.exp((r-0.5*sigma**2)*T+sigma*np.sqrt(T)*Z)
       Vcall=dFac(r,T)*(np.maximum(Sfinal-E,0))


       # Step 3. Compute sample mean and standard deviation
       aMcall=np.mean(Vcall); bMcall=np.std(Vcall);
```

Now we already have a MC valuation for the option price $\mathbb{E}[e^{-rT}\Lambda(S(T))]$, stored in the variable `aMcall`. However since this is a statistical estimate it is of little value without an accompanying confidence interval. We will do this next, at the 95% confidence level.

```
In []: # Step 4. Print 95
       confCall=aMcall+np.array([1,-1])*1.96*bMcall/np.sqrt(M)
       print(confCall)
```

By modifying only the payoff function, this method can be applied to any option with a payoff function of the form $V_T = \Lambda(S(T))$ that depends only on $S(T)$. Such options are called either *binary* (when Λ depends only on the direction of $S(T)$ relative to E) or *digital* (when Λ depends on both the direction and distance of $S(T)$ to E). If S follows the Black-Scholes model then the value of such an option with payoff $\Lambda(S(T))$ at expiry T may be expressed either as an improper integral:

$$V(S_0) = \frac{e^{-rT}}{\sigma\sqrt{2\pi T}} \int_0^{\infty} \Lambda(\tilde{u}) \exp\left[-\frac{\left(\log(\tilde{u}/S) - (r - \sigma^2/2)T\right)^2}{2\sigma^2 T} \right] \frac{d\tilde{u}}{\tilde{u}},$$

which may or may not resolve itself into a closed form expression depending on the form of Λ, or as an expectation:

$$V(S_0) = \mathbb{E}_{\mathbb{Q}}\left[e^{-rT} \Lambda\left(S_0 e^{(r-\sigma^2/2)T + \sigma W^{\mathbb{Q}}(T)} \right) \right], \tag{4.3}$$

where $W^{\mathbb{Q}}$ is a standard Brownian motion under the risk-neutral measure $\mathbb{Q}$.

4.2   A Regression-Based Method for Options with Early Exercise

Consider an American put option written on an underlying asset S with strike price E and expiry time T. It is possible to characterise the time $t = 0$ value of this option as

$$\sup_{\tau \in [0,T]} \mathbb{E}_{\mathbb{Q}} \left[e^{-r(\tau - t)} \left(E - S(\tau) \right)^+ \right], \tag{4.4}$$

where τ is a stopping time with respect to the natural filtration of S. If we think of each τ as representing an early exercise strategy then the observed value of each τ determines the exercise time of the option. The value of the American put option corresponds to the value generated by the best of these strategies. A naive approach to the Monte Carlo valuation of the RHS would involve characterising all possible early exercise strategies as stopping times, producing an estimate of the expectation on the RHS of (4.4) for each, and selecting the largest value thus computed. However even if it were possible to capture all strategies in this way, the computational cost would be prohibitive.

Instead, consider the value of the American put option via the dynamic programming representation given by (3.13) in Sect. 3.5 of Chap. 3. Just as for Step 3 of the Binomial Tree algorithm, we seek to assign an option value to each trajectory of S, proceeding from expiry backwards in time, working out whether and when to exercise according to the known optimal strategy in that setting. Since the continuation value $C_i(s) = e^{-r\delta t}\mathbb{E}_{\mathbb{Q}}[V_i(S(t_i))|S(t_{i-1}) = s]$ is an expectation, it seems reasonable that we try to use Monte Carlo at each step of the dynamic programming representation algorithm. Nonetheless, we cannot directly sample from the distribution of $V_i(S_i)$ since the closed form is unknown to us.

One way around this is to approximate the continuation value as a linear combination of elementary functions of the current asset price, and apply least-squares regression to estimate the coefficients of the approximating sum. This approximation, considered as a function of the current asset price, has a distribution that can be estimated by Monte Carlo.

There are several independently developed variations on this technique, and here we will demonstrate one that is commonly used in practice, due to Longstaff and Schwarz [48].

4.2.1   Approximating the Continuation Value

At expiry, the value of an option with payoff Λ is ($t = t_N = T$)

$$V_N(S(t_N)) = \Lambda(S(t_N)).$$

Using the notation $S_i := S(t_i)$, at the previous potential exercise date $(t = t_{N-1})$, the continuation value is

$$C_{N-1}(s) = \mathbb{E}_{\mathbb{Q}}\left[e^{-r\delta t} V_N(S_N) \,\Big|\, S_{N-1} = s \right].$$

Our strategy is to approximate this via a truncated series of R basis functions $(\{\psi_r\}_{r=1}^{R}$ (a set of elementary functions[1] that can be used in linear combination to approximate a given function) as

$$\widehat{C}_{N-1}(s) = \sum_{r=1}^{R} \beta_r \psi_r(s). \tag{4.5}$$

It is computationally convenient to choose $(\psi_i)_{i=1}^{\infty}$ to be a sequence of orthogonal polynomials such as the Legendre polynomials, the first few terms of which are

$$\psi_1(s) = 1, \quad \psi_2(s) = s, \quad \psi_3(s) = \frac{1}{2}(3s^2 - 1), \dots,$$

and whose general form is given by

$$\psi_n(s) = \sum_{k=0}^{n} \binom{n}{k} \binom{n+k}{k} \left(\frac{x-1}{2}\right)^k.$$

Estimating the coefficients $\{\beta_r\}_{r=1}^{R}$ is equivalent to an estimation of $\widehat{C}_{N-1}$, by (4.5). We can do this by least-squares. Minimise

$$\mathbb{E}_{\mathbb{Q}}\left[\left(\mathbb{E}_{\mathbb{Q}}\left[e^{-r\delta t} V_N(S_N) | S_{N-1} \right] - \widehat{C}_{N-1}(S_{N-1}) \right)^2 \right]$$

with respect to $\beta = (\beta_1, \dots, \beta_R)^T$ by differentiating with respect to each β_u and set equal to zero. This yields the system of equations

$$\mathbb{E}_{\mathbb{Q}}\left[\left(\mathbb{E}_{\mathbb{Q}}[e^{-r\delta t} V_N(S_N) | S_{N-1}] - \widehat{C}_{N-1}(S_{N-1}) \right) \psi_u(S_{N-1}) \right] = 0, \quad u = 1, \dots, R,$$

---

[1] Elementary functions are functions of a single variable that take sums, products, roots and compositions of finitely many polynomial, rational, trigonometric, hyperbolic, and exponential functions, including, potentially, the inverse functions.

which we want to solve for $\beta \in \mathbb{R}^R$. For each value of u, splitting the conditional expectation and then rearranging gives

$$\mathbb{E}_Q[e^{-r\delta t} V_N(S_N)\psi_u(S_{N-1})] = \mathbb{E}_Q\left[\widehat{C}_{N-1}(S_{N-1})\psi_u(S_{N-1})\right]$$

$$= \mathbb{E}_Q\left[\sum_{r=1}^{R} \beta_r \psi_r(S_{N-1})\psi_u(S_{N-1})\right]$$

$$= \sum_{r=1}^{R} \mathbb{E}_Q[\psi_r(S_{N-1})\psi_u(S_{N-1})]\,\beta_r, \qquad (4.6)$$

for $u = 1, \ldots, R$. The set of equations defined by (4.6) can be written in matrix-vector form as

$$A\boldsymbol{\beta} = B,$$

where $\boldsymbol{\beta} \in \mathbb{R}^R$, and $A \in \mathbb{R}^{R \times R}$ and $B \in \mathbb{R}^R$ are defined termwise as

$$(A)_{r,u} = \mathbb{E}_Q[\psi_r(S_{N-1})\psi_u(S_{N-1})]; \quad r, u = 1, \ldots, R;$$

$$(B)_r = \mathbb{E}_Q[e^{-r\delta t} V_N(S_N)\psi_r(S_{N-1})], \quad r = 1, \ldots, R.$$

Therefore

$$\beta = A^{-1}B. \qquad (4.7)$$

In the Monte Carlo approximation, each of the expectations in $A^{-1}B$ is replaced by an average over the values from M samples. For example

$$\mathbb{E}_Q[\psi_r(S_{N-1})\psi_u(S_{N-1})] \approx \frac{1}{M}\sum_{\omega=1}^{M} \psi_r(S_{N-1}(\omega))\psi_u(S_{N-1}(\omega)),$$

using the notation $S_{N-1}(\omega)$ to denote a simulated observation on S_{N-1}.

4.2.2   A Simple Example: Fitting the Legendre Polynomials by Least-Squares

Approximate the continuation value using (4.5) with Legendre polynomials truncated at the second term, so that $R = 2$ and $\psi_1(s) = 1$, $\psi_2(s) = s$:

$$\widehat{C}_{N-1}(s) = \beta_1 + \beta_2 s.$$

Then

$$A = \begin{pmatrix} 1 & \mathbb{E}_{\mathbb{Q}}[S_{N-1}] \\ \mathbb{E}_{\mathbb{Q}}[S_{N-1}] & \mathbb{E}_{\mathbb{Q}}[S_{N-1}^2] \end{pmatrix}; \quad B = e^{-r\delta t} \begin{pmatrix} \mathbb{E}_{\mathbb{Q}}[V_N(S_N)] \\ \mathbb{E}_{\mathbb{Q}}[V_N(S_N)S_{N-1}] \end{pmatrix}, \qquad (4.8)$$

and (4.7) can be written as

$$\begin{pmatrix} \beta_1 \\ \beta_2 \end{pmatrix} = e^{-r\delta t} \begin{pmatrix} 1 & \mathbb{E}_{\mathbb{Q}}[S_{N-1}] \\ \mathbb{E}_{\mathbb{Q}}[S_{N-1}] & \mathbb{E}_{\mathbb{Q}}[S_{N-1}^2] \end{pmatrix}^{-1} \begin{pmatrix} \mathbb{E}_{\mathbb{Q}}[V_N(S_N)] \\ \mathbb{E}_{\mathbb{Q}}[V_N(S_N)S_{N-1}] \end{pmatrix}.$$

Clearly, we must estimate the following expectations:

$$\mathbb{E}_{\mathbb{Q}}[S_{N-1}]; \quad \mathbb{E}_{\mathbb{Q}}[S_{N-1}^2]; \quad \mathbb{E}_{\mathbb{Q}}[V_N(S_N)]; \quad \mathbb{E}_{\mathbb{Q}}[V_N(S_N)S_{N-1}]. \qquad (4.9)$$

To do this we need to generate M path simulations of the asset model, each over the timeset $\{0 = t_0, t_1, \ldots, t_N = T\}$, using the function `bsEnsemble()` developed in Exercise 1.11 in Chap. 1.

```
In []: # Option and simulation parameters
       S0=100; E=110; T=1; r=0.05; sig=0.3; N=100
       dt=T/N; M=10**4

       # Generate ensemble
       S=bsEnsemble(S0,r,sig,T,M,N,rng)
```

The result is an $M \times (N + 1)$ array variable S, giving us everything we need to estimate the first two terms in (4.9) by Monte Carlo. To estimate the last two terms in (4.9) we need to compute $V_N(S_N)$ on each trajectory of the ensemble before averaging. Since N corresponds to the expiry time we do this by applying the payoff directly to our simulated values of S_N

```
In []: VN=np.maximum(S[:,N]-E,0)
```

Now we can produce the estimators

```
In []: EVS1=np.mean(VN); EVS2=np.mean(VN*S[:,N])
```

Now set up the values of the matrix A and vector B according to (4.8)

```
In []: ES1=np.mean(S[:,N-1])
       ES2=np.mean(S[:,N-1]**2)
       A=np.array([[1,ES1],[ES1,ES2]])
       B=np.array([EVS1,EVS2])
```

The `scipy.linalg` package allows us to invert A and hence compute β using the relation (4.7)

```
In []: Ainv=linalg.inv(A)
       beta=Ainv.dot(B)
```

Finally, we compute the approximate continuation value $\widehat{C}(S(t_{N-1}))$ on each trajectory in the ensemble by

```
In []: beta[0]+beta[1]*S[:,N-1]
```

In practice we would not wish to do this manually at every step. For example we may wish to choose use a higher-order polynomial approximation of the continuation value, and even to vary the order at each timestep. Fortunately the `numpy.polynomial` package automates polynomial fitting for us. After computing VN in the procedure above, fit a Legendre polynomial of degree 1 by

```
In []: p=Legendre.fit(S[:,N],VN,deg=1)
```

We present an implementation of the method in full next.

4.2.3   Using Continuation Estimates to Value an American Put Option

Start with the same setup from the previous section, with an additional variable R representing the desired order of polynomial approximation for the continuation value:

```
In []: # Option and simulation parameters
       S0=100; E=110; T=1; r=0.05; sig=0.3; N=100
       dt=T/N; M=10**4; R=2

       # Generate ensemble and instantiate
       # array to hold option values
       S=bsEnsemble(S0,r,sig,T,M,N,rng)
       V=np.zeros((M,N))

       # Payoff at expiry
       V[:,N-1]=np.maximum(E-S[:,N],0)
```

The Longstaff-Schwarz method requires that we carry out following procedure for $n = N - 1, N - 2, \ldots, 1$ to value the option. For each observed trajectory ω_i, $i = 1, \ldots, M$, perform the following iterative procedure over $n = N-1, N-2, \ldots, 1$:

- If $\Lambda\left(S(t_n, \omega_i)\right) > 0$ and $\widehat{C}_n\left(S(t_n, \omega_i)\right) < \Lambda\left(S(t_n, \omega_i)\right)$, exercise the option on that trajectory:

$$V_n(\omega_i) = \Lambda\left(S(t_n, \omega_i)\right), \quad i = 1, \ldots, M.$$

- Otherwise set

$$V_n(\omega_i) = e^{-r\delta t} V_{n+1}(\omega_i), \quad i = 1, \ldots, M.$$

At a given step n, we can see from the second point of the procedure above that it is not required to compute an estimation of $\widehat{C}_n\left(S(t_n, \omega_i)\right)$ for trajectories ω_i where $\Lambda\left(S(t_n, \omega_i)\right) = 0$. This makes sense intuitively since, if the intrinsic value of the option at a time prior to expiry is zero, we simply would not exercise at that time. We will use conditional indexing for NumPy arrays and the `np.where()` call, introduced in Sect. 1.3.1.4 of Chap. 1, to achieve this.

```
In []: # Longstaff-Schwarz valuation of American put option
       for k in range(N,2,-1):

           # Extract asset and intrinsic values considering
           # only ITM trajectories
           indKeep=np.where(S[:,k-1]<E)
           S1=S[indKeep,k-1]
           V1=dFac(r,dt)*V[indKeep,k-1]
           S1=S1[0,:]; V1=V1[0,:]

           # Fit Legendre polynomial to S1 and Y1
           p=Legendre.fit(S1,V1,deg=R)

           # Discount option values one step
           V[:,k-2]=dFac(r,dt)*V[:,k-1]

           # Use continuation value to check for early
           # exercise on each trajectory and overwrite
           # those values with intrinsic value
           earlyEx=(E-S[:,k-1])>p(S[:,k-1])
           for i in np.where(earlyEx):
               V[i,k-2]=np.maximum(E-S[i,k-1],0)
```

We used a degree $R = 2$ approximation of the continuation values here, and this is easily varied. The $t = 0$ value of the option is given by $\frac{1}{M}\sum_{i=1}^{M} V_0(\omega_i)$, the arithmetic average of observed values of $V_0 = e^{-r\delta t} V_1$:

```
In []: dFac(r,dt)*np.mean(V[:,1])
```

```
Out[]: 15.608008723753507
```

Exercise 4.4 asks you to compare the output for this example to the binomial tree valuation of the same option.

4.2.3.1 Computational Complexity of the Longstaff-Schwarz Method

If the number of samples M is greater than the number of basis functions, the main computational cost at each step is $O(MR^2)$. In [48], the authors give examples using 5–20 basis functions (the number can vary from step to step). If (4.5) is an exact series expansion of $C_{N-1}(s)$, then this method converges as $M \to \infty$. However, if (4.5) is a truncated estimate of $C_{N-1}(s)$ and R is fixed, the method yields an underestimate no matter how large we take M.

4.3 Monte Carlo for Exotic Options

Exotic options can be recognised by a payoff that depends on the history of the underlying asset $\{S(t); 0 \le t \le T\}$ and the fact that they are generally not quoted on an exchange. The holder may or may not be allowed to exercise early. This path dependence again leads us to simulate an ensemble of trajectories of the asset price process using `bsEnsemble()`. We will illustrate the approach with two exotic options that have known formulae against which we can benchmark our output: the down-and-out barrier call where the knock-out barrier is continuously monitored, and the geometric average Asian call.

4.3.1 Barrier Options

4.3.1.1 The Down-and-Out Call Option with Continuous or Discrete Monitoring

Recall from Sect. 3.6 that a down-and-out call option on an underlying asset S with strike price E, barrier $B < E$, and expiry time T, has value at time $t \in [0, T]$ given by

$$V(S, t) = C(S, t) - \left(\frac{S}{B}\right)^{-(k-1)} C\left(\frac{B^2}{S}, t\right), \qquad (4.10)$$

where $C(S, t)$ is the value of a European call option on the same underlying asset with the same strike and expiry time.

Note that (4.10) holds if the asset is continuously monitored for breaches of the barrier. If this monitoring only takes place at certain times then more trajectories would be expected to finish in-the-money and (4.10) is an underestimate of the value of the option. Figure 4.1 illustrates this. The Monte Carlo approach to valuation of such options allows us to sample from the distribution of the underlying asset only at the monitoring times and check for breaches of the barrier there, leading to an unbiased estimator.

Fig. 4.1 One trajectory of the underlying for a down-and-out call option with strike E and barrier $0 < B < E$ monitored at times M_i for $i = 1, 2, 3$. Since the barrier is achieved between M_1 and M_2, but the asset price recovers before detection, the payoff remains active and the option finishes in-the-money for this trajectory. If the barrier were continuously monitored, the payoff associated with this trajectory would be zero

If instead we wish to approximate the value of a down-and-out call option where the barrier is continuously monitored, the Monte Carlo estimator will tend to overestimate the benchmark value, and for the same reason. This bias can be mitigated by increasing the number and frequency of monitoring times.

4.3.1.2 Python Implementation

Because we will check each trajectory individually to see if the barrier has been breached before calculating the payoff, we must generate the entirety of the ensemble of asset price values S. As in Sect. 1.3.3 in Chap. 1, set up an array variable S with dimension $M \times (N + 1)$ to hold an ensemble with initial value $S_0 = 1$ and for which we have sampled at each of N equidistant monitoring times. Next, set up a 1-dimensional array to hold the payoff values and check each trajectory in turn to see if the barrier was breached at one of the monitoring times. If not, apply the payoff.

```
In []: V=np.zeros((M,1))
          for i in range(0,M):
```

```
            Smin=np.amin(S[i,:])
            if Smin>B:
                V[i]=dFac(r,T)*np.maximum(S[N,:]-E,0)
```

Notice that the default value for each component of V is zero and so if the barrier is breached the code doesn't need to do anything. Finally value the option by taking expectation and output a 95% confidence interval as before:

```
In []: aM=np.mean(V); bM=np.std(V)
       conf = aM+np.array([-1,1])*1.96*bM/np.sqrt(M)
       print(aM); print(conf);
```

Exercise 4.7 asks you to use this code to compare the resulting value to that produced by the formula (4.10) for various values of N, and the binomial tree valuation in Sect. 3.6 of Chap. 3. Exercise 4.5 asks you to rewrite the code making use of conditional indexing rather than a `for` loop.

4.3.2   Asian Options

Asian options have a payoff determined by the average behaviour of $S(t)$ over $[0, T]$. For example an *average price Asian call option* has payoff at expiry T given by

$$\max\left(\frac{1}{T}\int_0^T S(\tau)d\tau - E, 0\right).$$

In practice the integral component of the payoff must be computed from a finite set of observations, for example corresponding to the closing prices of the asset at the end of each trading day. The way this is done will have an effect on the price so it is possible to distinguish between geometric average Asian options, where we substitute

$$\frac{1}{T}\int_0^T S(\tau)d\tau \longrightarrow \left(\prod_{j=1}^n S(t_j)\right)^{1/n},$$

and arithmetic average Asian options, where we substitute

$$\frac{1}{T}\int_0^T S(\tau)d\tau \longrightarrow \frac{1}{n}\sum_{j=1}^n S(t_j).$$

The option value using the former is known precisely. The value for the latter is not.

4.3.2.1 The Geometric Average Asian Call Option

Theorem 4.1 *Consider a geometric average Asian call option on an underlying asset S with strike price E, expiry time T, and payoff given by*

$$
\max\left(\left(\prod_{j=1}^{n} S(t_j)\right)^{1/n} - E, 0\right),
$$

where $0 < t_1 < t_2 < \cdots < t_n = T$ are equally spaced, so that $t_i = i\delta t$, where $\delta t = T/n$. This option has value at time $t = 0$ given by

$$
S_0 e^{(\bar{r}-r)T} N(d_1) - E e^{-rT} N(d_2) \tag{4.11}
$$

where

$$
d_{1,2} = \frac{\log(S_0/E) + (\bar{r} \pm \frac{1}{2}\bar{\sigma}^2)T}{\bar{\sigma}\sqrt{T}}; \tag{4.12}
$$

$$
\bar{\sigma}^2 = \frac{(n+1)(2n+1)}{6n^2}\sigma^2; \tag{4.13}
$$

$$
\bar{r} = \frac{1}{2}\bar{\sigma}^2 + \left(r - \frac{1}{2}\sigma^2\right)\frac{(n+1)}{2n}. \tag{4.14}
$$

Proof By writing the product in the payoff as the telescoping product

$$
\prod_{j=1}^{n} S(t_j) = \frac{S(t_n)}{S(t_{n-1})}\left(\frac{S(t_{n-1})}{S(t_{n-2})}\right)^2 \times \cdots \times \left(\frac{S(t_2)}{S(t_1)}\right)^{n-1}\left(\frac{S(t_1)}{S(t_0)}\right)^n S_0^n,
$$

and then taking a logarithm

$$
Y := \ln\left(\left(\prod_{j=1}^{n} S(t_j)\right)^{1/n}/S_0\right) = \frac{1}{n}\sum_{j=1}^{n} j \cdot \underbrace{\ln\left(\frac{S(t_{n+1-j})}{S(t_{n-j})}\right)}_{=:X_j},
$$

we can express the logarithm of the geometric average as a linear combination of the log-returns

$$
X_j \sim N((r - \sigma^2/2)\delta t, \sigma^2\delta t), \quad j = 1, \ldots, n-1.
$$

The affine property of the normal distribution allows us to conclude that the geometric average $\left(\prod_{j=1}^{n} S(t_j)\right)^{1/n}$ has a lognormal distribution. The mean and variance can be computed.

We see that

$$\mathbb{E}[Y] = \mathbb{E}\left[\frac{1}{n}\sum_{j=1}^{n} j\, X_j\right] = \frac{1}{n}\sum_{j=1}^{n} j\, \mathbb{E}[X_j]$$

$$= \frac{1}{n}(r - \sigma^2/2)\delta t \sum_{j=1}^{n} j$$

$$= (r - \sigma^2/2)T\frac{(n+1)}{2n},$$

where we have used the fact that $\delta t n = T$ and $\sum_{j=1}^{n} j = n(n+1)/2$. In Exercise 4.10, we similarly show that

$$\mathrm{Var}[Y] = \sigma^2 T\frac{(n+1)(2n+1)}{6n^2}.$$

Since the value of the option is given by (4.1), and the only difference between the payoff of a European call and the geometric average Asian call is that we replace the log-normally distributed random variable $S(T)$ with the log-normally distributed random variable $\left(\prod_{j=1}^{n} S(t_j)\right)^{1/n}$, the formula given by (4.11) can be arrived at by making the following substitutions in the Black-Scholes formula for a European call:

$$r - \frac{\sigma^2}{2} \longrightarrow \left(r - \frac{\sigma^2}{2}\right)\frac{(n+1)}{2n} =: \bar{r} - \frac{\sigma^2}{2}$$

$$\sigma^2 \longrightarrow \sigma^2\frac{(n+1)(2n+1)}{6n^2} =: \bar{\sigma}^2$$

$$S_0 \longrightarrow S_0 e^{(\bar{r}-r)T}.$$

The last of these substitutions is motivated by the fact that Y has drift $\bar{r}t$, the value of S_0 at time $t = T$ is $S_0 e^{\bar{r}T}$. We then discount this back to time $t = 0$ at the risk-free rate. The statement of the theorem then follows. $\square$

4.3.2.2 The Arithmetic Average Asian Call Option

The value of an arithmetic average Asian call has no such convenient formula and must be approximated. Nonetheless, it seems reasonable to suppose that its value should be similar, though not identical, to its geometric average counterpart. In fact, the Arithmetic-Geometric Mean Inequality, which says that

$$\frac{1}{n}\sum_{j=1}^{n} S(t_j) \geq \left(\prod_{j=1}^{n} S(t_j)\right)^{1/n},$$

tells us that, of the two Asian call options, the arithmetic average version will have a higher (or at least not lower) value, since we would expect a larger proportion of trajectories to finish ITM. For a put option, the reverse is true.

In both the arithmetic and geometric average cases, once we have created the ensemble variable S and instantiated a payoff variable V, the payoff can be computed straightforwardly. Indeed we can do this simultaneously across all trajectories for the arithmetic average Asian call as follows:

```
In []: arithAvg=np.mean(S,axis=1)
       V=dFac(r,T)*np.maximum(arithAvg-E,0)
```

Then average to produce a Monte Carlo valuation as usual. For the geometric average Asian call it is better to recast the average as

$$\left(\prod_{i=1}^{N} S(t_i) \right)^{1/N} = \exp\left(\frac{1}{N} \sum_{i=1}^{N} \ln(S(t_i)) \right),$$

in order to avoid overflow errors when computing a potentially large product of asset values that may each be significantly greater than one. So the payoff computation looks like

```
In []: geoAvg=np.exp(np.mean(np.log(S),axis=1));
       V=dFac(r,T)*np.maximum(geoAvg-E,0)
```

Various other exotic options can be valued with a similar modification of the payoff computation. For example lookback options have a payoff that depends on either the maximum or minimum value of $S(t)$ over the interval $[0, T]$.

4.4 Variance Reduction

We return to the problem of computational inefficiency. Recall that if we use M samples to estimate the expectation of a random variable X with mean a and variance b^2 then the estimator $a_M \sim \mathcal{N}(a, b^2/M)$, and so the confidence interval for the estimate is proportional to $1/\sqrt{M}$. This is the weakest aspect of the Monte Carlo method across all areas of application.

There exist variance reduction techniques that work by identifying an alternative estimator with the same mean, but smaller variance. This allows us to generate estimates with shorter confidence intervals while avoiding a penalty on computational cost. In this section we will introduce three distinct strategies for variance reduction: by control variate, by antithetic variate, and by importance sampling.

4.4.1 Control Variates

Suppose we wish to estimate $\mathbb{E}[X]$. Our approach here is to use another random variable Y, called the control variate, where $\mathbb{E}[Y]$ is known. Then, for some $c \in \mathbb{R}$, define

$$Z = X + c(Y - \mathbb{E}[Y]).$$

Taking expectations on both sides,

$$\mathbb{E}[Z] = \mathbb{E}[X] + c\mathbb{E}[Y - \mathbb{E}[Y]] = \mathbb{E}[X].$$

Now we can estimate $\mathbb{E}[X]$ as the sample average of independent draws from the distribution of Z. However, this is only useful if $\mathrm{Var}[Z] < \mathrm{Var}[X]$. It turns out that the more closely correlated X and Y, the better the variance reduction.

We can show this as follows.

$$\mathrm{Var}[Z] = \mathrm{Var}[X] + c^2\mathrm{Var}[Y] + 2c\mathrm{Cov}[X, Y]. \tag{4.15}$$

The RHS of (4.15) is a quadratic in c and we can minimize $\mathrm{Var}[Z]$ by differentiating with respect to c, setting equal to zero, yielding

$$\bar{c} = -\frac{\mathrm{Cov}[X, Y]}{\mathrm{Var}[Y]}.$$

By making the substitution $c \to \bar{c}$, we have

$$\min_{c \in \mathbb{R}} \mathrm{Var}[Z] = \mathrm{Var}[X] + \frac{\mathrm{Cor}[X, Y]^2}{\mathrm{Var}[Y]^2} \cdot \mathrm{Var}[Y] - 2\frac{\mathrm{Cov}[X, Y]^2}{\mathrm{Var}[Y]}$$

$$= \mathrm{Var}[X]\left(1 - \frac{\mathrm{Cov}[X, Y]^2}{\mathrm{Var}[X]\mathrm{Var}[Y]}\right)$$

$$= \mathrm{Var}[X]\left(1 - \rho(X, Y)^2\right),$$

where $\rho(X, Y)$ denotes the correlation of X and Y, and therefore $1 - \rho(X, Y)^2 \in [0, 1]$. So it is certainly the case that $\min_{c \in \mathbb{R}} \mathrm{Var}[Z] \leq \mathrm{Var}[X]$, and we see that the variance reduction will be better the closer $\rho(X, Y)^2$ is to 1.

There is no general method to identify the optimal control variate for a particular estimation problem and so one may have to apply either wisdom or experience. Fortunately for those of us with little of either, a suitable control variate sometimes suggests itself.

4.4.1.1 A Python Example of the Control Variate Technique for Asian Options

Consider a path dependent option without an explicit value formula, such as the arithmetic average Asian option, but where there is an option with known formula which should have a similar though not identical value, such as the geometric average Asian option.

Let's try this. In context, $\mathbb{E}[X]$ is the value of the arithmetic average Asian call that we wish to estimate, and $\mathbb{E}[Y]$ is the value of the geometric average Asian option given by (4.11)–(4.14) in Sect. 4.3.2.1. Define parameters for the option. Then compute the precise value of $\mathbb{E}[Y]$ using the function `geoAsian()` function constructed in Exercise 4.6 and store it as a variable named `geo`:

```
In []: S0=4; E=4; sigma=0.25; r=0.03; T=1; N=100
        geo=geoAsian(S0,E,sigma,r,T,N)
```

Choosing $c = 1$ as an illustration, we will now proceed with a comparative Monte Carlo estimation of the arithmetic average Asian call. Compute an ensemble of M Black-Scholes asset price trajectories using the function `bsEnsemble` developed in Exercise 1.11, and store in a variable `Spath`. Now compute the arithmetic and geometric average of asset prices along each trajectory

```
In []: arithAvg=np.mean(Spath,axis=1)
        geoAvg=np.exp((1/(N+1))*np.sum(np.log(Spath),axis=1))
```

Compute the payoff of each call option along each trajectory:

```
In []: pArith=dFac(r,T)*np.maximum(arithAvg-E,0)
        pGeo=dFac(r,T)*np.maximum(geoAvg-E,0)
```

Construct the values of $Z = X + (Y - \mathbb{E}[Y])$ along each trajectory:

```
In []: z=pArith+geo-pGeo
```

Finally, we will compute values for the arithmetic average Asian option using both a brute-force approach

```
In []: pMean=np.mean(pArith)
        pStd=np.std(pArith)
        bruteCI=pMean+np.array([-1,1])*1.96*pStd/np.sqrt(M)
```

and a control variate approach

```
In []: zMean=np.mean(z)
        zStd=np.std(z)
        cvCI=zMean+np.array([-1,1])*1.96*zStd/np.sqrt(M)
```

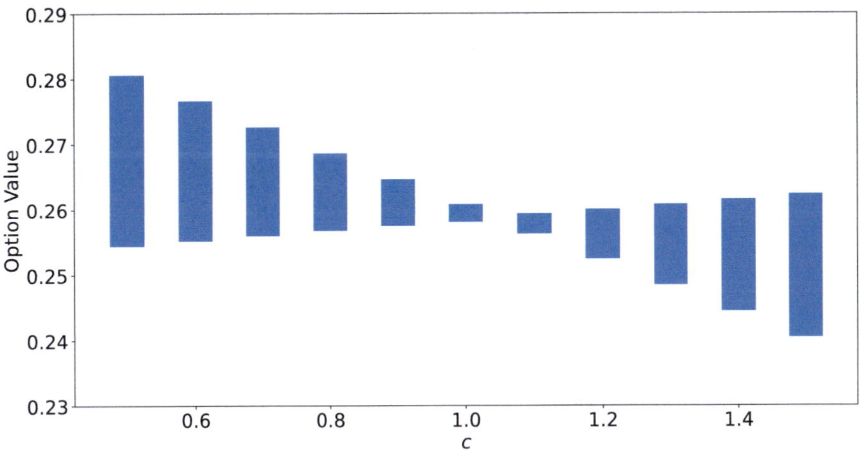

Fig. 4.2 95% confidence intervals for the Monte Carlo valuation of an arithmetic Asian call using a geometric average Asian call as a control variate for a range of values of c, using the parameter values $S_0 = E = 4$, $\sigma = 0.25$, $r = 0.03$, $T = 1$, $N = 100$, and $M = 1000$. For this example, and reusing a common ensemble of observed trajectories, the variance is minimized by choosing $c \approx 1$

It is not always obvious how to choose c in order to maximise the variance reduction achieved, but one can demonstrate that its value has an effect: see Fig. 4.2.

4.4.2 Antithetic Variates

An approach using *variance reduction by antithetic variate* to value a binary/digital option requires that we statistically estimate

$$\mathbb{E}\left[e^{-rT} \left(\frac{\Lambda\left(S_0 e^{(r-\sigma^2/2)T + \sigma\sqrt{T}Z} \right) + \Lambda\left(S_0 e^{(r-\sigma^2/2)T - \sigma\sqrt{T}Z} \right)}{2} \right) \right] \tag{4.16}$$

instead of (4.3), reusing the sampled value of Z in both evaluations of the payoff. Here, Z and $-Z$ have the same $\mathcal{N}(0, 1)$ distribution.

We will illustrate the effectiveness of this technique via the related *toy problem* of estimating the expected future value of a Black-Scholes asset model that has been discounted back to time $t = 0$:

$$\mathbb{E}[e^{-rT} S(T)] = \mathbb{E}\left[e^{-rT} S_0 e^{(r-\sigma^2/2)T + \sigma\sqrt{T}Z} \right] = \mathbb{E}\left[S_0 e^{-\sigma^2 T/2 + \sigma\sqrt{T}Z} \right]$$

$$= S_0 e^{-\sigma^2 T/2 + \sigma^2 T/2} = S_0,$$

where we have used the mean of a lognormal random variable at the penultimate step, and similarly we can calculate the variance as

$$\mathrm{Var}[e^{-rT}S(T)] = S_0^2(e^{\sigma^2 T} - 1). \tag{4.17}$$

As in Sect. 4.4.1, the goal is to find an alternative random variable with the same mean and smaller variance. Define the random variable

$$X = \frac{S_0}{2}\left(e^{-\sigma^2 T/2 + \sigma\sqrt{T}Z} + e^{-\sigma^2 T/2 - \sigma\sqrt{T}Z}\right).$$

One can show that X has the following mean and variance (see Exercise 4.12)

$$\mathbb{E}[X] = S_0; \quad \mathrm{Var}[X] = \frac{S_0^2}{2}\left(e^{\sigma^2 T} - 1\right) + \frac{S_0^2}{2}\left(1 - e^{\sigma^2 T}\right). \tag{4.18}$$

Since $T > 0$ we have $1 - e^{\sigma^2 T} < 0$, and it is clear by comparison with (4.17) that it is always the case that $\mathrm{Var}[X] < \frac{1}{2}\mathrm{Var}[e^{-rT}S(T)]$. This represents a considerable variance reduction of at least 50%, allowing us to efficiently estimate $\mathbb{E}[e^{-rT}S(T)]$ by sampling from X instead.

More generally, if we wish to estimate $\mathbb{E}[f(Z)]$, where f is some function of the standard normal random variable Z then we can make use of the fact that the antithetic variate $-Z$ has the same distribution as Z, since both are standard normal random variables, and

$$\mathbb{E}[f(Z)] = \frac{\mathbb{E}[f(Z)] + \mathbb{E}[f(-Z)]}{2} = \mathbb{E}\left[\frac{f(Z) + f(-Z)}{2}\right].$$

We can guarantee a variance reduction by sampling from the distribution of the secondary random variable $(f(Z) + f(-Z))/2$ when $\mathrm{Cov}[f(Z), f(-Z)]$ is negative, since

$$\mathrm{Var}\left[\frac{f(Z) + f(-Z)}{2}\right] = \frac{1}{4}\left(\mathrm{Var}[f(Z)] + \mathrm{Var}[f(-Z)] + 2\mathrm{Cov}[f(Z), f(-Z)]\right)$$

$$= \frac{1}{2}\mathrm{Var}[f(Z)] + \frac{1}{2}\mathrm{Cov}[f(Z), f(-Z)].$$

Since Z is symmetrically distributed on $\mathbb{R}$, then this will generally be the case for functions f that are monotone or close to monotone.[2] For example in (4.16) we have

$$f(z) = e^{-rT}\Lambda\left(S_0 e^{(r - \sigma^2/2)T + \sigma\sqrt{T}z}\right).$$

---

[2] A function f is monotone non-decreasing (resp. non-increasing) if, for all $x \leq y$, we have $f(x) \leq f(y)$ (resp. $f(x) \geq f(y)$).

The parameters S_0 and σ are positive, and so the inner exponential part of f is certainly monotone increasing in z, and the extent of the variance reduction is therefore influenced by the effect of the payoff function Λ on that monotonicity. In practice the inner exponential means that the form of the payoff Λ is likely to have very little effect.

4.4.2.1 An Antithetic Variate Example for Power Options

Let's try applying this technique to the valuation of an option with a payoff that is polynomial in the underlying asset. Suppose we wish to estimate the value of a call option with strike E, where $\Lambda(S(T)) = (S(T)^n - E)^+$, for some choice of $n = 1, 2, \dots$. See Fig. 4.3. When $n > 1$, power options allow the holder to benefit from leverage while transferring the risk to the writer and so they will tend to be more expensive than their European counterparts, which correspond to the $n = 1$ case. They are then also referred to as leveraged options.

If S is governed by the Black-Scholes model then a power call has a value given by the formula

$$V_{\text{pow}}(S_0, 0) = S_0^n e^{((n-1)(r+n\sigma^2/2))T} N(d_1) - E e^{-rT} N(d_2),$$

where

$$d_1 = \frac{\ln(S_0/E^{1/n}) + (r + (n - 1/2)\sigma^2)T}{\sigma\sqrt{T}}; \quad d_2 = d_1 - n\sigma\sqrt{T}.$$

In practice, power options may be written with an interest rate or volatility as the underlying asset and so this formula may not produce an appropriate valuation. Sampling from the distribution of SDE models for interest rates and stochastic volatility processes will be covered in Chaps. 7 and 8.

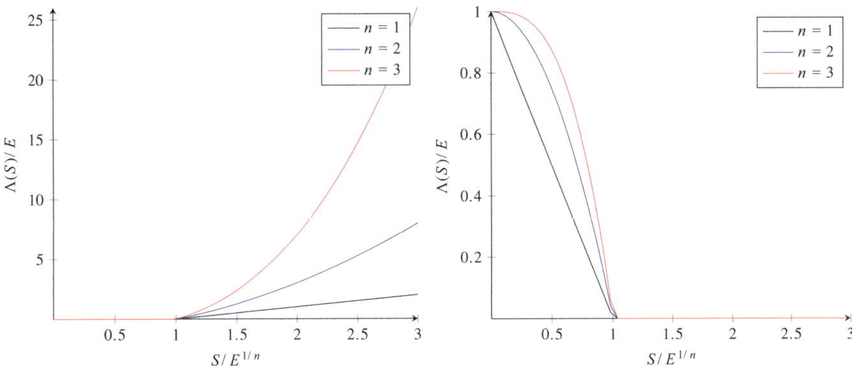

Fig. 4.3 A strike-independent representation of the payoff functions Λ of power call (on the left) and power put (on the right) options for $n = 1, 2, 3$

Suppose $n = 1$ (so that we are working with a European call). In Python, we modify Step 2 of the code provided in Sect. 4.1 to incorporate an antithetic variate as follows. First, generate the sample of observations on the asset model:

```
In []: # Set up option and simulation parameters
       S0=4; E=4; sig=0.25; r=0.03;T=1
       M=10**3

       # Sample from asset price and antithetic counterpart
       Z=rng.normal(0,1,M)
       S=S0*np.exp((r-0.5*sig**2)*T+sig*np.sqrt(T)*Z)
       Santi=S0*np.exp((r-0.5*sig**2)*T-sig*np.sqrt(T)*Z)

       # Compute payoffs
       Vcall=dFac(r,T)*(np.maximum(S-E,0))
       VcallAnti=dFac(r,T)*(np.maximum(Santi-E,0))
```

Step 3 then becomes

```
In []: Vcall=0.5*(Vcall+VcallAnti)
       aManti=np.mean(Vcall); bManti=np.std(Vcall)
       CI=aManti+np.array([-1,1])*1.96*bManti/np.sqrt(M)
       print(CI)
```

For path-dependent cases such as the barrier or Asian options, we need to generate the ensemble of trajectories as in Sect. 1.3.3 along with a second ensemble array variable containing their antithetic counterparts. Start by generating the $M \times (N + 1)$ array containing the standard Brownian motion ensemble named B. Then construct S and Santi:

```
In []:  S=S0*np.exp((r-0.5*sig**2)*T+sig*B)
        Santi=S0*np.exp((r-0.5*sig**2)*T-sig*B)
```

Note that at each step the ensemble of values for B is generated once, with the same observations used for both asset price arrays. The only difference is the sign in front of sig in each case. See Fig. 4.4 for a comparison of trajectories with their antithetic counterparts. Exercise 4.9 asks you to explore the use of variance reduction by antithetic variate for power options with $n > 1$.

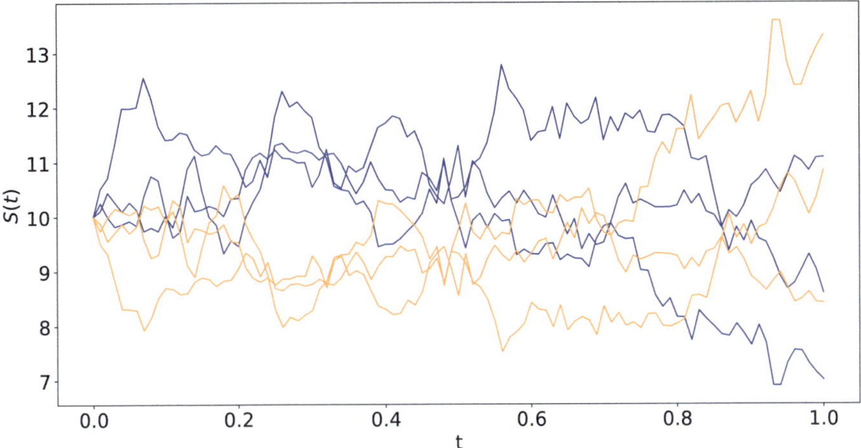

Fig. 4.4 Five asset price trajectories in blue and their antithetic counterparts in orange

4.4.3 Importance Sampling

Change-of-measure can be applied to reduce the computational cost of pricing an option where a significant number of sampled trajectories taken in a brute-force approach will be zero-valued, and therefore will not contribute materially to the estimation of the option value.

4.4.3.1 The Example of a European Option far Out-of-the-Money

Consider a European put option with a short time to expiry, moderate risk-free rate and volatility, and a strike price that is significantly in excess of the the present value of the underlying. Intuitively we would expect that in order to estimate the option value with some reasonable level of precision, our sample size M should be high enough that we include in our sample some of the relatively rare samples where the asset finishes ITM, and enough of these that the sample average reflects the true contribution of this subset of the sample space to the overall value of the option. The idea of importance sampling is to simulate the option value as though the underlying was closer to the strike, but to use a change of measure to ensure that the expectation of observations in our sample remains that taken under the risk-neutral measure.

For example, choose

```
In []: S0=50; E=10; T=0.25; r=0.01; sig=0.2; M=100
```

Carry out Step 2 of the algorithm for brute-force Monte Carlo from Sect. 4.1 and inspect the contents of the variable Vput. The majority of values will be zero and, in order to capture a characteristic sample of trajectories that finish ITM, we will need to choose M large. Intuitively we might expect that if we were to bring the value of the underlying asset closer to the strike, then a higher proportion of trajectories would finish ITM.

To see this, consider that, under the risk-neutral measure $\mathbb{Q}$, we estimate

$$\mathbb{E}_{\mathbb{Q}}\left[e^{-rT}(E - S(T))^{+}\right],$$

where $Y := \ln(S(T)/S_0) = (r - \sigma^2/2)T + \sigma B(T) \sim \mathcal{N}((r - \sigma^2/2)T, \sigma^2 T)$. If we change to a measure $\mathbb{P}$ where

$$Y \sim \mathcal{N}(\ln(E/S_0) - \sigma^2 T/2, \sigma^2 T), \tag{4.19}$$

then

$$\mathbb{E}_{\mathbb{P}}[S(T)] = S_0 e^{\ln(E/S_0) - \sigma^2 T/2 + \sigma^2 T/2} = E.$$

This is in practice the same as replacing rT with $\ln(E/S_0)$ in the risk-neutral form of the Black-Scholes model. If the risk neutral measure is denoted $\mathbb{Q}$ and the importance sampling measure is denoted $\mathbb{P}$, the value of the option is given by

$$\mathbb{E}_{\mathbb{P}}\left[e^{-rT}(E - S_0 e^{Y})^{+}\frac{d\mathbb{Q}}{d\mathbb{P}}\right],$$

where $d\mathbb{Q}/d\mathbb{P}$ is the appropriate Radon-Nikodym derivative. Under importance sampling the mean of the underlying asset at expiry is E, so we should expect a higher proportion of our sample call values to finish in-the-money. See Fig. 4.5 for an illustration of the difference between risk-neutral sampling and importance sampling in the context of a European put option. The same idea may be applied

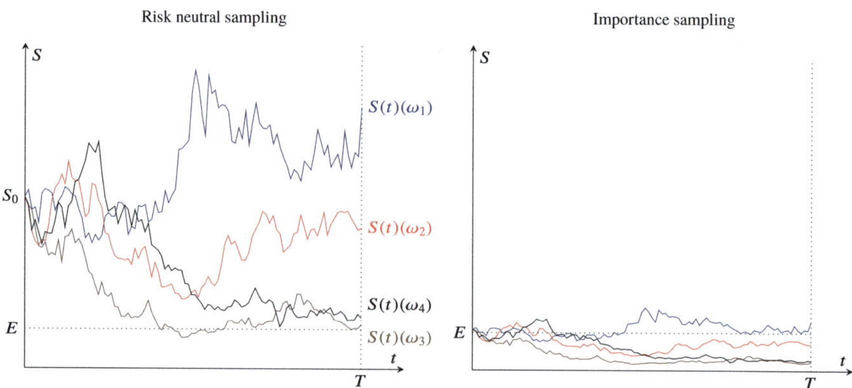

Fig. 4.5 Four trajectories of the underlying asset for Monte Carlo estimation of the value of a European put option with strike E. On the left, we see the trajectories generated for a given Brownian ensemble using risk-neutral sampling. None of the sampled trajectories finish ITM. On the right we see the same trajectories under importance sampling. Corresponding trajectories are indicated by matching colour, and three out of four trajectories finish ITM. The relative scale of the horizontal axis and vertical axis is the same in each case

(for example) to barrier options in order to reduce the overall number of samples required to capture a characteristic sample of trajectories that breach the barrier.

Now we must determine the form of the Radon-Nikodym derivative. Let X be a real-valued random variable, and let $\mathbb{P}_1$ and $\mathbb{P}_2$ be probability measures associated with PDFs f_1 and f_2 respectively taking positive values over all of $\mathbb{R}$ such that

$$\mathbb{E}_{\mathbb{P}_1}[X] = \int_{\mathbb{R}} x f_1(x) dx; \quad \mathbb{E}_{\mathbb{P}_2}[X] = \int_{\mathbb{R}} x f_2(x) dx.$$

Then the ratio of PDFs (referred to as a *likelihood ratio*) is denoted $L(x) := f_1(x)/f_2(x)$. When taking the random variable X as its argument, $L(X)$ corresponds a.s. to the Radon-Nikodym derivative and may be used to effect the change between probability measures:

$$\mathbb{E}_{\mathbb{P}_1}[X] = \int_{\mathbb{R}} x f_1(x) dx = \int_{\mathbb{R}} x L(x) f_2(x) dx = \mathbb{E}_{\mathbb{P}_2}[L(X)X].$$

For this importance sampling application, the likelihood ratio is given by

$$L(s) = \frac{\varphi\left(\ln(s); rT - \sigma^2 T/2, \sigma\sqrt{T}\right)}{\varphi\left(\ln(s); \ln(E/S_0) - \sigma^2 T/2, \sigma\sqrt{T}\right)}, \tag{4.20}$$

for $s \in [0, T]$, and where φ is the density function of a normal random variable:

$$\varphi(x; m, v) = \frac{1}{\sqrt{2\pi}\,v} e^{-\frac{(x-m)^2}{2v^2}}, \quad x \in \mathbb{R}.$$

We can use the `norm.pdf()` call from the `scipy.stats` module to access this in code, which takes the same three arguments x, m, and v.

In order to estimate the value of the option we sample from the distribution of

$$e^{-rT}\left(E - S_0 e^Y\right)^+ \frac{\varphi\left(Y; rT - \sigma^2 T/2, \sigma\sqrt{T}\right)}{\varphi\left(Y; \ln(E/S_0) - \sigma^2 T/2, \sigma\sqrt{T}\right)}, \tag{4.21}$$

where

$$Y = \ln(E/S_0) - \sigma^2 T/2 + \sigma\sqrt{T} Z,$$

and $Z \sim \mathcal{N}(0, 1)$.

4.4.3.2 Implementation in Python

The major change to the Monte Carlo algorithm described in Sect. 4.1 is in Step 2, where we create a vector holding call payoffs for each random sample. Start by first sampling M observations on Y according to (4.19):

```
In []: LSE=np.log(E/S0)
       Y=LSE-0.5*T*sig**2+sig*rng.normal(0,T,M)
```

Now compute the corresponding values of the Radon-Nikodym derivative according to (4.20):

```
In []: num=norm.pdf(Y,r*T-0.5*T*sig**2,sig*np.sqrt(T))
       denom=norm.pdf(Y,LSE-0.5*T*sig**2,sig*np.sqrt(T))
       radNik=num/denom
```

Finally compute the vector of call values according to (4.21):

```
In []: Vcall=dFac(r,T)*np.maximum(S0*np.exp(Y)-E,0)*radNik
```

Step 3, where we compute sample mean and standard deviation using the values in `Vcall`, proceeds as before.

4.5   Estimating the Greeks: Bump-and-Revalue

The lessons of Sect. 4.4 are particularly relevant for the Monte Carlo estimation of the *Greeks*, or option price sensitives, which are expressed as first- and second-order partial derivatives of the option price: see Definition 2.11 in Chap. 2. Two important examples are the first- and second-order sensitivities of option prices to changes in the underlying asset value, denoted Δ and Γ respectively.

$$\Delta = V_S(S, t); \quad \Gamma = V_{SS}(S, t).$$

Using the bump-and-revalue technique, there are two approximation steps involved in estimating either Δ or Γ. First, use a finite difference approximation of the partial derivative, then use Monte Carlo to estimate the individual option values contained therein. So, if we expand V as a Taylor series to first order in the S direction around the point $(S, t) \in \mathbb{R}^+ \times [0, T]$, then[3]

$$V(S + \delta S, t) = V(S, t) + V_S(S, t)\delta S + O\left((\delta S)^2\right).$$

---

[3] We write $f(x) = O(g(x))$ as $x \to 0$ iff $\limsup_{x \to 0} |f(x)|/g(x) < \infty$. For notational convenience, we often omit the expression "as $x \to 0$".

and we can rearrange to get

$$V_S(S, t) = \frac{V(S + \delta S, t) - V(S, t)}{\delta S} + O(\delta S). \tag{4.22}$$

The first term on the RHS of (4.22) is referred to as the *forward difference approximation* of V_S at the point (S, t). If we truncate the remainder terms that are $O(\delta S)$ and above, we can write down the estimator

$$\widehat{V_S}(S, t) = \frac{\hat{V}(S + \delta S, t) - \hat{V}(S, t)}{\delta S},$$

where $\hat{V}$ denotes the Monte Carlo estimated value of the option value at each of the points $(S + \delta S, t)$ and (S, t). By now the reason we refer to this as the *bump-and-revalue* method should be clear: we must value the option at (S, t), then perturb (bump) the initial asset value by an amount δS and revalue the option at $(S + \delta S, t)$.

Let's consider the variance of this estimator:

$$\text{Var}\left[\widehat{V_S}(S, t)\right]$$

$$= \frac{1}{(\delta S)^2} \text{Var}\left[\hat{V}(S + \delta S, t) - \hat{V}(S, t)\right]$$

$$= \frac{1}{(\delta S)^2} \left(\text{Var}\left[\hat{V}(S + \delta S, t)\right] + \text{Var}\left[\hat{V}(S, t)\right] - 2\text{Cov}\left[\hat{V}(S + \delta S, t), \hat{V}(S, t)\right]\right).$$

We will make the simplifying assumption that

$$\text{Var}\left[\hat{V}(S + \delta S, t)\right] \approx \text{Var}\left[\hat{V}(S, t)\right], \tag{4.23}$$

which is reasonable if $\text{Var}[V(S, t)]$ is a continuous function of S and δS is small. Then we can write

$$\text{Var}\left[\widehat{V_S}(S, t)\right] \approx \frac{2}{(\delta S)^2} \text{Var}\left[\hat{V}(S, t)\right] \left(1 - \rho(\hat{V}(S + \delta S, t), \hat{V}(S, t))\right), \tag{4.24}$$

where ρ denotes the correlation of $\hat{V}(S + \delta S, t)$ and $\hat{V}(S, t)$. There are three factors on the RHS of (4.24). We established in Sect. 4.1 that the second of these, the variance of the Monte Carlo estimator $\hat{V}(S, t)$, is $O(1/M)$, where M is the number of samples drawn from the distribution of $V(S, t)$ and so this dependence is also inherited by the variance of the Δ estimator $\widehat{V_S}$. Variance reduction strategies such as those described in Sect. 4.4 can help to mitigate this.

The first factor on the RHS of (4.24) shows an inverse dependence on the square of our choice of discretisation step $(\delta S)^2$. This presents a new and somewhat paradoxical challenge, since it implies that we cannot choose δS very small without inflating the variance of the estimator, but in spite of this we somehow need to choose it small in order to avoid large truncation error as given by the remainder terms on the RHS of (4.22). There is some hope contained in the third factor on the RHS, which tells us that if $\rho(\hat{V}(S+\delta S,t),\hat{V}(S,t))$ can be brought close to one by our choice of δS then this source of deviation might be offset.

We will discuss two approaches of contrasting effectiveness here. The first involves estimating $\hat{V}(S+\delta S,t)$ and $\hat{V}(S,t)$ using separate and independent draws from the distribution of $V(S,t)$. For example, if V represents the value of a digital option,

$$\Delta \approx \frac{e^{-r(T-t)}}{\delta S}\left(\mathbb{E}\left[\Lambda\left((S+\delta S)e^{(r-\sigma^2/2)(T-t)+\sigma\sqrt{(T-t)}Z}\right)\right]\right.$$
$$\left.-\mathbb{E}\left[\Lambda\left(Se^{(r-\sigma^2/2)(T-t)+\sigma\sqrt{(T-t)}Z}\right)\right]\right), \qquad (4.25)$$

and we estimate both expectations on the RHS by making two independent samples from the distribution of Z. In general for the bump-and-revalue approach the best we can say is that

$$\mathrm{Var}\left[\hat{V}_S(S,t)\right]=\frac{1}{(\delta S)^2}\mathrm{Var}\left[\hat{V}(S+\delta S,t)-\hat{V}(S,t)\right]$$
$$\approx\frac{2}{(\delta S)^2}\mathrm{Var}\left[\hat{V}(S,t)\right]=O\left(\frac{1}{(\delta S)^2 M}\right).$$

where we have relied again on the simplifying assumption (4.23). Due to the $(\delta S)^2$ in the denominator of the order term, the potential for blow-up exists if we revalue using an entirely new sample. Moreover it is not clear how one might choose δS to avoid this.

The second approach instead uses the same set of observations to estimate both $\hat{V}(S+\delta S,t)$ and $\hat{V}(S,t)$ in a bid to maximise their correlation. It can be shown that in this case, as long as the discounted payoff of the option is almost everywhere a continuous function of S, then $\mathrm{Var}\left[\widehat{V_S}(S,t)\right]=O(1/M)$, so that the negative influence of δS has been mitigated. Going back to the example of a digital option, we would rewrite the RHS of (4.25) as a single expectation:

$$\Delta \approx \frac{e^{-r(T-t)}}{\delta S}\left(\mathbb{E}\left[\Lambda\left((S+h)e^{(r-\sigma^2/2)(T-t)+\sigma\sqrt{(T-t)}Z}\right)\right.\right.$$
$$\left.\left.-\Lambda\left(Se^{(r-\sigma^2/2)(T-t)+\sigma\sqrt{(T-t)}Z}\right)\right]\right), \qquad (4.26)$$

so that we estimate by averaging M samples from the distribution of

$$\Lambda\left((S+\delta S)e^{(r-\sigma^2/2)(T-t)+\sigma\sqrt{(T-t)}Z}\right)-\Lambda\left(Se^{(r-\sigma^2/2)(T-t)+\sigma\sqrt{(T-t)}Z}\right),$$

using the same observation on Z for both instances of its appearance.

In practice however, many options that require a Monte Carlo approach will have a discounted payoff that is not almost everywhere continuous in S (this is the case for barrier options that are monitored at discrete times, for example). Nonetheless, if we reuse our random sample, then $\mathrm{Var}\left[\widehat{V_S}(S, t)\right] = O(1/(M\delta S))$ and the effect of the approximating step can be partially mitigated.

Variations on this approach are available. For example if we instead expand V as a Taylor series to first order in the S direction around the point $(S - \delta S)$ over a step of length $2\delta S$ then we arrive at the *central difference approximation*

$$\widehat{V_S}(S, t) = \frac{\hat{V}(S + \delta S, t) - \hat{V}(S - \delta S)}{2(\delta S)}$$

which is computationally more expensive in that we now need to recompute the estimation of V by shifting our computation by δS in two directions, but can improve the rate of convergence of the bias of $\hat{V}(S, t)$. The additional computations can be re-used to estimate Γ via the *symmetric central difference approximation*

$$V_{SS}(S, t) \approx \frac{V(S + \delta S, t) - 2V(S, t)) + V(S - \delta S, t)}{(\delta S)^2}$$

In Chap. 5 we will take a more detailed look at finite difference approximation in the context of numerically solving the Black-Scholes PDE.

4.6   Further Reading

Those interested in a specialist treatment of Monte Carlo methods and their applications in finance which includes such detail may refer to the monographs of Glasserman [25] and Jaeckel [37]. Indeed both give treatments of the estimation of Greeks which include the bump-and-revalue method described in Sect. 4.5, and several other methods which aren't discussed here. Readers will find other applications for importance sampling there too, including to the valuation of barrier options. See also the textbook of Oosterlee and Grzelak [58].

As we proceed through the book we have introduced various options with closed-form valuation formulae to be used as benchmarks. In this chapter alone we saw barrier options with a continuously monitored barrier, geometric average asian options, and power options. A useful reference in this regard is Haug [27].

Regression-based methods for American and Bermudan options are covered in more detail in Glasserman [25], including the method of Longstaff and Schwarz [48] presented here, but also the similar method of Tsitsiklis and van Roy [65], and the broader class of stochastic mesh methods that such techniques fall into. See Jain and Oosterlee [38] for a more recent development along these lines that is effective for multidimensional Bermudan options and fast approximation of the associated Greeks.

4.6.1 Exercises

4.1 Run the code in Sect. 4.1 for $M = 10, 10^2, 10^3, 10^4, 10^5$ and compare the lengths of the resulting confidence intervals. Does the Black-Scholes value given by `euroCallBenchmark()` lie on each confidence interval?

4.2 Making use of (4.3), modify Step 2 of the Monte Carlo algorithm given in Sect. 4.1 to value a Cash-or-Nothing call with payoff

$$\Lambda(S(T)) = \begin{cases} A, & S(T) > E; \\ \frac{A}{2}, & S(T) = E; \\ 0, & S(T) < E. \end{cases}$$

You may use the same parameter values as before, and take $A = 10$. This type of option is effectively a bet for a fixed amount A that the asset will finish above the strike, and is an example of a binary, rather than digital, option.

4.3 Compare the variance of the estimator for brute force Monte Carlo to that of Monte Carlo with importance sampling for the example implemented in Sect. 4.4.3.

4.4

1. Compare the value for an American put option given by the Longstaff-Schwarz method in Sect. 4.2.3 to that provided by a binomial tree valuation using the function `amerPutTree()` from Exercise 3.7 in Chap. 3.
2. What happens to the value as the order of the polynomial approximation to the continuation values changes?
3. By reference to the online documentation at https://numpy.org/doc/stable/reference/routines.polynomials.html, try replacing the Legendre polynomials with an alternative set of basis polynomials in this algorithm. Do you notice a significant difference in the output?

4.5 In Sect. 4.3.1.2 we presented a Python implementation for the Monte Carlo valuation of a down-and-out call option with N equidistant monitoring times. Rewrite this code without using a `for` loop.

4.6 Write a Python function `geoAsian()` that takes values for S_0, E, T, r, σ, n and implements the valuation formula for a geometric average Asian option defined by (4.11)–(4.14) in Sect. 4.3.2.1.

4.7 In Sect. 3.6 of Chap. 3 we used a Cox-Ross-Rubenstein binomial tree to value a down-and-out call option written on an underlying asset with initial value $S_0 = 40$, risk free rate $r = 0.04$, volatility $\sigma = 0.4$, with strike price $E = 40$, continuously monitored barrier $B = 20$, and time to expiry $T = 1$. We found that a tree valuation with 439 timesteps was within a tolerance of $\varepsilon = 0.0036$ of the benchmark value of the continuous-time option.

1. For a brute-force Monte Carlo valuation of the same option, set $N = 439$ and, by experimentation, identify a sample size M to produce an estimate with a confidence interval of length at most 2ε.
2. Using the sample size M computed in part 1 of this exercise, produce Monte Carlo interval estimates of the value of the option for $N = 2^3, 2^6, 2^9, 2^{12}$ and comment.

4.8

1. Use the function `geoAsian()` from Exercise 4.6 to compute exact values for a geometric average Asian call option with strike $E = 100$, expiring in 6 months time on an asset whose underlying price is currently $S_0 = 99$ if $r = 0.01$, $\sigma = 0.4$, and the payoff at expiry is determined by averaging the asset price over $n = 2^4$ equally spaced observations.
2. Compare this exact value to the confidence interval of your Monte Carlo valuation for $M = 10^2, 10^4, 10^6$.

4.9 Modify the code in Sect. 4.4.2 to value power options for user-selected n, (otherwise keep the option and simulation parameters the same) and perform a comparison of the resulting confidence intervals with and without the use of antithetic variates for $n = 1, 2, 3$. Repeat for power puts.

4.10 Complete the proof of Theorem 4.1 by showing that

$$\mathrm{Var}\left[\ln\left(\left(\prod_{j=1}^{n} S(t_j)\right)^{1/n} / S_0\right)\right] = \sigma^2 T \frac{(n+1)(2n+1)}{6n^2}.$$

where $0 < t_1 < t_2 < \cdots < t_n = T$ are equally spaced, so that $t_i = i\delta t$, where $\delta t = T/n$. You may use the elementary series identity $\sum_{i=1}^{n} i^2 = n(n+1)(2n+1)/6$.

4.11 In Sect. 4.4.1.1, code was presented implementing a control variate applied to the Monte Carlo valuation of an arithmetic average Asian option.

1. Working from this code, calculate the ratio of the length of the confidence intervals for the brute force versus control variate Monte Carlo methods. Does use of the control variate lead to an improvement?
2. By reusing a single common ensemble of asset price trajectories, reproduce the output in Fig. 4.2 and hence confirm that $c \approx 1$ minimises the variance reduction for this example.

4.12 Prove (4.18) in Sect. 4.4.2.

4.13 Let X be a real-valued random variable and suppose that $\mathbb{P}_1$ and $\mathbb{P}_2$ are probability measures associated with density functions

$$u_1(x) = \frac{1}{\sqrt{2\pi}}e^{-\frac{1}{2}x^2}; \quad u_2(x) = \frac{1}{\sqrt{2\pi}}e^{-\frac{1}{2}(x-\mu)^2}$$

for some $\mu \in \mathbb{P}$. Compute the likelihood function $L(x)$ that corresponds a.s. to the Radon-Nikodym derivative $d\mathbb{P}_1/d\mathbb{P}_2$ when evaluated at X.

4.14 Implement both approaches to the Monte Carlo estimation of Δ described in Sect. 4.5. Try varying δS. What happens?

Finite Difference Methods 5

In Chaps. 3 and 4 we examined several ways of numerically estimating the value of an option as the discounted expected payoff under the risk-neutral measure. Now we will take a slightly different approach, instead looking at methods of numerically solving the Black-Scholes-Merton partial differential equation (PDE). These methods are useful for options where the Black-Scholes-Merton PDE has no closed-form solution and can accommodate variable model parameters such as local volatilities. However they experience the same complexity issues as binomial tree models for multi-asset options and are less-well suited than Monte Carlo methods for high dimensional financial problems as a result. In practice, there are Python packages available that work as PDE solvers (for example `py-pde`); this chapter is intended to give the reader an understanding of the fundamental issues that must be considered when designing and implementing a numerical scheme of this kind.

We will begin by reviewing the form of the Black-Scholes-Merton PDE, the general form of its boundary conditions, and a transformation that converts the PDE into the canonical heat equation form. Next, we construct an explicit discretisation scheme over a uniform mesh based upon finite difference approximations of the partial derivatives in the heat equation. We met some of these approximations already in Sect. 4.5 in the context of Monte Carlo estimation of option price sensitivities. We investigate the stability of this scheme and describe how errors, produced locally at each point on the mesh as an unavoidable consequence of the approximation, are propagated over time. We find that stability depends on maintaining a particular relationship between the size of the discretisation steps in time and in space. Moreover we find that the local errors themselves have a quadratic relationship with the stepsize in space but only a linear relationship with the stepsize in time.

C. Kelly, *Computation and Simulation for Finance*,
Springer Undergraduate Texts in Mathematics and Technology,
https://doi.org/10.1007/978-3-031-60575-8_5

Seeking to improve matters, we develop an implicit scheme and confirm its unconditional stability. Next, we present the Crank-Nicolson approximation method, which is both unconditionally stable and for which the local errors produced at each step are quadratic in both the space and time steps, ensuring greater computational efficiency for the method.

We then demonstrate how these methods, having been developed in quite a general setting, may be applied to the valuation of European options and barrier options. We also describe how these methods may be adapted in order to numerically approximate the optimal exercise boundary for options that allow early exercise, and hence to value American and Bermudan options. Finally, a link to the binomial methods treated in Chap. 3 is made explicit.

Python Setup To follow the Python code in this chapter, you will need to import

```
In []: import numpy as np
        import scipy.linalg as linalg
```

as well as, for plotting,

```
In []: %matplotlib notebook
        import matplotlib.pyplot as plt;
        from mpl_toolkits.mplot3d import axes3d
```

The `%matplotlib notebook` line enables interactive plotting features when invoked before importing the PyPlot package. In this chapter we don't require any user-defined functions from Chaps. 1–4.

5.1   The Black-Scholes-Merton PDE

5.1.1   The General Form of the PDE

Recall from Chap. 2 that, under the Black-Scholes framework, if $V(s, t)$ is the value of an option at time $t \in [0, T]$ when the underlying asset takes value $S(t) = s$, then V satisfies the partial differential equation

$$V_t(s, t) + \frac{1}{2}\sigma^2 s^2 V_{ss}(s, t) + rs V_s(s, t) - rV(s, t) = 0. \tag{5.1}$$

Indeed, (5.1) must hold for any option on the asset S whose value can be expressed as a sufficiently smooth function V. The state (or asset-price) domain of this PDE, which we denote $\mathcal{D}$, will vary depending on the type of option to be valued, and in this chapter will be $\mathcal{D} = (0, \infty)$, or some subinterval thereof with endpoints $0 \leq a < b \leq \infty$. This PDE is both linear in V and parabolic (see Sect. 5.1.3 below). To ensure the existence of a unique solution, two boundary conditions in s

and one final time condition are required, taking the form

$$
V(s, t) = \begin{cases} V_T(s) & \text{when} \quad t = T; \\ V_a(t) & \text{when} \quad s = a; \\ V_b(t) & \text{when} \quad s = b. \end{cases}
$$

These conditions, together with the choice of state domain $\mathcal{D}$, characterise the option to be solved. It may be the case where the domain interval $\mathcal{D}$ is open at one or both of its endpoints. Here the boundary conditions should be understood to hold as $s \to a^+$ at the left endpoint, and as $s \to b^-$ at the right endpoint.

In this chapter we learn how to value options in the Black-Scholes framework by discretising (5.1), and to do so we must take into consideration how to approximate such boundary conditions as well as how to approximate the partial derivatives. This can be done by applying finite difference methods directly to the Black-Scholes PDE (5.1), or indirectly by transforming (5.1) into the heat equation (given in the next section as (5.6)). We will focus on the latter approach, since the simple canonical form of the heat equation allows us to directly investigate such notions as stability, consistency and convergence of candidate numerical schemes.

5.1.2 Asset Price Domain and Boundary Conditions for European and Barrier Options

For all options written on the continuous-time asset model, the time domain is $t \in [0, T]$. However the state domain may change depending on the nature of the option, along with the final-time and boundary conditions. Consider the following as examples.

5.1.2.1 European Options

Recall from Chap. 2 that, for a European call option, the asset price domain of solutions of the Black-Scholes PDE (5.1) is given by $\mathcal{D} = (0, \infty)$ and $t \in [0, T]$ and the final-time and boundary conditions take the form

$$
V(s, T) = \max(s - E, 0); \quad \lim_{s \to 0^+} V(0, t) = 0; \quad V(s, t) \sim s \quad \text{as } s \to \infty.
$$

$$\tag{5.2}$$

For a European put option we also have $\mathcal{D} = (0, \infty)$ and

$$
V(s, T) = \max(E - s, 0); \quad \lim_{s \to 0^+} V(s, t) = Ee^{-r(T-t)}; \quad \lim_{s \to \infty} V(s, t) = 0.
$$

$$\tag{5.3}$$

5.1.2.2 Barrier Options

For a down-and-out barrier call option with barrier $B < E$ the asset-price domain of (5.1) changes to $\mathcal{D} = [B, \infty)$, and the boundary conditions become

$$V(s, T) = \Lambda(s); \quad V(B, t) = 0; \quad V(s, t) \sim s \quad \text{as } s \to \infty. \tag{5.4}$$

For an up-and-out barrier call option with barrier $B > E$ the asset-price domain is $\mathcal{D} = (0, B]$ and our boundary conditions are

$$V(s, T) = \Lambda(s); \quad \lim_{s \to 0^+} V(s, t) = 0; \quad V(B, t) = 0. \tag{5.5}$$

In each case Λ is the payoff for a European call or put option, depending on whether the barrier option is a call or a put.

The corresponding knock-in options can be valued via in-out parity, as described in Sect. 3.6.3 of Chap. 3.

5.1.3　Parabolic PDEs and the Heat Equation

Disregarding boundary conditions, a general linear PDE of second order is of the form

$$a u_{xx}(x, t) + b u_{xt}(x, t) + c u_{tt} + d u_x(x, t) + e u_t(x, t) + f u(x, t) = g$$

where a, b, c, d, e, f, g may depend on x and t, but not on their derivatives. When $b^2 - 4ac = 0$, the PDE is referred to as *parabolic*.

The canonical form of a parabolic PDE is the *heat equation*, which may be written with solutions defined on a strip $(x, \tau) \in \mathbb{R} \times [0, \bar{T}]$, for some $0 < \bar{T} < \infty$, as

$$u_\tau(x, \tau) = u_{xx}(x, \tau); \quad x \in \mathbb{R}, \quad \tau \in [0, \bar{T}]. \tag{5.6}$$

In this case we can ensure the existence of a unique solution with an initial condition at $\tau = 0$ and boundary conditions capturing the limit behaviour of u as $x \to \pm\infty$.

When designing computational approaches to the solution of (5.1), we can glean valuable intuition by considering the action of (5.6). See Fig. 5.1 for an illustration of how (5.6) evolves a solution over time, and a hint as to why this equation is sometimes referred to as the diffusion equation. The diffusive dynamic, where the action of the PDE is to dampen and spread out the solution, is characteristic of parabolic PDEs. Since any numerical approximation method applied to (5.6) will produce local errors, by which we mean errors produced at each point of approximation, the tendency of the PDE will be to damp these errors out over time. Therefore if we see errors propagating in an uncontrolled way in our numerical

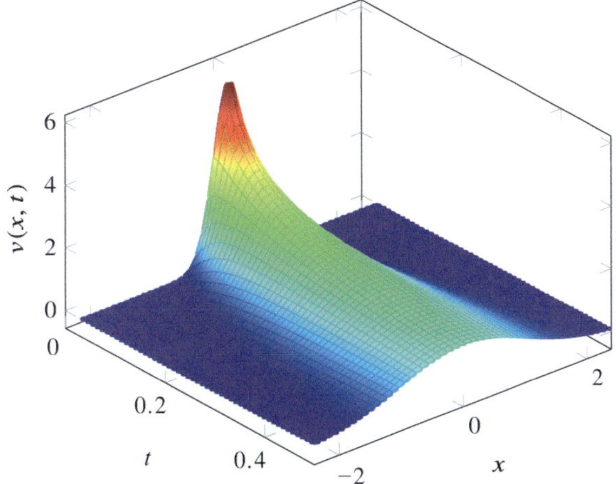

Fig. 5.1 The fundamental solution of the heat equation $v(x, \tau)$ (see (5.48) in Exercise 5.1) in the (x, τ) plane

approximation, this is numerical instability, and will be the fault of the scheme that we have designed. An important goal in this chapter is to see how we can avoid this.

5.1.4   Transforming the Black-Scholes-Merton PDE to the Heat Equation

Suppose that (5.1) holds for $s \in \mathcal{D}$ and $t \in [0, T]$. Apply the following set of transformations to (5.1):

$$x = \log(S/E), \quad \tau = \frac{1}{2}\sigma^2(T - t), \tag{5.7}$$

and

$$V(S, t) = E e^{\alpha x + \beta \tau} u(x, \tau), \tag{5.8}$$

where

$$\alpha = -\frac{1}{2}(k - 1), \quad \beta = -\frac{1}{4}(k + 1)^2, \quad k = \frac{2r}{\sigma^2}. \tag{5.9}$$

Then the transformed function u satisfies the diffusion equation

$$u_\tau(x, \tau) = u_{xx}(x, \tau), \quad x \in \mathcal{D}', \quad \tau \in \left[0, \bar{T}\right], \tag{5.10}$$

where $\bar{T} = \sigma^2 T/2$ and $\mathcal{D}' = \log(\mathcal{D}/E)$, the latter to be understood in the following sense:

$$\log(\mathcal{D}/E) := \{\log(s/E) \text{ for all } s \in \mathcal{D}\}.$$

For example in the case where $\mathcal{D} = (0, \infty)$, we would have $\mathcal{D}' = (-\infty, \infty)$.

5.1.5   A Uniform Finite Difference Mesh Over a Rectangular Domain

5.1.5.1 Construction of the Mesh, and Notation

In order to compute approximate values of $u(x, \tau)$ over a finite mesh, we will need to truncate the spatial dimension of the domain $\mathcal{D}' \times [0, \bar{T}]$, unless $\mathcal{D}'$ is already a finite interval. So we restrict x to the closed interval $[L_1, L_2]$, for some $-\infty < L_1 < L_2 < \infty$, and look later at a suitable way to choose L_1 and L_2 based upon our boundary conditions. We will construct a mesh with N_x steps in space and N_τ steps in time. Since the part of u that we are interested in is a surface defined on the rectangle $[L_1, L_2] \times [0, \bar{T}]$, this leads to stepsizes

$$\delta x = \frac{L_2 - L_1}{N_x}, \quad \delta \tau = \frac{\bar{T}}{N_\tau}.$$

At each point on the mesh we are going to numerically solve (5.10) to obtain the approximation

$$u(L_1 + n\delta x, m\delta \tau) \approx U_n^m; \quad n = 0, \dots, N_x, \quad m = 0, \dots, N_\tau, \tag{5.11}$$

by replacing the partial derivatives with approximations based on truncated Taylor series expansions. The values U_n^0, for $n = 1, \dots, N_x - 1$, are already determined by the initial condition and the values U_0^m and $U_{N_x}^m$, for $m = 1, \dots N_\tau$, will be determined by boundary conditions at the truncated points $x = L_1$ and $x = L_2$ respectively. So our scheme must give us a way to compute the values of U_n^m on the interior of the mesh (i.e. for $n = 1, \dots, N_x - 1$ and $m = 1, \dots, N_\tau$).

See Fig. 5.2 for an illustration of this mesh, with an indication of which mesh points are associated with a boundary/initial condition, and which must be computed via a numerical scheme approximating the action of (5.10). Although in previous chapters we represented time along the horizontal axis for asset price trajectories, in the context of PDEs it is traditional to transpose the axes so that the direction of movement in the time direction τ is up.

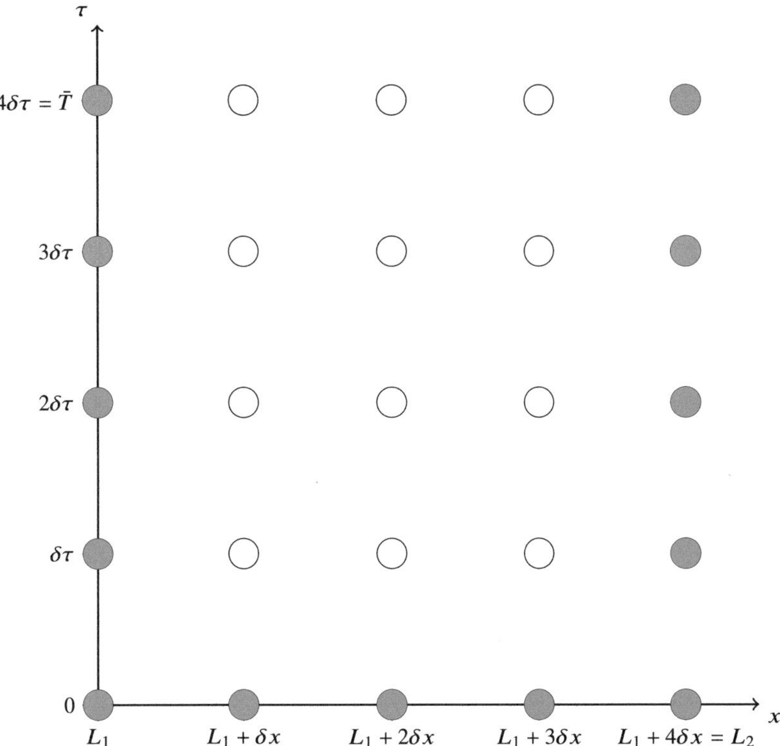

Fig. 5.2 The finite difference mesh over the domain $[L_1, L_2] \times [0, \bar{T}]$, when $N_x = N_\tau = 4$. Filled circles show meshpoints associated with predetermined boundary conditions. Empty circles show meshpoints at which the solution will be computed by the finite difference scheme

5.1.5.2 Python Representation of the Mesh Using `np.meshgrid()`

Suppose we take $N_x = N_\tau = 4$, $\bar{T} = 1$ and $[L_1, L_2] = [0, 1]$. Set up the stepsize in each direction:

```
In []: Nx=4; Ntau=5
       dx=1/Nx; dt=1/Ntau
```

Now define two array variables each containing all possible values of x and τ to be used in our computation

```
In []: x=np.linspace(0,1,Nx+1)
       tau=np.linspace(0,1,Ntau+1)
```

A computational Cartesian product of the array variables `x` and `tau` can then be created using `np.meshgrid()`:

```
In []: xx,tt = np.meshgrid(x,tau)
```

This structures data in a way that is a little non-intuitive for those seeing it for the first time, so let's examine `np.meshgrid()` in more detail. Invoking **xx** (respectively **tt**) then returns a 6×5 array where each row (respectively column) is the vector of values in **x** (respectively tau). For this example, if we inspect the contents of **x** followed by **xx**, we will see the output

```
In []: print(x)

Out[]: [0. 0.25 0.5 0.75 1. ]

In []: print(xx)

Out[]: [[0. 0.25 0.5 0.75 1. ]
        [0. 0.25 0.5 0.75 1. ]
        [0. 0.25 0.5 0.75 1. ]
        [0. 0.25 0.5 0.75 1. ]
        [0. 0.25 0.5 0.75 1. ]
        [0. 0.25 0.5 0.75 1. ]]
```

Similarly, the output for `tau` and `tt` looks like

```
In []: print(tau)

Out[]: [0. 0.2 0.4 0.6 0.8 1. ]

In []: print(tt)

Out[]: [[0.  0.  0.  0.  0. ]
        [0.2 0.2 0.2 0.2 0.2]
        [0.4 0.4 0.4 0.4 0.4]
        [0.6 0.6,0.6 0.6 0.6]
        [0.8 0.8 0.8 0.8 0.8]
        [1.  1.  1.  1.  1. ]]
```

This is useful if, for example, we wish to generate a matrix containing a linear function of (x, τ) (say $f(x, \tau) = 3x + 5\tau$) applied simultaneously at all points on the mesh $(n\delta x, m\delta \tau)$:

```
In []: print(3 * xx + 5 * tt)

Out[]: [[0.   0.75 1.5  2.25 3.  ]
        [1.   1.75 2.5  3.25 4.  ]
        [2.   2.75 3.5  4.25 5.  ]
        [3.   3.75 4.5  5.25 6.  ]
        [4.   4.75 5.5  6.25 7.  ]
        [5.   5.75 6.5  7.25 8.  ]]
```

We will do this when reversing the transformation defined by (5.7)–(5.9) for the purposes of plotting a Black-Scholes surface of option values in Sect. 5.5.

5.2 Construction of an Explicit Discretisation Scheme

5.2.1 Taylor series Expansion of Partial Derivatives

For Eq. (5.10), let's begin with the first order partial derivative in time u_τ. Just as we did for the estimation of $V_S(S, t)$ in Sect. 4.5, we can develop the forward difference approximation of u_τ at the point (x, τ) by expanding u as a Taylor series to first order in the τ direction around the point $(x, \tau) \in \mathbb{R} \times \left(0, \bar{T}\right)$:

$$u(x, \tau + \delta\tau) = u(x, \tau) + u_\tau(x, \tau)\delta\tau + O\left((\delta\tau)^2\right),$$

then rearrange to get

$$u_\tau(x, \tau) = \frac{u(x, \tau + \delta\tau) - u(x, \tau)}{\delta\tau} + O(\delta\tau). \tag{5.12}$$

What about the second order partial derivative in space u_{xx}? For this, we start with the following Taylor series expansions in the x direction, but make explicit all terms up to third order:

$$u(x + \delta x, \tau) = u(x, \tau) + u_x(x, \tau)\delta x + u_{xx}(x, \tau)\frac{(\delta x)^2}{2}$$

$$+ u_{xxx}(x, \tau)\frac{(\delta x)^3}{6} + O\left((\delta x)^4\right);$$

$$u(x - \delta x, \tau) = u(x, \tau) - u_x(x, \tau)\delta x + u_{xx}(x, \tau)\frac{(\delta x)^2}{2}$$

$$- u_{xxx}(x, \tau)\frac{(\delta x)^3}{6} + O\left((\delta x)^4\right).$$

Add both sides of both equations:

$$u(x + \delta x, \tau) + u(x - \delta x, \tau) = 2u(x, \tau) + u_{xx}(x, \tau)(\delta x)^2 + O\left((\delta x)^4\right),$$

and solve for $u_{xx}(x, \tau)$:

$$u_{xx}(x, \tau) = \frac{u(x + \delta x, \tau) - 2u(x, \tau) + u(x - \delta x, \tau)}{(\delta x)^2} + O\left((\delta x)^2\right). \tag{5.13}$$

The first term on the RHS of (5.13) is the symmetric central difference approximation of u_{xx} at (x, τ), the symmetry in question being invariance under the substitution $\delta x \to -\delta x$.

5.2.2   Building the FTCS Scheme

If we substitute the expressions (5.12) and (5.13) into (5.10) then we have

$$\frac{u(x, \tau + \delta\tau) - u(x, \tau)}{\delta\tau} + O(\delta\tau)$$

$$= \frac{u(x + \delta x, \tau) - 2u(x, \tau) + u(x - \delta x, \tau)}{(\delta x)^2} + O\left((\delta x)^2\right),$$

for $x \in [L_1 + \delta x, L_2 - \delta x]$ and $\tau \in [0, \bar{T} - \delta\tau]$. This is still an exact expression of (5.10), because all of the higher order terms in the Taylor series expansions have been rolled into the $O(\delta\tau)$ and $O\left((\delta x)^2\right)$ terms. If we disregard these in the hope that the resulting inaccuracy can be controlled by a suitable choice of $(\delta x, \delta\tau)$, then we switch notation to that defined in (5.11) in order to reflect the approximation, and write

$$\frac{U_n^{m+1} - U_n^m}{\delta\tau} = \frac{U_{n+1}^m - 2U_n^m + U_{n-1}^m}{(\delta x)^2}, \tag{5.14}$$

for $n = 1, \ldots, N_x - 1$ and $m = 0, \ldots, N_\tau - 1$. Since the superscript is the time index, we can rearrange the numerical scheme in (5.14) to write U_n^{m+1} explicitly in terms of U_{n-1}^m, U_n^m, and U_{n+1}^m, resulting in the following definition.

Definition 5.1   The *Forward in Time and Centred in Space (FTCS) finite difference scheme* for the heat Eq. (5.6) is given by the difference equation

$$U_n^{m+1} = \alpha U_{n+1}^m + (1 - 2\alpha)U_n^m + \alpha U_{n-1}^m, \tag{5.15}$$

for $n = 1, \ldots, N_x - 1$ and $m = 0, \ldots, N_\tau - 1$, where we define the stepsize ratio parameter $\alpha := \delta\tau/(\delta x)^2$.

A schematic guide to the use of information at each step of the scheme, called a stencil, is provided in Fig. 5.3.

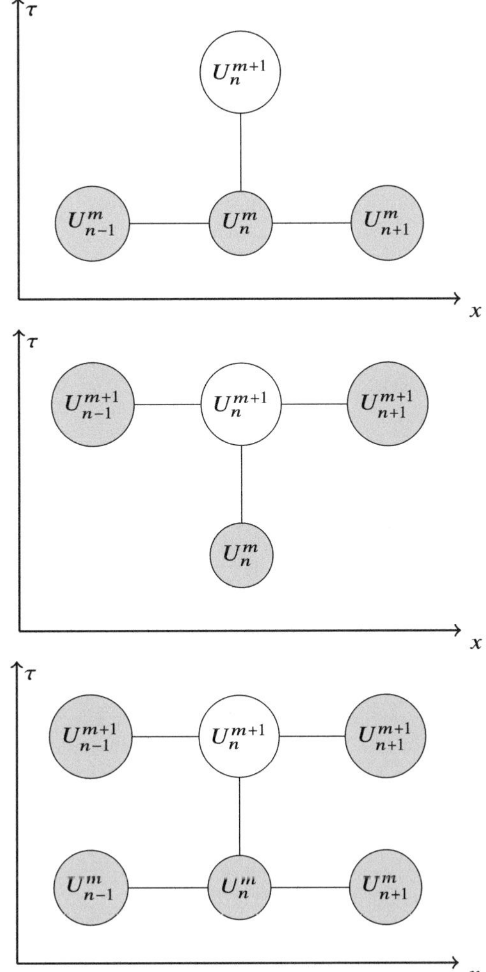

Fig. 5.3 Stencils for the FTCS (top), BTCS (middle) and Crank-Nicolson (bottom) schemes. The value we wish to compute, U_n^{m+1} is represented by an unfilled node. Grey nodes are the values that each scheme needs in order to compute U_n^{m+1} in each case

5.2.3   Python Implementation of the FTCS Scheme and Numerical Instability

Some preliminary computation will illustrate the use, and limitations, of the FTCS scheme. Suppose that $L_1 = 0$, $L_2 = 1$, $\bar{T} = 1$, so that we are approximating the solution for points $(x, \tau) \in [0, 1] \times [0, 1]$, and take $N_x = N_\tau = 4$, so that $\delta x = \delta \tau = 0.25$ and $\alpha = 0.25/(0.25)^2 = 4$:

```
In []: Nx = 4; Ntau = 4
       dx = 1/Nx; dt = 1/Ntau
       alpha = dt/(dx**2)
```

We will choose the initial condition

$$u_0(x) = -4\left(x - \frac{1}{2}\right)^2 + 1, \quad x \in [0, 1],$$

and boundary conditions satisfying

$$u(0, \tau) = u(1, \tau) = 0, \quad \tau \in [0, 1].$$

The mesh will contain $(N_x + 1) \times (N_\tau + 1)$ points, and so we set up an array variable of these dimensions to contain our approximate solution, and define the initial profile:

```
In[]:  U = np.zeros((Nx+1,Ntau+1))
       U[:,0] = -4*(np.linspace(0,1,Nx+1)-0.5)**2+1
```

The zero boundary conditions are automatically in place from the start, by construction of the variable U. By substituting $\alpha = 4$ into (5.15), we can generate the interior values of the solution using the update rule

$$U_n^{m+1} = 4U_{n+1}^m - 7U_n^m + 4U_{n-1}^m, \quad m = 0, 1, 2, 3, \quad n = 1, 2, 3.$$

Let's do this line by line, with a view to investigating the behaviour of the scheme. Generate the values at the first step (so that $m = 0$) by

```
In[]:   for n in range(1,Nx):
            U[n,1] = 4*U[n+1,0]-7*U[n,0]+4*U[n-1,0]
```

Repeat this loop for each of $m = 1, 2, 3$ to generate the solution vectors U[:,2], U[:,3], U[:,4]. The output of the computation is contained in Table 5.1, and we clearly see that the solution is displaying oscillations and blowup. This is inconsistent with the dynamic we expect from the action of the heat equation, illustrated (albeit for a problem with a different initial profile) in Fig. 5.1, and

Table 5.1 Numerical instability observed in an application of the FTCS scheme with $\alpha = 4$. Each vector of values $U^m = [U_0^m, \ldots, U_4^m]$ for $m = 0, \ldots, 4$ is represented on a row

x	0	0.25	0.5	0.75	1
U^0	0	0.75	1	0.75	0
U^1	0	−1.25	−1	−1.25	0
U^2	0	4.75	−3	4.75	0
U^3	0	−45.25	59	−45.25	0
U^4	0	552.75	−775	552.75	0

is evidence of a numerically unstable scheme. Such a scheme has a tendency to magnify errors that accumulate at each step, leading to the observed blowup.

Clearly, we should not trust this output. But how can we tell a priori if a scheme is reliable? We must understand two things: how errors are generated by the scheme and, once generated, how they are propagated.

5.3 Error Control for the FTCS Scheme

5.3.1 Global Error of a Numerical Scheme

Definition 5.2 For the finite difference approximation of a solution of (5.10), the *global error at the point* $(L_1 + n\delta x, m\delta\tau)$ is given by

$$e_n^m = u(L_1 + n\delta x, m\delta t) - U_n^m, \tag{5.16}$$

for every $n = 0, \ldots, N_x$ and $m = 0, \ldots, N_\tau$.

Our overarching goal is to construct a scheme which is convergent, in the sense that the maximum error across all meshpoints can be made arbitrarily small by choosing suitably small stepsizes in the x and τ directions:

$$\lim_{\delta x, \delta\tau \to 0} \left(\max_{\substack{n=1,\ldots,N_x, \\ m=1,\ldots,N_\tau}} e_n^m \right) = 0.$$

In our setting, the convergence of a scheme has two crucial aspects. First, truncation errors are unavoidably generated at each meshpoint by the use of a finite difference approximation rather than a partial derivative. If we can control the size of these errors by increasing the resolution of our mesh via the stepsizes δx and $\delta\tau$, we say that the scheme is *consistent*. Second, we must be able to ensure that those errors, along with any errors introduced by the use of finite precision arithmetic, will be damped out over time by repeated iterations of the scheme itself. The latter is ensured if our scheme is *stable*.

5.3.2 Stability and Consistency Together Imply Convergence

An important result called the *Lax Equivalence Theorem* tells us that a scheme which is both consistent and stable will be convergent. While we don't state the theorem in its general form here, we can illustrate it by a careful decomposition of the global error for the FTCS scheme as follows.

Rearrange (5.16) and write

$$U_n^m = u(L_1 + n\delta x, m\delta\tau) - e_n^m. \tag{5.17}$$

If we substitute the RHS of (5.17) into the FTCS update Eq. (5.15) we obtain (setting $\bar{n} := (n + L_1/\delta x)$)

$$\begin{aligned}
u(\bar{n}\delta x, (m+1)\delta\tau) &- e_n^{m+1} \\
&= \alpha\left[u((\bar{n}+1)\delta x, m\delta\tau) - e_{n+1}^m\right] + (1-2\alpha)\left[u(\bar{n}\delta x, m\delta\tau) - e_n^m\right] \\
&\quad + \alpha\left[u((\bar{n}-1)\delta x, m\delta\tau) - e_{n-1}^m\right] \\
&= \alpha u((\bar{n}+1)\delta x, m\delta\tau) + (1-2\alpha)u(\bar{n}\delta x, m\delta\tau) + \alpha u((\bar{n}-1)\delta x, m\delta\tau) \\
&\quad - \left[\alpha e_{n+1}^m + (1-2\alpha)e_n^m + \alpha e_{n-1}^m\right].
\end{aligned} \tag{5.18}$$

The error introduced if we take the true solution at the start of the timestep and apply a single iteration of the scheme (5.15) is known as the *local accuracy* of the FTCS scheme, and we define it

$$\begin{aligned}
R_n^m := \ &u(\bar{n}\delta x, (m+1)\delta\tau) \\
&- \left[\alpha u((\bar{n}+1)\delta x, m\delta\tau) + (1-2\alpha)u(\bar{n}\delta x, m\delta\tau) + \alpha u((\bar{n}-1)\delta x, m\delta\tau)\right].
\end{aligned} \tag{5.19}$$

Notice that if the scheme were to perfectly reproduce the action of the heat Eq. (5.10) over a single step, then no local error would be generated and $R_n^m = 0$. Substituting (5.19) into (5.18) and rearranging yields the relation

$$\underbrace{e_n^{m+1}}_{\text{Global error}} = \underbrace{R_n^m}_{\text{Local accuracy}} + \underbrace{\left[\alpha e_{n+1}^m + (1-2\alpha)e_n^m + \alpha e_{n-1}^m\right]}_{\text{Propagated error}}. \tag{5.20}$$

This decomposition of the global error at $(\bar{n}\delta x, (m+1)\delta\tau)$, possible because of the linearity of (5.15), allows us to study the consistency and stability of the scheme separately.

In Sect. 5.3.5 we will investigate the behaviour of the local accuracy R_n^m as $\delta x, \delta\tau \to 0$, showing that it vanishes and therefore the FTCS scheme is consistent. Stability can be considered in isolation if we eliminate local accuracy as a factor by setting $R_n^m = 0$ in (5.20), and study the asymptotic behaviour of solutions of

$$e_n^{m+1} = \alpha e_{n+1}^m + (1-2\alpha)e_n^m + \alpha e_{n-1}^m, \tag{5.21}$$

as $m \to \infty$. We will do this first, taking two distinct approaches.

5.3.3   Stability in the Sense of von Neumann

We don't know what form e_n^m might take, but under certain conditions it may admit a Fourier series representation and can be broken down into a (potentially infinite) sum of terms of the form $\lambda^m e^{in\omega}$, for some $\omega \in [-\pi, \pi]$ and $\lambda \in \mathbb{R}$. We can substitute this into (5.21) and observe the effect of the scheme on the amplitude λ^m of a prototypical term in this Fourier series expansion as m increments. The interpretation here is that for any given ω, if $|\lambda| > 1$ then we would expect to see $|e_n^m| \to \infty$ as $m \to \infty$ and errors are magnified by the action of the scheme. If $|\lambda| < 1$ then we would expect to see $|e_n^m| \to 0$ as $m \to \infty$ and errors are damped out by the scheme. Finally, if $|\lambda| = 1$ then errors will remain bounded.

Definition 5.3 We say that the FTCS finite difference method generating approximations U_n^m is *Fourier stable*, or *stable in the sense of von Neumann* if, ignoring initial and boundary conditions, any solution of (5.21) of the form

$$e_n^m = \lambda^m \exp(in\omega) \tag{5.22}$$

satisfies $|\lambda| \le 1$ for all $\omega \in [-\pi, \pi]$.

Let's check the conditions under which the FTCS scheme can be said to be stable in the sense of this definition. By substitution of (5.22) into (5.21),

$$e_n^{m+1} = \alpha e_{n+1}^m + (1 - 2\alpha)e_n^m + \alpha e_{n-1}^m$$

$$\Leftrightarrow \ \lambda^{m+1} e^{in\omega} = \alpha \lambda^m e^{i(n+1)\omega} + (1 - 2\alpha)\lambda^m e^{in\omega} + \alpha \lambda^m e^{i(n-1)\omega}$$

$$\Leftrightarrow \ \lambda = \alpha e^{i\omega} + (1 - 2\alpha) + \alpha e^{-i\omega}.$$

Using $e^{i\omega} + e^{-i\omega} = 2\cos(\omega)$ we obtain, after some manipulation,

$$\lambda = 1 - 4\alpha \sin^2(\omega/2).$$

Since $\sin^2(\omega/2) \in [0, 1]$ for all $\omega \in [-\pi, \pi]$, we will have $|\lambda| \le 1$ if and only if $0 < \alpha \le 1/2$. We will expect to see stable output if and only if we choose N_x, N_τ (and therefore $\delta x, \delta \tau$) such that $\alpha = \delta\tau/(\delta x)^2 \le 1/2$, and as a result we refer to FTCS as a *conditionally stable* scheme. This also explains the results seen in Table 5.1, which were generated by a scheme with $\alpha = 4 > 1/2$.

5.3.4   Matrix Stability

5.3.4.1 The Matrix-Vector Form of the FTCS Scheme

An alternative way to understand the propagation of error in (5.15) is to express the scheme in matrix-vector form. This form is also the most suitable for a structured implementation in code. Consider the FTCS finite difference approximation (5.15) for the heat Eq. (5.10) with solutions defined on $[L_1, L_2] \times [0, \bar{T}]$ and with boundary conditions

$$u(x, 0) = u_0(x); \quad u(L_1, \tau) = a(\tau); \quad u(L_2, \tau) = b(\tau),$$

where the scheme is given by (5.15) for $n = 1, \ldots, N_x - 1$ and $m = 0, \ldots, N_\tau - 1$. The vector of interior values $U^m = [U_1^m, \ldots, U_{N_x-1}^m]^T$ then satisfies the linear equation

$$U^{m+1} = FU^m + p^m, \tag{5.23}$$

where $F \in \mathbb{R}^{(N_x-1) \times (N_x-1)}$ and $p^m \in \mathbb{R}^{N_x-1}$ are defined to be

$$F = \begin{pmatrix} 1 - 2\alpha & \alpha & 0 & \cdots & \cdots & 0 \\ \alpha & 1 - 2\alpha & \alpha & 0 & \cdots & \vdots \\ 0 & \ddots & \ddots & \ddots & \ddots & \vdots \\ \vdots & \ddots & \ddots & \ddots & \ddots & 0 \\ \vdots & & \ddots & \ddots & \alpha & 1 - 2\alpha & \alpha \\ 0 & \cdots & \cdots & 0 & \alpha & 1 - 2\alpha \end{pmatrix}; \quad p^m = \begin{pmatrix} \alpha a(m\delta\tau) \\ 0 \\ \vdots \\ 0 \\ \alpha b(m\delta\tau) \end{pmatrix}.$$

Exercise 5.4 asks you to confirm that (5.23) corresponds to the update Eq. (5.15).

5.3.4.2 Characterising Stability in Terms of the Matrix F

Again we consider the action of the update Eq. (5.23) on the propagated error component of our numerical solution alone, disregarding the vector p^m containing boundary conditions. At each iteration we will premultiply the vector of interior errors by the matrix F, and so the scheme will be stable if F is a convergent matrix, i.e. if

$$\lim_{i \to \infty} F^i = 0,$$

where the RHS is understood as the zero matrix of dimension $(N_x - 1) \times (N_x - 1)$. Therefore, we are motivated to give the following definition.

Definition 5.4 The FTCS scheme given by (5.23) is *matrix stable* if and only if $|\zeta| \leq 1$, where ζ is any eigenvalue of F.

Since F is a tridiagonal and symmetric square matrix, all of its eigenvalues are real. Moreover we have the following useful result for computing the eigenvalues:

Theorem 5.5 *An $(N-1) \times (N-1)$ tridiagonal matrix with constant entries on each descending diagonal*

$$A = \begin{pmatrix} a & c & 0 & \cdots & 0 \\ b & a & c & \ddots & \vdots \\ 0 & b & a & \ddots & 0 \\ \vdots & \ddots & \ddots & \ddots & c \\ 0 & \cdots & 0 & b & a \end{pmatrix}$$

has eigenvalues given by

$$\lambda_s = a + 2\sqrt{bc}\cos\left(\frac{s\pi}{N}\right), \quad s = 1, \ldots N-1.$$

Proof The eigenvalues λ and eigenvectors $\underline{v}$ of A satisfy

$$A\underline{v} = \lambda\underline{v}, \quad \underline{v} = [v_1, \ldots, v_{N-1}]^T.$$

This corresponds to

$$bv_{i-1} + av_i + cv_{i+1} = \lambda v_i, \quad i = 1, \ldots, N-1; \quad v_0 = v_N = 0,$$

which can be written as the second-order difference equation

$$bv_{i-1} + (a - \lambda)v_i + cv_{i+1} = 0, \quad i = 1, \ldots, N-1; \quad v_0 = v_N = 0, \quad (5.24)$$

with auxiliary equation given by

$$cm^2 + (a - \lambda)m + b = 0. \tag{5.25}$$

To see this substitute $v_i = m^i$ into (5.24). Solutions of (5.24) have the general form

$$v_j = \begin{cases} \bar{a}m_1^j + \bar{b}m_2^j, & m_1 \neq m_2; \\ (\bar{a} + \bar{b}j)m_1^j, & m_1 = m_2, \end{cases}$$

where m_1 and m_2 are the (possibly complex) roots of (5.25) and $\bar{a}, b$ are real constants to be determined.

m_1 and m_2 satisfy

$$m_1 m_2 = b/c; \quad m_1 + m_2 = \frac{\lambda - a}{c}.$$

So we have

$$v_j = \bar{a}m_1^j + \bar{b}m_2^j \;\Rightarrow\; v_0 = \bar{a} + \bar{b} = 0 \;\Leftrightarrow\; \bar{a} = -\bar{b},$$

which implies that neither $\bar{a}$ nor $\bar{b}$ are zero-valued (since the eigenvector $\underline{v}$ cannot be $\underline{0}$). Therefore

$$v_{N+1} = \bar{a}(m_1^N - m_2^N) = 0 \;\Leftrightarrow\; \left(\frac{m_1}{m_2}\right)^N = 1,$$

which gives

$$\frac{m_1}{m_2} = e^{\frac{2\pi i s}{N}}, \quad s = 1, \ldots N - 1.$$

Combine with $m_1 m_2 = b/c$ to get

$$m_1 m_2 = m_2^2 e^{\frac{2\pi i s}{N}} = \frac{b}{c}.$$

It follows that

$$m_2 = \sqrt{\frac{b}{c}}\, e^{\frac{-\pi i s}{N}}, \quad m_1 = \sqrt{\frac{b}{c}}\, e^{\frac{\pi i s}{N}},$$

and (since $e^{ix} + e^{-ix} = 2\cos(x)$)

$$m_1 + m_2 = \sqrt{\frac{b}{c}}\left[e^{\frac{\pi i s}{N}} + e^{\frac{-\pi i s}{N}}\right] = 2\sqrt{\frac{b}{c}} \cos\left(\frac{\pi s}{N}\right), \quad s = 1, \ldots, N - 1.$$

Since $m_1 + m_2 = (\lambda - a)/c$, we arrive at

$$\lambda_s = a + 2c\sqrt{\frac{b}{c}} \cos\left(\frac{\pi s}{N}\right), \quad s = 1, \ldots, N - 1,$$

from which the statement of the theorem follows. $\qquad\square$

Let's now check the conditions under which the FTCS scheme is matrix stable. Applying Theorem 5.5 to F with $b = c = \alpha$ and $a = 1 - 2\alpha$, we see that the eigenvalues of F are given by

$$\lambda_i = 1 - 2\alpha + 2\alpha \cos\left(\frac{s\pi}{N_x - 1}\right), \quad s = 1, \ldots N_x - 1, \tag{5.26}$$

and therefore $|\lambda_s| \leq 1$ for all $s = 1, \ldots, N_x - 1$ if and only if $0 < \alpha \leq 1/2$. This again indicates conditional stability, where the condition on the mesh ratio is identical to that given by the von Neumann characterisation of stability.

5.3.5   Local Accuracy and Consistency of the FTCS Scheme

Since $\alpha = \delta\tau/(\delta x)^2$, the local accuracy of the FTCS scheme, given by (5.19), may be written

$$R_n^m = (\delta\tau)^{-1}\left[u\left(\bar{n}\delta x, (m+1)\delta\tau\right) - u\left(\bar{n}\delta x, m\delta\tau\right)\right]$$

$$- (\delta x)^{-2}\left[u\left((\bar{n}+1)\delta x, m\delta\tau\right) - 2u\left(\bar{n}\delta x, m\delta\tau\right) + u\left((\bar{n}-1)\delta x, m\delta\tau\right)\right].$$
$$(5.27)$$

Our goal is to show that R_n^m vanishes as $\delta x, \delta\tau \to 0$.

Returning to the Taylor series expansion from Sect. 5.2, we can write

$$u\left(\bar{n}\delta x, (m+1)\delta\tau\right) - u\left(\bar{n}\delta x, m\delta\tau\right) = (\delta\tau)u_\tau + \frac{(\delta\tau)^2}{2}u_{\tau\tau} + O\left((\delta\tau)^3\right),$$

where the partial derivatives on the RHS are evaluated at the point $(\bar{n}\delta x, m\delta\tau)$, but we suppress this for notational clarity. Similarly

$$u\left((\bar{n}+1)\delta x, m\delta\tau\right) - 2u\left(\bar{n}\delta x, m\delta\tau\right) + u\left((\bar{n}-1)\delta x, m\delta\tau\right)$$

$$= (\delta x)^2 u_{xx} - \frac{(\delta x)^4}{12}u_{xxxx} + O\left((\delta x)^6\right).$$

Making the substitution back into (5.27), we see that

$$R_n^m = u_\tau - u_{xx} + \frac{\delta\tau}{2}u_{\tau\tau} + \frac{(\delta x)^2}{12}u_{xxxx} + O(\delta\tau) + O\left((\delta x)^4\right).$$

By (5.10), $u_\tau - u_{xx} = 0$ for all $(x, \tau) \in [L_1, L_2] \times [0, \bar{T}]$, and the leading order terms of what remains on the RHS tell us that

$$R_n^m = O(\delta\tau) + O\left((\delta x)^2\right).$$

Note that this relies upon the existence of a uniform bound on $u_{\tau\tau}$ and u_{xxxx} over $[L_1, L_2] \times [0, \bar{T}]$.

The FTCS method is therefore *consistent of order 1 in $\delta\tau$ and order 2 in δx*, and by the Lax Equivalence Theorem our method will be convergent as long as a mesh ratio of $0 < \alpha \leq 1/2$ is maintained. This is unsatisfactory, since it binds the resolution of the mesh in both dimensions in a way that leads to computational inefficiency. For example if we wish to increase the resolution of our mesh in the x direction by a factor of 10 we will need to increase it in the τ direction by a factor of 100 in order to maintain the same mesh ratio, and the overall computational complexity of our calculation will be 1000 times higher.

5.4   Two Unconditionally Stable Schemes

It is possible to remove this constraint on α, moving from the conditional stability of the FTCS to an unconditionally stable scheme by modifying the way we approximate the partial derivative u_τ. We will show this here, and then demonstrate the application of the new scheme to the option pricing problem.

5.4.1   Construction and Stability of an Implicit Scheme

Instead of taking a forward difference in our Taylor series approximation, as we did in Sect. 5.2, we could write

$$u(x, \tau - \delta\tau) = u(x, \tau) - u_\tau(x, \tau)\delta\tau + O\left((\delta\tau)^2\right),$$

and rearrange to get

$$u_\tau(x, \tau) = \frac{u(x, \tau) - u(x, \tau - \delta\tau)}{\delta\tau} + O(\delta\tau). \tag{5.28}$$

This first term on the RHS of (5.28) is called the *backward difference approximation* of u_τ and we will use it to construct an implicit scheme. As before, we can use the symmetric central difference approximation for u_{xx}. Equating both these approximations gives the scheme

$$\frac{U_n^m - U_n^{m-1}}{\delta\tau} = \frac{U_{n+1}^m - 2U_n^m + U_{n-1}^m}{(\delta x)^2},$$

for $n = 1, \ldots, N_x - 1$ and $m = 1, \ldots, N_\tau$. For notational consistency with the FTCS scheme (5.14), we shift the time index forwards by one step, to obtain

$$\frac{U_n^{m+1} - U_n^m}{\delta\tau} = \frac{U_{n+1}^{m+1} - 2U_n^{m+1} + U_{n-1}^{m+1}}{(\delta x)^2},$$

for $n = 1, \ldots, N_x - 1$ and $m = 0, \ldots, N_\tau - 1$. Rearranging so that all $(m + 1)$-indexed terms are on the left and all m-indexed terms are on the right leads us to the following definition.

Definition 5.5 The *Backward in Time and Centred in Space (BTCS) finite difference scheme* for the heat Eq. (5.6) is given by the difference equation

$$-\alpha U_{n+1}^{m+1} + (1 + 2\alpha)U_n^{m+1} - \alpha U_{n-1}^{m+1} = U_n^m. \tag{5.29}$$

for for $n = 1, \ldots, N_x - 1$ and $m = 0, \ldots, N_\tau - 1$, and $\alpha = \delta\tau/(\delta x)^2$.

The stencil is illustrated in Fig. 5.3. Notice that (5.29) does not explicitly describe how to update a vector of internal values at the mth step in order to generate values at the $(m + 1)^{th}$ step, and so we refer to this as an implicit scheme.

5.4.1.1 The Matrix-Vector Form of the BTCS Scheme

Although Exercise 5.6 confirms that the BTCS scheme is unconditionally stable in the sense of von Neumann, the implicit characterisation of (5.29) means that it is difficult to do the kind of quick-and-dirty iterative computations seen in Sect. 5.2 for the FTCS scheme to confirm what we actually see in practice. Indeed implementation requires that we express the scheme in matrix-vector form. Consider the BTCS finite difference approximation for the heat Eq. (5.10) with solutions defined on $[L_1, L_2] \times [0, \bar{T}]$ and with boundary conditions

$$u(x, 0) = u_0(x); \quad u(L_1, \tau) = a(\tau); \quad u(L_2, \tau) = b(\tau),$$

where the scheme is given by (5.29) for $n = 1, \ldots, N_x - 1$ and $m = 0, \ldots, N_\tau - 1$. The vector of interior values $U^m = [U_1^m, \ldots, U_{N_x-1}^m]^T$ then satisfies the linear equation

$$GU^{m+1} = U^m + q^m, \tag{5.30}$$

where $G \in \mathbb{R}^{(N_x-1)\times(N_x-1)}$ and $q^m \in \mathbb{R}^{N_x-1}$ are defined to be

$$G = \begin{pmatrix} 1+2\alpha & -\alpha & 0 & \cdots & & \cdots & 0 \\ -\alpha & 1+2\alpha & -\alpha & 0 & & \cdots & \vdots \\ 0 & \ddots & \ddots & \ddots & \ddots & & \vdots \\ \vdots & & \ddots & \ddots & \ddots & \ddots & 0 \\ \vdots & & & \ddots & -\alpha & 1+2\alpha & -\alpha \\ 0 & \cdots & & \cdots & 0 & -\alpha & 1+2\alpha \end{pmatrix}; \quad q^m = \begin{pmatrix} \alpha a((m+1)\delta\tau) \\ 0 \\ \vdots \\ 0 \\ \alpha b((m+1)\delta\tau) \end{pmatrix}. \tag{5.31}$$

To compute U^{m+1} from U^m, we need G^{-1}. Rewrite (5.30) as

$$U^{m+1} = G^{-1}U^m + G^{-1}q^m. \tag{5.32}$$

Since (5.32) also shows the effect of iterating the scheme on the vector of internal errors, we can expect stability if G^{-1} is a convergent matrix and so the BTCS scheme is matrix stable if and only if all eigenvalues ξ of G^{-1} satisfy $|\xi| \leq 1$.

We finish with a point on the computational cost of implicit methods. Since (5.30) is a linear system, we can write the update equation exactly as (5.32), invert G once and then store its value. Therefore there is little additional computation involved. If, for example, we wish to value the option with time-varying volatilities then we may need to invert this matrix at each step, which will increase the computational cost accordingly. The worst case scenario for implicit methods is when the numerical scheme is nonlinear, in which case it is necessary to iteratively solve a nonlinear system at each timestep. We showed how this can be done using `fsolve()` from the `scipy.optimize` package Sect. 2.3 of Chap. 2. Fortunately we are never faced with this possibility when solving the Black-Scholes-Merton PDE.

5.4.2   The Crank-Nicolson Finite Difference Approximation

So far we have seen that both the FTCS and BTCS schemes are of local accuracy $O(\delta\tau) + O\left((\delta x)^2\right)$, but that only the BTCS scheme is unconditionally stable. With another choice of finite difference approximation of u_τ we can improve the local accuracy in time to second order, giving $O\left((\delta\tau)^2\right) + O\left((\delta x)^2\right)$ without sacrificing stability. The idea is to combine central difference approximations over steps of length $\delta\tau/2$, while the scheme still takes steps of length $\delta\tau$.

5.4.2.1 Taylor Series Expansion of Partial Derivatives

If we expand u to second order around $(x, \tau + \delta\tau/2)$ we can write

$$u(x, \tau + \delta\tau) = u(x, \tau + \delta\tau/2) + u_\tau(x, \tau + \delta\tau/2)\frac{\delta\tau}{2}$$

$$+ u_{\tau\tau}(x, \tau + \delta\tau/2)\frac{(\delta\tau)^2}{8} + O\left((\delta\tau)^3\right);$$

$$u(x, \tau) = u(x, \tau + \delta\tau/2) - u_\tau(x, \tau + \delta\tau/2)\frac{\delta\tau}{2}$$

$$+ u_{\tau\tau}(x, \tau + \delta\tau/2)\frac{(\delta\tau)^2}{8} + O\left((\delta\tau)^3\right).$$

Subtracting the second equation from the first and solving for u_τ gives

$$u_\tau(x, \tau + \delta\tau/2) = \frac{u(x, \tau + \delta\tau) - u(x, \tau)}{\delta\tau} + O\left((\delta\tau)^2\right). \tag{5.33}$$

the first term on the RHS is the central difference approximation of u_τ. We cannot use this by itself to approximate u_τ, since that would lead to an unstable scheme, but it can used as a stepping stone to better things.

Now consider the Taylor expansions of $u_{xx}(x, \tau + \delta\tau)$ and $u_{xx}(x, \tau)$ around the point $(x, \tau + \delta\tau/2)$:

$$u_{xx}(x, \tau + \delta\tau) = u_{xx}(x, \tau + \delta\tau/2) + \frac{\delta\tau}{2}u_{xx\tau}(x, \tau + \delta\tau/2) + O\left((\delta\tau)^2\right);$$

$$u_{xx}(x, \tau) = u_{xx}(x, \tau + \delta\tau/2) - \frac{\delta\tau}{2}u_{xx\tau}(x, \tau + \delta\tau/2) + O\left((\delta\tau)^2\right).$$

Sum both sides to get

$$u_{xx}(x, \tau + \delta\tau) + u_{xx}(x, \tau) = 2u_{xx}(x, \tau + \delta\tau/2) + O\left((\delta\tau)^2\right),$$

then rearrange:

$$u_{xx}(x, \tau + \delta\tau/2) = \frac{u_{xx}(x, \tau + \delta\tau) + u_{xx}(x, \tau)}{2} + O\left((\delta\tau)^2\right).$$

Via the heat Eq. (5.10), the RHS can be equated to $u_\tau(x, \tau + \delta\tau/2)$:

$$u_\tau(x, \tau + \delta\tau/2) = \frac{u_{xx}(x, \tau + \delta\tau) + u_{xx}(x, \tau)}{2} + O\left((\delta\tau)^2\right).$$

Now substitute (5.33) for $u_\tau(x, \tau + \delta\tau/2)$ on the LHS:

$$\frac{u(x, \tau + \delta\tau) - u(x, \tau)}{\delta\tau} + O\left((\delta\tau)^2\right)$$
$$= \frac{1}{2}u_{xx}(x, \tau + \delta\tau) + \frac{1}{2}u_{xx}(x, \tau) + O\left((\delta\tau)^2\right).$$

Finally, replace the u_{xx} terms with their symmetric central difference approximations (see (5.13) in Sect. 5.2) to get

$$\frac{u(x, \tau + \delta\tau) - u(x, \tau)}{\delta\tau}$$
$$= \frac{u(x + \delta x, \tau + \delta\tau) - 2u(x, \tau + \delta\tau) + u(x - \delta x, \tau + \delta\tau)}{2(\delta x)^2}$$
$$+ \frac{u(x + \delta x, \tau) - 2u(x, \tau) + u(x - \delta x, \tau)}{2(\delta x)^2} + O\left((\delta x)^2\right) + O\left((\delta\tau)^2\right),$$

for $x \in [L_1 + \delta x, L_2 - \delta x]$ and $\tau \in [0, \bar{T} - \delta\tau]$.

5.4.2.2 The Difference Equation of the Numerical Scheme

As before, set $\alpha = \delta\tau/(\delta x)^2$ and replace the true values with approximations to get the numerical scheme

$$U_n^{m+1} - U_n^m = \frac{1}{2}\alpha\left(U_{n+1}^{m+1} - 2U_n^{m+1} + U_{n-1}^{m+1}\right) + \frac{1}{2}\alpha\left(U_{n+1}^m - 2U_n^m + U_{n-1}^m\right),$$

for $n = 1, \ldots, N_x - 1$ and $m = 0, \ldots, N_\tau - 1$, which has local accuracy corresponding to the truncated remainder terms $O\left((\delta x)^2\right) + O\left((\delta\tau)^2\right)$. Rearrange so that present terms are gathered on the RHS and future terms on the LHS and we arrive at the following definition:

Definition 5.6 The *Crank-Nicolson finite difference approximation* of the heat Eq. (5.6) is given by the difference equation

$$(1 + \alpha)U_n^{m+1} - \frac{1}{2}\alpha\left(U_{n+1}^{m+1} + U_{n-1}^{m+1}\right) = (1 - \alpha)U_n^m + \frac{1}{2}\alpha\left(U_{n+1}^m + U_{n-1}^m\right),$$

(5.34)

for $n = 1, \ldots, N_x - 1$ and $m = 0, \ldots, N_\tau - 1$ and $\alpha = \delta t/(\delta x)^2$.

The stencil is displayed in Fig. 5.3.

5.4.2.3 The Matrix-Vector Form of the Numerical Scheme

Since it is clear from (5.34) that Crank-Nicolson is an implicit scheme, we will again turn to the matrix-vector formulation as the best approach to implementation. As before, specify the boundary conditions for (5.10) to be

$$u(x, 0) = u_0(x); \quad u(L_1, \tau) = a(\tau); \quad u(L_2, \tau) = b(\tau),$$

and consider the Crank-Nicolson scheme given by (5.34) for $n = 1, \ldots, N_x - 1$ and $m = 0, \ldots, N_\tau - 1$. The vector of interior values $U^m = \left[U_1^m, \ldots, U_{N_x-1}^m\right]^T$ then satisfies the linear equation

$$\widehat{B}U^{m+1} = \widehat{F}U^m + r^m,$$

(5.35)

where $\widehat{B}, \widehat{F} \in \mathbb{R}^{(N_x-1)\times(N_x-1)}$ and $r^m \in \mathbb{R}^{N_x-1}$ are defined to be

$$\widehat{B} = \begin{pmatrix} 1+\alpha & -\frac{1}{2}\alpha & 0 & \cdots & \cdots & 0 \\ -\frac{1}{2}\alpha & 1+\alpha & -\frac{1}{2}\alpha & 0 & \cdots & \vdots \\ 0 & \ddots & \ddots & \ddots & \ddots & \vdots \\ \vdots & \ddots & \ddots & \ddots & \ddots & 0 \\ \vdots & & \ddots & \ddots & -\frac{1}{2}\alpha & 1+\alpha & -\frac{1}{2}\alpha \\ 0 & \cdots & \cdots & 0 & -\frac{1}{2}\alpha & 1+\alpha \end{pmatrix},$$

$$\widehat{F} = \begin{pmatrix} 1-\alpha & \frac{1}{2}\alpha & 0 & \cdots & \cdots & 0 \\ \frac{1}{2}\alpha & 1-\alpha & \frac{1}{2}\alpha & 0 & \cdots & \vdots \\ 0 & \ddots & \ddots & \ddots & & \vdots \\ \vdots & & \ddots & \ddots & \ddots & 0 \\ \vdots & & \ddots & \ddots & \frac{1}{2}\alpha & 1-\alpha & \frac{1}{2}\alpha \\ 0 & \cdots & \cdots & 0 & \frac{1}{2}\alpha & 1-\alpha \end{pmatrix},$$

and

$$
r^m = \begin{pmatrix} \frac{1}{2}\alpha\left(a(m\delta\tau) + a((m+1)\delta\tau)\right) \\ 0 \\ \vdots \\ 0 \\ \frac{1}{2}\alpha\left(b(m\delta\tau) + b((m+1)\delta\tau)\right). \end{pmatrix}.
$$

To compute U^{m+1} from U^m, rewrite (5.35) as

$$
U^{m+1} = \left(\widehat{B}\right)^{-1} \widehat{F} U^m + \left(\widehat{B}\right)^{-1} r^m. \tag{5.36}
$$

It makes sense then to define matrix stability for this scheme in terms of eigenvalues of the matrix

$$
\left(\widehat{B}\right)^{-1} \widehat{F} = \left(\widehat{B}\right)^{-1} \left(2\mathbb{I} - \widehat{B}\right) = 2\left(\widehat{B}\right)^{-1} - \mathbb{I},
$$

where we have used the fact that $\widehat{B} + \widehat{F} = 2\mathbb{I}$, with $\mathbb{I}$ the $(N_x - 1) \times (N_x - 1)$-identity matrix.

In implementation, we can still follow the code scheme developed in Sect. 5.5, making changes only to Step 3, where we set up the iteration matrix $(\widehat{B})^{-1}\widehat{F}$ instead of G; Step 6, where the array of boundary conditions q is replaced by r; and Step 7, to reflect (5.36) rather than (5.32).

We now have everything we need to generate reliable approximate solutions of the Black-Scholes-Merton PDE by numerical discretisation of the heat equation.

5.5 Python Valuation of Options Using an Implicit Scheme

In this section we first walk through the implementation of the BTCS scheme for a European call option with parameters $E = 5$, $T = 1$, $r = 0.01$, $\sigma = 0.4$. We then describe what modifications are required, first for a barrier call option with knock-out barrier at $B = 2$, and then for a cash-or-nothing call option.

5.5.1 Implementation of the BTCS Scheme for a European Call Option

5.5.1.1 Step 1: Parameters Are Defined for the Option, then Transformed

Set up parameter values for the European option:

```
In []: E=5.; T=1; r=0.01; sigma=0.4
```

Notice that we don't specify an initial value S_0. This is because we will solve for option prices over a rectangular domain of possible values (s, t). Although the underlying asset S can take any value $s \in (0, \infty)$ we must truncate this interval at both ends. Because of the transformation $x = \ln(s/E)$ it is not possible to include $s = 0$ in our computation, nor would it be financially realistic to do so, and so we choose a minimum value for s, denoted s_{low}, that is very small but nonetheless positive. Similarly we choose a maximum value for s, denoted s_{high}, that is large but finite.

```
In []: sLow=np.exp(-8); sHigh=np.exp(8)
```

These values are chosen heuristically; some of the consequences of taking `sHigh` too small are illustrated in Exercise 5.8.

Next, we transform these values from the model PDE to the transformed PDE using (5.7) and (5.9):

$$x = \ln(s/E),\ \tau = \sigma^2 T/2,\ k = 2r/\sigma^2,\ \alpha = -(k-1)/2,\ \beta = -(k+1)^2/4.$$

In code, this looks like

```
In []: xLow=np.log(sLow/E)
       xHigh=np.log(sHigh/E)
       Tbar=0.5*(sigma**2)*T
       k=2*r/sigma**2;
       alphaT=-0.5*(k-1); betaT=-0.25*(k+1)**2
```

Notice that we name the transformed parameters `alphaT` and `betaT` to avoid confusion with the mesh ratio variable `alpha`. Moreover, the variables `xLow` and `xHigh` represent our choice for the values L_1 and L_2 in the boundary of the domain for (5.10).

5.5.1.2 Step 2: Parameters Are Defined for the Scheme

Once we choose N_x and N_τ, the number of steps in each direction, we can immediately compute δx, $\delta \tau$, and the mesh ratio α:

```
In []: Nx=500; dx=(xHigh-xLow)/Nx
       Nt=500; dt=Tbar/Nt
       alpha=dt/(dx**2)
```

This allows us to set up array variables containing all values of x and τ for which we will compute an approximation, and their pre-transformed counterparts S and t:

```
In []: x=np.linspace(xLow,xHigh,Nx+1);
       tau=np.linspace(0,Tbar,Nt+1);
       S=E*np.exp(x);
       t=T-(2/sigma**2)*tau;
```

5.5.1.3 Step 3: Set Up the Iteration Matrix G and Invert

The structure of G, tridiagonal with constant entries along each of the main, sub- and super-diagonals, lends itself to use of the `np.eye()` and `np.ones()` commands. Note the shift in the second argument of the latter used to produce off-diagonal entries:

```
In []: subDiagG=-alpha*np.diag(np.ones(Nx-2),-1)
       diagG=(1+2*alpha)*np.eye(Nx-1)
       superDiagG=-alpha*np.diag(np.ones(Nx-2),1)
       G=superDiagG+diagG+subDiagG
```

The inversion of large sparse matrices is associated with a large body of research, partly because of its usefulness in this kind of finite difference scheme. Fortunately the `scipy.linalg` package is quite capable of inverting large square, tridiagonal, real-valued matrices such as G, so for European options at least we can leave the heavy lifting to that package with a clear conscience.

```
In []: Ginv=linalg.inv(G);
```

For American options, we cannot be so cavalier as to use a "black-box" approach to inverting G. As we will see, direct methods for inverting G are insufficient when using implicit methods in that context, and so in Sect. 5.6 we will look more closely at iterative procedures for matrix inversion, and specifically the method of successive over-relaxation (SOR).

5.5.1.4 Step 4: Set Up an Array to Contain the Solution Values

We will append the boundary conditions separately, so only include $N_x - 1$ entries in the x direction at first. In the τ direction we make room for the initial vector, so include $N_\tau + 1$ entries there:

```
In []: U = np.zeros((Nx-1,Nt+1))
```

Note the use of double brackets in the `np.zeros()` call.

5.5.1.5 Step 5: Define the Initial Profile and Boundary Conditions

Since, by (5.8),

$$u(x, \tau) = \frac{1}{E} e^{-\alpha x - \beta \tau} V(s, t),$$

we can apply the boundary conditions from (5.2) as follows.

1. When $s = 0$, $x = -\infty$. In Step 1 we chose the smallest value of s to be $s_{\text{low}} = e^{-8} = 0.0003$, which corresponds to $x_{\text{low}} = -11.951$. The associated boundary condition is $V(0, t) = 0$. So we set

$$a(\tau) = u(x_{\text{low}}, \tau) = \frac{1}{E} e^{-\alpha x_{\text{low}} - \beta \tau} \cdot 0 = 0.$$

In code, this could be constructed as an array of length $N_\tau + 1$

```
In []: a=0*np.ones(Nt+1)
```

2. When $s = \infty$, $x = \infty$. In Step 1 we chose the largest value of s to be $s_{\text{high}} = e^{8} = 2981$, which corresponds to $x_{\text{high}} = 4.0488$. The associated boundary condition is $V(s, t) \sim s$ as $s \to \infty$. So we set (since $s = E e^x$)

$$b(\tau) = u(x_{\text{high}}, \tau) = \frac{1}{E} e^{-\alpha x_{\text{high}} - \beta \tau} s = e^{-\alpha x_{\text{high}} - \beta \tau} e^{x_{\text{high}}}$$

$$= e^{(1-\alpha)x_{\text{high}} - \beta \tau}.$$

In code, this could be written, again as an array of length $N_\tau + 1$,

```
In []: b=np.exp((1-alphaT)*xHigh-betaT*tau)
```

3. The final-time condition for (5.1) corresponding to the payoff for a European call option is transformed by (5.7)–(5.9) in Sect. 5.1.4 to the initial profile

$$u(x, 0) = u_0(x) = \max\left(e^{\frac{1}{2}(k+1)x} - e^{\frac{1}{2}(k-1)x}, 0 \right). \tag{5.37}$$

This is implemented by direct modification to the first column of U:

```
In []: xVals=np.linspace(xLow+dx,xHigh-dx, Nx-1)
        ic=np.exp(0.5*(k+1)*xVals)-np.exp(0.5*(k-1)*xVals)
        U[:,0]=np.maximum(ic,0)
```

Note that since U does not yet contain rows for the boundary conditions, we are careful to omit xLow and xHigh from the computation.

5.5.1.6 Step 6: Set Up the Collection of Vectors q Containing the Boundary Conditions

Since there is a different vector q at each timestep, we set it up as an array variable:

```
In []: q=np.zeros((Nx-1,Nt+1))
```

where each column corresponds to a particular vector q. Then populate the first and last rows:

```
In []: q[0,:]=alpha*a; q[Nx-2,:]=alpha*b
```

5.5.1.7 Step 7: Compute the Interior Solution Values via the BTCS Scheme

Using the update form (5.32) we have

```
In []: for i in range(0,Nt):
           U[:,i+1]=Ginv.dot(U[:,i])+Ginv.dot(q[:,i+1])
```

Once all interior values are populated we can append the boundary conditions to the solution matrix.

```
 > U=np.r_[[a],U,[b]]
```

5.5.1.8 Step 8: Transform Back to the Black-Scholes Variables

We can use the `np.meshgrid()` command to produce an array of all (x, τ) pairs on our grid and then reverse the transformation (5.8) at each individual coordinate (x, τ) using elementwise multiplication:

```
In []: xx,tt=np.meshgrid(x,tau)
       V=E*np.multiply(np.exp(alphaT*xx+betaT*tt).T,U)
```

Notice the transposition of the array variable in the second line.

5.5.1.9 Step 9: Plot the Surfaces U and V

First, we display the direct solution of the diffusion equation stored in U:

```
In []: fig=plt.figure()
       ax=plt.axes(projection='3d')
       ax.plot_surface(tt,xx,U,cmap='Spectral')
       plt.show();
```

The parameter `cmap` allows us to choose a colour scheme (or map) for the surface plot. Next, plot the solution of the Black-Scholes-Merton PDE (5.1) over the full computational domain.

```
In []: # Set up axis object
       fig=plt.figure()
       ax=plt.axes(projection = '3d')

       # Reset state and time vectors
       tVals=np.linspace(T,0,Nt+1)
       xVals=np.linspace(xLow,xHigh,Nx+1)
```

```
# Transform mesh values back to original coordinates
X,Y=np.meshgrid(tVals,E*np.exp(xVals))

# Plot zoomed-out view of Black-Scholes surface
ax.plot_surface(X,Y,V,cmap='Spectral')
plt.show();
```

Note that we constructed a uniform mesh to approximate the function $u(x, \tau)$. Since $x = \log(s/E)$, the mesh points after transforming back to V will not be uniformly distributed in the s direction.

Finally, zoom in on the part of the Black-Scholes surface showing option values for $s \in [0, 10]$:

```
In []: # Mystery variable. What does it do?
       xr=m.ceil((np.log(10/E)-xLow)/dx);

       # Set up new axis object and transformed mesh values
       ax2=plt.axes(projection = '3d')

       # Reset state and time vectors
       tVals=np.linspace(T,0,Nt+1)
       xVals=np.linspace(xLow,xr*dx+xLow,xr+1)

       # Transform mesh values back to original coordinates
       X,Y=np.meshgrid(tVals,E*np.exp(xVals))

       # Plot zoomed-in view of Black-Scholes surface
       ax2.plot_surface(X,Y,V[range(0,xr+1),:]);
       plt.show();
```

Exercise 5.10 asks you to explain the role of the variable **xr** in this code segment. Exercise 5.11 asks you to modify Step 5 so that the finished code values a European put.

The Black-Scholes surface for a European call and put using the code developed in this section for the BTCS scheme with $N_x = N_\tau = 500$ is displayed in Fig. 5.4. We can compare the array of time $t = 0$ option values thus produced to the benchmark values returned by the relevant Black-Scholes formula. The maximum error for the European call values is (to 4 significant digits) 2.817×10^{-4}; for the European put values the maximum error is 3.3546×10^{-4}. Exercise 5.12 asks you to modify the algorithm so that it uses the Crank-Nicolson scheme instead of the BTCS scheme and recompute the maximum error in each case.

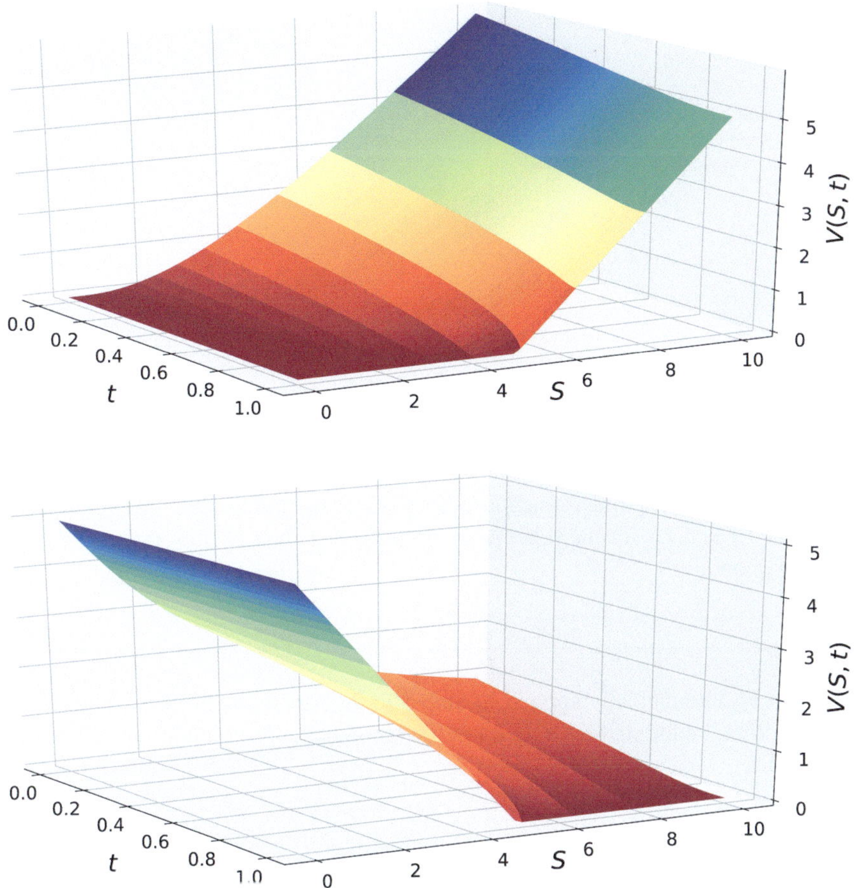

Fig. 5.4 Black-Scholes surfaces for European options with $T = 1$, $E = 5$, $r = 0.01$, and $\sigma = 1$. On the top we show the value surface for a call, and on the bottom for a put

5.5.2   Modification for a Down-and-Out Barrier Call Option

This algorithm can be straightforwardly modified to value (for example) a down-and-out barrier call option with parameters $E = 5$, $B = 2$, $T = 1$, $r = 0.01$, $\sigma = 0.4$. The primary change is to restrict the state domain for the Black-Scholes-Merton PDE to $S \in [B, \infty)$ and impose the final-time and boundary conditions

- $V_T(s) = (s - E)^+$;
- $V(B, t) = 0$;
- $V(s, t) \sim s$ as $s \to \infty$.

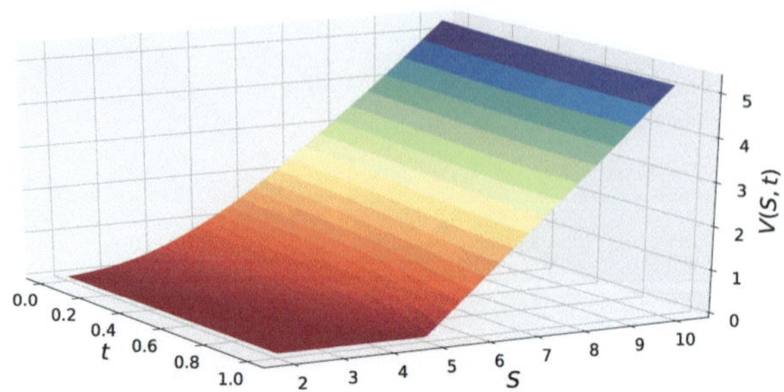

Fig. 5.5 Black-Scholes surface for a down-and-out barrier call option with $B = 2$, $T = 1$, $E = 5$, $r = 0.01$, and $\sigma = 0.5$

This means that in Step 1 from Sect. 5.5.1.1, we redefine the limit `sLow` to reflect the knock-out barrier, leaving `sHigh` the same.

```
In []: sLow=2.; sHigh=np.exp(8)
```

These values are then transformed as before into `xLow` and `xHigh`, representing our choice of L_1 and L_2 for the scheme. Without further modification to the code, the boundary condition

$$a(\tau) = \frac{1}{E} e^{-\alpha x_{\text{low}} - \beta \tau} \cdot 0 = 0$$

will then be automatically applied in Step 5 to the new value of `xLow`. The rest of the algorithm, including construction of the second boundary condition

$$b(\tau) = e^{(1-\alpha)x_{\text{high}} - \beta \tau},$$

and the initial profile $u_0(x)$, proceeds unmodified. Figure 5.5 shows the Black-Scholes surface that results. Notice the zero boundary at $s = 2$. Exercise 5.13 asks you to compute the maximum error at time $t = 0$ by comparing these approximate values with those produced by the benchmark valuation formula.

5.5.3   Modification for a Cash-or-Nothing Call Option

Consider a cash-or-nothing call option with payoff at expiry time $t = T$ given by

$$\Lambda(s) = \begin{cases} D, & s > E; \\ \frac{D}{2}, & s = E; \\ 0, & s < E, \end{cases}$$

for some $D > 0$ and any $S(T) = s$. Here, the asset price domain is $\mathcal{D} = (0, \infty)$, and so Step 1 of the algorithm for a European call described above is unchanged.

The final-time and boundary conditions for this option may be written

- $V_T(s) = D\mathcal{H}(s - E)$;
- $\lim_{s \to 0^+} V(s, t) = 0$;
- $\lim_{s \to \infty} V(s, t) = B$,

where

$$\mathcal{H}(x) = \begin{cases} 1, & x > 0; \\ \frac{1}{2}, & x = \frac{1}{2}; \\ 0, & x < 0, \end{cases}$$

is the Heaviside step function. So under the transformation (5.7)–(5.9), the initial profile will be

$$u_0(x) = \frac{1}{E} e^{-\alpha x} D\mathcal{H}(s - E)$$

$$= \frac{1}{E} e^{-\alpha x} D\mathcal{H}(Ee^x - E)$$

$$= \frac{1}{E} e^{-\alpha x} D\mathcal{H}(e^x - 1)$$

$$= \frac{1}{E} e^{-\alpha x} D\mathcal{H}(x), \quad x \in \left[x_{\text{low}}, x_{\text{high}} \right].$$

Following the same procedure as described in Step 5, the transformed boundary conditions will be, for $\tau \in [0, \bar{T}]$,

$$a(\tau) = u(x_{\text{low}}, \tau) = \frac{1}{E} e^{-\alpha x_{\text{low}} - \beta \tau} \cdot 0 = 0$$

and

$$b(\tau) = u(x_{\text{high}}, \tau) = \frac{1}{E} e^{-\alpha x_{\text{high}} - \beta \tau} \times De^{-r(T-t)} = \frac{D}{E} e^{-\alpha x_{\text{high}} - (\beta + k)\tau}.$$

We can modify the code in Step 5 accordingly, using the `np.heaviside()` function call to define the initial profile. Exercise 5.14 asks you to do this. Figure 5.6 shows the Black-Scholes surface that results. Notice that, since the payoff for this option is capped at $D = 22$, the value of the option at earlier times $t < T$ can never exceed $e^{-r(T-t)} D$.

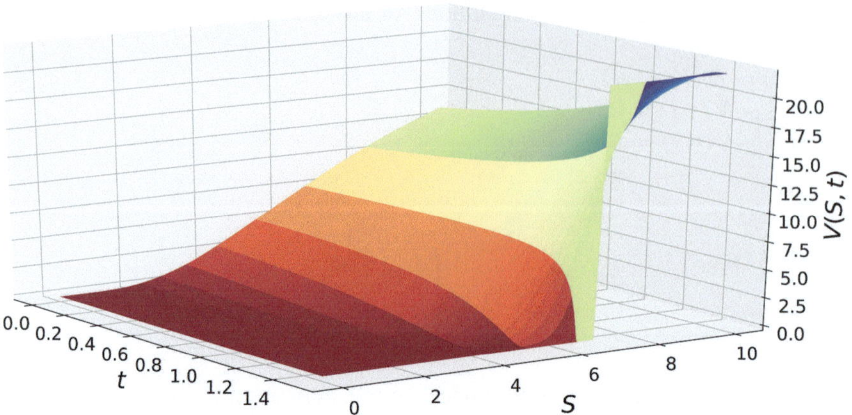

Fig. 5.6 Black-Scholes surface for a cash-or-nothing call option with $D = 22$, $T = 1.5$, $E = 7$, $r = 0.04$, and $\sigma = 0.5$

5.6   The Finite Difference Approach for American Options

Consider an American put option with strike E and expiry T. Denote its value at time t for a given asset price $S(t) = s$ by $P^{\mathrm{Am}}(s, t)$, and its intrinsic value $\Lambda(s) = \max(E - s, 0)$. Recall from Sect. 3.5 in Chap. 3 that there exists a non-decreasing function $S_f(t)$ defined for all $t \in [0, T]$ satisfying $S_f(T) = E$ known as the optimal exercise boundary, and a schematic illustration of this boundary was shown in Fig. 3.6.

5.6.1   The Free Boundary Formulation of an American Put

When $s > S_f(t)$ at some time $t \in [0, T)$, the value associated with continuing to hold the option is greater than that gleaned from immediate exercise and so the holder of the option should not exercise at time t. In this case the value of the option is governed by the Black-Scholes-Merton PDE and we have (suppressing arguments in the PDE)

$$P^{\mathrm{Am}}(s, t) > E - s; \qquad \frac{\partial P^{\mathrm{Am}}}{\partial t} + \frac{1}{2}\sigma^2 s^2 \frac{\partial^2 P^{\mathrm{Am}}}{\partial s^2} + rs\frac{\partial P^{\mathrm{Am}}}{\partial s} - rP^{\mathrm{Am}} = 0. \qquad (5.38)$$

When $s \leq S_f(t)$ it is optimal for the holder of the option to exercise at time t. In this case the value of the option satisfies

$$P^{\mathrm{Am}}(s, t) = E - s; \qquad \frac{\partial P^{\mathrm{Am}}}{\partial t} + \frac{1}{2}\sigma^2 s^2 \frac{\partial^2 P^{\mathrm{Am}}}{\partial s^2} + rs\frac{\partial P^{\mathrm{Am}}}{\partial s} - rP^{\mathrm{Am}} < 0. \qquad (5.39)$$

In the first case, when it is not optimal to exercise, (5.38) tells us that the option can be valued by solving the Black-Scholes-Merton PDE. The boundary conditions for this at the optimal exercise boundary are

$$P^{\text{Am}}(S_f(t), t) = (E - S_f(t))^+; \qquad P_s^{\text{Am}}(S_f(t), t) = -1.$$

In the second case, when it is optimal to exercise, the value of the option is simply $E - S(t)$ at time $t \in [0, T]$, and the partial differential inequality in the second part of (5.39) can be confirmed by making the substitution $P^{\text{Am}}(s, t) = E - s$ into the LHS, so that

$$\frac{\partial P^{\text{Am}}}{\partial t} + \frac{1}{2}\sigma^2 s^2 \frac{\partial^2 P^{\text{Am}}}{\partial s^2} + rs\frac{\partial P^{\text{Am}}}{\partial s} - rP^{\text{Am}} = -rE < 0. \tag{5.40}$$

This reflects the fact that if one sets up a dynamically hedged portfolio for the American put option of the kind discussed in Sect. 2.2.4 for a European option, it is not possible to hedge away the risk that the option will be exercised early.

Taken together, (5.38) and (5.39) are called the free boundary formulation of the option value, and a more detailed derivation may be found, for example, in Shreve [62].

The free boundary formulation is hard to solve directly. To approximate the value of $P^{\text{Am}}(s, t)$ on a discretisation of the domain $\mathbb{R}^+ \times [0, T]$, we must identify where each point (s, t), at which we wish to compute an option value, is relative to $S_f(t)$. If $S(t) > S_f(t)$ we can in principle use the methods developed in this chapter to solve the Black-Scholes-Merton PDE with boundary conditions (5.40). If $S(t) < S_f(t)$, we can in principle set the value equal to the intrinsic value at that point. However, the location of the boundary S_f is not known, so in practice we must compare the value predicted by the Black-Scholes-Merton PDE at each point to the intrinsic value and select the larger of the two. This is analogous to the comparison step with which we valued American options using binomial trees in Chap. 3, and the Longstaff-Schwarz method in Chap. 4.

5.6.2 The Compact Linear Complementarity Form of the Problem

We seek to incorporate this comparison step into the finite difference approach. As we have transformed the Black-Scholes-Merton PDE (5.1) into the heat Eq. (5.10), we will also need to transform the free boundary formulation using the change of variables given by (5.1)–(5.9). The intrinsic value $\Lambda(s) = \max\{E - s, 0\}$ under the new coordinate system becomes

$$g(x, \tau) = e^{(k+1)^2 \tau/2} \left(e^{(k-1)x/2} - e^{(k+1)x/2} \right)^+.$$

Let's set up a function for later use:

```
In []: def g(x,tau,k):
           innerTerm=np.exp(0.5*x*(k-1))-np.exp(0.5*x*(k+1))
           outerterm=np.exp(0.5*tau*(k+1)**2)
           return(outerTerm*np.maximum(innerTerm,0))
```

The option value can then be characterised by

$$u(x, \tau) > g(x, \tau), \quad u_\tau(x, \tau) = u_{xx}(x, \tau),$$

corresponding to (5.38), and

$$u(x, \tau) = g(x, \tau), \quad u_\tau(x, \tau) > u_{xx}(x, \tau),$$

corresponding to (5.39). The optimal exercise boundary, denoted $x_f(\tau)$ in this coordinate system, is such that

$$u(x_f(\tau), \tau) = g(x_f(\tau), \tau) \text{ and } u(x, \tau) > g(x, \tau) \text{ for } x > x_f(\tau).$$

This characterisation may be recast in a form that makes no reference to the unknown boundary $x_f(\tau)$, the *compact linear complementarity form* (see Chapter 7.4 of Wilmott et al. [69]):

$$(u_\tau(x, \tau) - u_{xx}(x, \tau))\,(u(x, \tau) - g(x, \tau)) = 0;$$
$$u_\tau(x, \tau) \geq u_{xx}(x, \tau), \quad u(x, \tau) \geq g(x, \tau),$$

and in principle we are solving this problem for $(x, \tau) \in [L_1, L_2] \times [0, \bar{T}]$ in order to value the American put.

5.6.3 Valuing an American Put Using the Method of Projected Successive Over-Relaxation (PSOR)

Let's apply the BTCS scheme, which is described by the matrix-vector form of the update equation given by (5.30):

$$GU^{m+1} = U^m + q^m, \quad m = 0, \ldots, N_\tau - 1, \tag{5.41}$$

where G and each q^m are as given in (5.31). For a European option, we can simply multiply both sides by G^{-1} in order to see how to move from the solution vector U^m to U^{m+1}. We need to build an additional check into this step so that entries in the vector U^{n+1} can be set equal to the intrinsic value g if they fall at or under the optimal exercise boundary x_f. However, we have to be a little careful when doing this for an implicit method, since the value of approximation U_n^{m+1} depends on its neighbours U_{n-1}^{m+1} and U_{n+1}^{m+1} (see Fig. 5.3), and so at any given timestep, changes to one part of the solution must somehow be propagated to all other parts of the solution at the same timestep.

The PSOR method is designed to deal with this. Instead of directly inverting G and updating from timestep m to $m + 1$ in a single matrix-vector multiplication, we will modify Step 7 of the algorithm described in Sect. 5.5 so that the matrix inversion is carried out by the iterative method of *successive over-relaxation (SOR)*. At each iteration, we can adjust, or *project (P)*, any values for which the intrinsic value is higher than that predicted by the PDE, and therefore for which the value of the option should reflect early exercise at that point. Since the iteration that inverts G will continue until we decide it has converged, this has the desired effect of propagating any such adjustments.

5.6.3.1 Implementation Details

First, decompose G additively into a diagonal component and strictly lower and upper triangular components, written $G = LT + D + UT$. This leads to an adjustment of Step 3 of the algorithm detailed in Sect. 5.5, where recall that we defined the parts of G separately as:

```
In []: subDiagG=-alpha*np.diag(np.ones(Nx-2),-1)
       diagG=(1+2*alpha)*np.eye(Nx-1)
       superDiagG=-alpha*np.diag(np.ones(Nx-2),1)
```

The matrix-vector equation (5.41) can be written, for some relaxation constant $\omega > 1$,

$$(D + \omega LT)U^{m+1} = \omega(U^m + q^m) - [\omega UT + (\omega - 1)D]U^{m+1}.$$

Since U^m is already known, we can solve this expression for U^{m+1} by iterating the expression

$$U(k + 1) = L_\omega U(k) + c(U^m + q^m); \quad U(0) = 0, \tag{5.42}$$

where $U(k)$ is our approximation of U^{m+1} after k iterations, and

$$c = (D + \omega LT)^{-1}\omega; \quad L_\omega = -(D + \omega LT)^{-1}[\omega UT + (\omega - 1)D].$$

The rate of convergence of this procedure depends on our choice of ω, and for this demonstration we will set omega $= 1.8$. So set up c and L_ω:

```
In []: X=linalg.inv(diagG+omega*subDiagG)
       c=omega*X
       elOmega=-X.dot(omega*superDiagG+(omega-1)*diagG)
```

Now apply the iteration in (5.42), at each step projecting $U(k+1)$ onto $\max\{U(k+1), g\}$, where g represents the vector of intrinsic values corresponding to the current timestep, and the maximum is taken componentwise. We terminate the iteration when $\|U(k+1) - U(k)\|$ falls below some user defined tolerance parameter `tol`, where $\|\cdot\|$ denotes the Euclidean norm implemented in the call `linalg.norm()`. Now in Step 7, we would have the nested loop

```
In []: for i in range(0,Nt):
           u=np.zeros(Nx-2)
           convergence = False

           while (convergence==False):
               old=u;
               new=elOmega.dot(old)+c.dot(U[:,i]+q[:,i+1])
               u=np.maximum(new,g(x,tau[i+1]))
               if (linalg.norm(u-old)<tol):
                   convergence=True

           U[:,i+1]=u
```

The remainder of the algorithm can be applied exactly as described in Sect. 5.5.

5.7   The Link to Binomial Methods

Suppose that instead of applying the transformation (5.7)–(5.9) to the Black-Scholes-Merton PDE (5.1), we only transform the state variable and discount the value function, so that

$$y = \ln(s), \quad \text{and} \quad W(y,t) = e^{-rt}V(s,t). \tag{5.43}$$

Then by various applications of the chain rule for partial derivatives (Exercise 5.2 is good practice for this), (5.1) becomes the constant coefficient PDE

$$W_t(y,t) + \frac{\sigma^2}{2}W_{yy}(y,t) + \left(r - \frac{\sigma^2}{2}\right)W_y(y,t) = 0, \tag{5.44}$$

for all $(y,t) \in \log(\mathcal{D}) \times [0,T]$. Exercise 5.17 asks you to confirm this.

Now, we will apply an implicit finite difference approximation scheme to solutions of (5.44) on a truncated domain $[L_1, L_2] \times [0,T]$, using a forward difference approximation in time, a central difference approximation for the first order partial in space, and a symmetric central difference approximation for the second order partial in space. Let N_x be the number of steps on our mesh in the y direction and N_t the number of steps in the t direction, so that $\delta y = (L_2 - L_1)/N_y$ and $\delta t = T/N_t$. For $n = 1, \ldots, N_y - 1$ and $m = 1, \ldots, N_t$ denote $W_n^m \approx$

$W(n\delta y, m\delta t)$, and

$$W_t(y, t) \approx \frac{W_n^{m+1} - W_n^m}{\delta t};$$

$$W_y(y, t) \approx \frac{W_{n+1}^{m+1} - W_{n-1}^{m+1}}{2\delta y};$$

$$W_{yy}(y, t) \approx \frac{W_{n+1}^{m+1} - 2W_n^{m+1} + W_{n-1}^{m+1}}{(\delta y)^2}.$$

Substituting back into (5.44) gives the scheme

$$\frac{W_j^{i+1} - W_j^i}{\delta t} + \frac{\sigma^2}{2}\frac{W_{j+1}^{i+1} - 2W_j^{i+1} + W_{j-1}^{i+1}}{(\delta y)^2} + \left(r - \frac{\sigma^2}{2}\right)\frac{W_{j+1}^{i+1} - W_{j-1}^{i+1}}{2\delta y} = 0,$$

$$(5.45)$$

for $j = 1, \ldots, N_y$ and $i = 1, \ldots, N_t$. Now we chose the stepsizes so that $\alpha = \delta t/(\delta y)^2 = 1/\sigma^2$. Then (5.45) may be written

$$W_j^i = \bar{p}W_{j+1}^{i+1} + (1 - \bar{p})W_{j-1}^{i+1},$$

where

$$\bar{p} = \frac{1}{2}\left(1 + \frac{\sqrt{\delta t}}{\sigma}\left(r - \frac{\sigma^2}{2}\right)\right).$$

Making the conversion back to the original value process, we can make the substitution

$$W_j^i = e^{-ri\delta t}V_j^i; \quad W_{j+1}^{i+1} = e^{-r(i+1)\delta t}V_{j+1}^{i+1}; \quad W_{j-1}^{i+1} = e^{-r(i+1)\delta t}V_{j-1}^{i+1};$$

giving

$$V_j^i = e^{-r\delta t}\left[\bar{p}V_{j+1}^{i+1} + (1 - \bar{p})V_{j-1}^{i+1}\right], \qquad (5.46)$$

which we recognise as a structural mimic of Step 3 of the binomial algorithm for European options described in Sect. 3.2.

5.8 Further Reading

A general but accessible introduction to finite difference approximation methods, including error analysis and details on direct and iterative methods for matrix inversion, may be found in Strikwerda [63]. This chapter makes some use of matrix analysis and readers interested in more detail may consult the classic text of Horn and Johnson [35].

The transformation of the Black-Scholes-Merton PDE (5.1) and associated boundary conditions to the heat Eq. (5.10) was the basis of the original derivation of the Black-Scholes formula for the price of a European option in [8], and has become the standard textbook treatment of the derivation: see Wilmott et al [69] for example. That book also gives a description of the PSOR method for valuing American options, and describes an implementation that uses Crank-Nicolson on the heat equation.

In this book we choose to discretise the heat equation form of the Black-Scholes-Merton PDE in order to more directly analyse the local accuracy and stability of our numerical schemes. One may also to discretise the Black-Scholes-Merton PDE directly using finite difference approximations and to solve it backwards from the final time condition - this approach may be preferable, for example, for exotic options where transformation to the heat equation is not possible. A treatment of the direct approach using Matlab may be found in Higham [32]. A recent study that compares numerical approximations of the optimal exercise boundary found using the PSOR method to common analytic approximations may be found in Lauko and Ševčovič [44].

Our construction of the link between binomial methods and finite difference methods in Sect. 5.7 follows that of Chapter 24 of Higham [32], and in an exercise he uses an FTCS approximation of the heat equation

$$u_\tau(x, \tau) = \frac{\sigma^2}{2} u_{xx}(x, \tau), \qquad (5.47)$$

to show that the stepsize coupling $(\delta x)^2 = \sigma^2 \delta t$ places this method at the boundary of stability and instability in the sense of von Neumann for (5.47). Although not a direct stability analysis of the binomial method, since we have arrived at (5.46) by discretisation of (5.44) rather than (5.47), this gives some insight into the non-monotone nature of convergence observed in Sect. 3.4. The use of benchmark test equations, usually linear, for the stability analysis of numerical methods is standard practice and we will consider this in the context of numerical methods for SDE models in Chap. 8.

5.8.1 Exercises

5.1 Verify that

$$v(x, \tau) = \frac{1}{\sqrt{\tau}} e^{-\frac{x^2}{4\tau}}, \quad \tau \neq 0, \tag{5.48}$$

satisfies the heat Eq. (5.6), and identify the initial and boundary conditions associated with it.

5.2 Check that the transformation given by (5.7) and applied to the Black-Scholes-Merton PDE (5.1) leads to

1. the constant coefficient PDE

$$0 = \frac{1}{2}\sigma^2 f_{xx}(x, \tau) + (r - \sigma^2/2) f_x(x, \tau) - \frac{\sigma^2}{2} f_\tau(x, \tau) - r f(x, \tau),$$

   where $f(x, \tau) = EV(s, t)$;
2. the heat equation in (5.10), using the full transformation given in (5.8) and the constant definitions in (5.9).

Confirm that the final time conditions for a European call and put option contained in (5.2) and (5.3) are transformed to the initial conditions (5.37) and (5.49) respectively.

5.3 Review the computation in Sect. 5.2 which led to the output in Table 5.1. If we maintain the number of steps taken in the spatial direction $N_x = 4$, how many steps must we take in the time direction in order to ensure a stable scheme? Adjust your computation for this new value of N_τ and output a surface plot of the numerical solution. Does it work?

5.4 Setting $N_x = N_\tau = 4$, check that the matrix form of the FTCS scheme (5.23) corresponds to the update Eq. (5.15) for $n = 1, 2, 3$ by multiplying out the terms of (5.23). Identify for which values of n the boundary conditions a and b appear on the RHS of the update equation.

5.5 Confirm, using Theorem 5.5, that (5.26) holds for the eigenvalues of F, and show explicitly why this implies that $|\lambda_s| \leq 1$ for all $s = 1, \ldots, N_x - 1$ if and only if $0 < \alpha \leq 1/2$.

5.6 By adapting the computations from Sect. 5.3.3, demonstrate that the BTCS scheme characterised by (5.29) is unconditionally stable in the sense of von Neumann (i.e. stable for all $\alpha > 0$).

5.7 If $G \in \mathbb{R}^{N \times N}$ is an invertible symmetric matrix, then the eigenvalues of G^{-1} are the reciprocals of the eigenvalues of G. Use this fact, along with Theorem 5.5, to show that the BTCS scheme characterised by (5.30) is unconditionally matrix stable.

5.8 Modify this code so that, instead of approximating V over a large set of values for S ($[e^{-8}, e^8]$ in the current version of the code) and zooming in to the part of the Black-Scholes surface we are interested in (e.g. $[1, 100]$), V is approximated only over the smaller range. What do you observe? Can you explain why?

5.9 Confirm that the Crank-Nicolson scheme (5.34) is unconditionally stable in the sense of von Neumann.

5.10 What role is played by the variable `xr` in the last code snippet in Step 9 of the algorithm described in Sect. 5.5?

5.11 To produce a Black-Scholes surface for a European put, the main substantial change is to the setup of the initial and boundary conditions in Step 5. The final-time condition for (5.1) corresponding to the payoff for a European put option is transformed by (5.7)–(5.9) in Sect. 5.1.4 to the initial profile

$$u(x, 0) = u_0(x) = \max\left(e^{\frac{1}{2}(k-1)x} - e^{\frac{1}{2}(k+1)x}, 0\right). \tag{5.49}$$

Now consider the boundary condition at $S = 0$ from (5.3). We have $V(0, t) = Ee^{-r(T-t)}$ which gives

$$a(\tau) = u(x_{\min}, \tau) = \frac{1}{E}e^{-\alpha x_{\min} - \beta \tau} Ee^{-r(T-t)}$$

$$= e^{-\alpha x_{\min} - \beta \tau} e^{-r(T-t)}$$

$$= e^{-\alpha x_{\min} - \beta \tau} e^{-k\tau} = e^{-\alpha x_{\min} - (\beta + k)\tau}.$$

The second boundary condition is $b(\tau) = 0$ for all τ, and the initial profile is given by (5.49).

Modify Step 5 of the algorithm in Sect. 5.5 so that it produces a value for the European put.

5.12 Modify your code from Sect. 5.5 so that it implements a Crank-Nicolson approximation of the transformed Black-Scholes-Merton PDE and recompute the maximum error against the benchmark valuation formulae for the European call and put options.

5.13 In Sect. 5.5.2 it is described how to modify the BTCS valuation of a European call option to produce a Black-Scholes surface for a down-and-out barrier call option with parameters $E = 5$, $B = 2$, $T = 1$, $r = 0.01$, $\sigma = 0.4$.

1. Using the benchmark valuation formula given by (3.15) in Chap. 3, compute the maximum error at time $t = 0$ using BTCS.
2. Modify your code again to use Crank-Nicolson and recompute the maximum error.

5.14 Consider a cash-or-nothing call option with payoff

$$\Lambda(S) = B\mathcal{H}(s - E),$$

where $B = 20$, $E = 5$, $T = 1.5$, $r = 0.01$, $\sigma = 0.5$, and $\mathcal{H}$ denotes the Heaviside step function.

1. Modify the calculation of U[:,0] in Step 5 of your implementation code to reflect the initial function $u_0(x)$ associated with Λ.
2. By further reference to Sect. 5.5.3, modify the calculation of a, b in Step 5 of your implementation code to reflect suitable boundary conditions for this option.
3. Use either BTCS or Crank-Nicolson to produce a surface plot of prices for the cash-or-nothing call over the approximate range $S \in [0, 10]$. Comment on your output.

5.15 Implement the PSOR method for an American put option with $T = 1$, $E = 5$, $r = 0.01$, and $\sigma = 1$. Compare the resulting value surface with that given in Fig. 5.4 for a European put option.

5.16 How would you modify this algorithm to value a Bermudan option?

5.17 Verify that, under the change-of-variables given by (5.43), the Black-Scholes-Merton PDE (5.1) becomes the constant coefficient PDE (5.44).

Part III

Simulation Methods Beyond the Black-Scholes Framework

Simulation II: Modelling Multivariate Financial Data   6

We show how to construct multi-asset models within the Black-Scholes framework with a given correlation structure and normally distributed log-returns. This is necessary when valuing what are known as rainbow options that are characterised by a payoff depending on two or more assets trading in the same market with a single risk-free asset and multiple risky assets. It is also necessary when constructing larger-scale market models to study, for example, the effects of systemic risk.

Next, we investigate the real-world distribution of an equity asset pair and demonstrate that the log-returns are not, in fact, normal but instead display a bivariate distribution with marginal distributions that are heavy-tailed. This motivates an exploration of tools for sampling from multivariate non-normal distributions, both directly by using a multivariate Student's t_ν distribution, and by the use of copulas.

Copulas provide a characterisation of the dependence structure of a multivariate distribution by converting the distribution to a standardised form with uniform marginal distributions. They provide a set of tools that allow us to examine and compare dependencies in different multivariate distributions without worrying about the specifics of the marginal distributions. They also allow us to construct bespoke multivariate distributions by imposing a target dependence structure on arbitrary marginal distributions in order to match observed data.

Though copulas were developed as a tool in classical statistics, their ability to simplify the characterisation of dependence structure in high dimensional data makes them a useful step in the direction of supervised machine learning.

© The Author(s), under exclusive license to Springer Nature Switzerland AG 2024   183
C. Kelly, *Computation and Simulation for Finance*,
Springer Undergraduate Texts in Mathematics and Technology,
https://doi.org/10.1007/978-3-031-60575-8_6

Python Setup To follow the Python code in this chapter, you will need to import the NumPy, PyPlot, Stats, and Pandas packages.

```
In []: import numpy as np
       import matplotlib.pyplot as plt
       import scipy.stats as stats
       import pandas as pd
```

Pandas is used for real-world data analysis in Python, and is included in the Anaconda distribution. It depends on the NumPy package and may optionally make use of functionality in Scipy and Matplotlib. We will use it to manage financial data sets in order to compare real-world data to simulation, making particular use of the built-in `DataFrame` structure for representing tabular data.

The most convenient way to import data available online is to use a datareader package. We will use Pandas Datareader, which creates Pandas DataFrame objects from internet-accessible data sources. Documentation is available at the official site pandas-datareader.readthedocs.io. At the time of writing it is not part of the Anaconda distribution, and before it is used for the first time it must be installed using the `pip` command:

```
In []: %pip install pandas_datareader
```

If you have used Pandas Datareader before you can omit this step. The appropriate package can then be imported in the usual way:

```
In []: import pandas_datareader.data as web
```

It should be noted that the availability of data from internet sources can change over time, and some sources require sign up to an account or payment for API access. We will use stooq.pl, which at the time of writing gives free access to up to 5 years of historical data from major financial markets and is sufficient for learning the techniques presented here. Free online data sources should be used with caution, and we do not recommend that these sources be used for commercial applications.

For pseudo-random number generation throughout this chapter we will, as usual, instantiate and use a local RNG from the `numpy.random` module

```
In []: rng=np.random.default_rng()
```

6.1   Rainbow Options

Rainbow options are options whose payoff depends on a collection of $d > 1$ underlying risky assets labelled $S_1, \ldots, S_d$, modelled as a d-dimensional stochastic process with correlated noises. Option prices in closed form are only available for simple cases, and in general pricing requires the use of Monte Carlo or binomial tree methods.

There are many types of rainbow option. Consider the following (non-exhaustive) list of examples:

1. A *two-asset spread call* option on the (signed) difference between two assets S_1 and S_2 with expiry T and strike E has payoff

$$\Lambda(S_1, S_2) = ((S_1(T) - S_2(T)) - E)^+ .$$

   Notice that as defined here, if $S_1(T) < S_2(T)$ there is no possibility for a nonzero payoff.
2. A *basket call* option on a portfolio of d assets with expiry T and strike E has payoff

$$\Lambda(S_1, \ldots, S_d) = ((c_1 S_1(T) + c_2 S_2(T) + \cdots + c_d S_d(T)) - E)^+ ,$$

   for some weights $c_i, i = 1, \ldots d$.
3. An *outperformance call* option on the maximum of d asset holdings with expiry T and strike E has payoff

$$\Lambda(S_1, \ldots, S_d) = (\max\{c_1 S_1(T), c_2 S_2(T), \ldots, c_d S_d(T)\} - E)^+,$$

   again for some weights $c_i, i = 1, \ldots d$.
4. A *two-asset barrier put* option with expiry T, strike E and barrier B may have payoff

$$\Lambda(S_1, S_2) = I_{\{\min_{i=1,\ldots,n} S_2(t_i) < B\}} (E - S_1(T))^+,$$

   so that the option pays $(E - S_1(T))^+$ if and only if the second asset S_2 takes a value below B at one of the designated times t_i, for $i = 1, \ldots, n$.

Let us consider in more detail a *swaption*. This is an option written at time $t = 0$ giving the holder the right, but not the obligation, to swap their holding in an asset S_2 for a holding in asset S_1 at time $t = T$. Let $c_1, c_2 > 0$ represent the size of the holdings of S_1 and S_2 respectively. The payoff of the swaption is given by

$$\Lambda(S_1, S_2) = (c_1 S_1(T) - c_2 S_2(T))^+ . \tag{6.1}$$

Suppose that the pair of assets is modelled computationally as

$$dS_1(t) = r S_1(t)dt + \sigma_1 S_1(t)dW_1^{\mathbb{Q}}(t);$$

$$dS_2(t) = r S_2(t)dt + \sigma_2 S_2(t)dW_2^{\mathbb{Q}}(t), \quad t \in [0, T]. \tag{6.2}$$

Here we let $W_1^{\mathbb{Q}}, W_2^{\mathbb{Q}}$ be standard Brownian motions on a probability space $(\Omega, \mathcal{F}, \mathbb{Q})$ where the pair $[W_1^{\mathbb{Q}}, W_2^{\mathbb{Q}}]^T$ is adapted to a filtration denoted $(\mathcal{F}_t)_{t \in [0,T]}$, and the risk neutral measure $\mathbb{Q}$ is such that both of the discounted asset price processes $e^{-rt}S_1(t)$ and $e^{-rt}S_2(t)$ are $\mathcal{F}_t$-martingales. Finally, $\sigma = [\sigma_1, \sigma_2]^T$ is a constant vector of asset price volatilities such that $\sigma_1, \sigma_2 > 0$.

A more general form of the argument presented in Sect. 2.1 in Chap. 2 tells us that the value of the swaption at time $t \in [0, T]$ is given by

$$\mathbb{E}_{\mathbb{Q}}\left[e^{-r(T-t)}\left(c_1 S_1(T) - c_2 S_2(T)\right)^+ \Big| \mathcal{F}_t\right]. \qquad (6.3)$$

It is unreasonable to suppose that the pair of Brownian motions $W_1^{\mathbb{Q}}, W_2^{\mathbb{Q}}$ are independent. Consider, for example, two equity assets trading on the same exchange. The evolution of both assets will be influenced by similar though not identical flows of information, and should therefore have some correlation. A similar observation applies for problems where we must sample from the joint distribution of a vector of d asset price processes $[S_1(t), \ldots, S_d(t)]^T$ for $t \in [0, T]$.

6.2 Jointly Distributed Random Variables

If we wish to model a collection of d assets as random variables, then we should understand how to characterise their joint distribution. Let $X = [X_1, \ldots, X_d]^T$ be a vector of $\mathbb{R}$-valued random variables defined on the same probability space $(\Omega, \mathcal{F}, \mathbb{P})$.

The *joint cumulative distribution function (CDF)* of X is written

$$F_X(x_1, \ldots, x_d) := \mathbb{P}[X_1 \leq x_1, \ldots, X_d \leq x_d]$$

for all $x_1, \ldots, x_d \in \mathbb{R}$.

X is said to be *continuous* if there is a function $f_X(x_1, \ldots, x_d)$, called the *joint probability density function (PDF)* of X, such that the joint CDF can be written as

$$F_X(x_1, \ldots, x_d) = \int_{-\infty}^{x_d} \cdots \int_{-\infty}^{x_1} f_X(s_1, \ldots, s_d)ds_1 \cdots ds_d, \qquad (6.4)$$

for $x_1, \ldots, x_d \in \mathbb{R}$.

It follows from (6.4) that the joint PDF can be obtained from the joint CDF as

$$f_X(x_1, \ldots, x_d) = \frac{\partial^d}{\partial x_1 \cdots \partial x_d} F_X(x_1, \ldots, x_d), \quad x_1, \ldots, x_d \in \mathbb{R}.$$

Finally, X has d *marginal CDFs* defined in each case to be

$$F_{X_i}(x) := \lim_{\substack{x_j \to \infty; \\ i \neq j}} F_X(x_1, \ldots, x_{i-1}, x, x_{i+1}, \ldots, x_d) = \mathbb{P}[X_i \leq x], \quad x \in \mathbb{R},$$

for each $i = 1, \ldots, d$.

6.3   Joint Centred Normal Distributions

Recall from (1.2) in Definition 1.3 of Chap. 1 that the time-t values of a scalar standard Brownian motion have an $\mathcal{N}(0, t)$ distribution, with CDF defined by setting $m = 0$ and $v = t$ in

$$\Phi(x; m, v) := \int_{-\infty}^{x} \frac{1}{\sqrt{2\pi v}} e^{-\frac{(s-m)^2}{2v^2}} \, ds, \quad x \in \mathbb{R}. \tag{6.5}$$

Notationally, when $m = 0$ and $v = 1$ we write $\Phi(x) := \Phi(x; 0, 1)$. We can extend the definition of a normal random variable to the multivariate case as follows.

Definition 6.1 A real-valued random vector $X = [X_1, \ldots, X_d]^T$ has a *multivariate normal distribution* if every linear combination

$$Y = a_1 X_1 + \cdots + a_d X_d,$$

has a normal distribution. Define the covariance matrix

$$\Sigma := \begin{pmatrix} \mathrm{Var}(X_1) & \mathrm{Cov}(X_1, X_2) \cdots \mathrm{Cov}(X_1, X_d)) \\ \mathrm{Cov}(X_2, X_1) & \mathrm{Var}(X_2) & \mathrm{Cov}(X_2, X_d) \\ \vdots & \ddots & \vdots \\ \mathrm{Cov}(X_d, X_1) \, \mathrm{Cov}(X_d, X_2) \cdots & \mathrm{Var}(X_d) \end{pmatrix}. \tag{6.6}$$

Any such matrix $\Sigma \in \mathbb{R}^{d \times d}$ is symmetric and positive semi-definite,[1] so that all its eigenvalues are non-negative. If Σ is positive definite, so that all its eigenvalues are positive, it is invertible, and the joint PDF of X is given by

$$f_{\mu, \Sigma}(x_1, \ldots, x_d) = \frac{1}{\sqrt{(2\pi)^d \, \det \Sigma}} \exp\left(-\frac{1}{2}[x - \mu]^T \Sigma^{-1}[x - \mu]\right), \tag{6.7}$$

for some vector of means $\mu = [\mu_1, \ldots, \mu_d]^T \in \mathbb{R}^d$ and all $[x_1, \ldots, x_d]^T \in \mathbb{R}^d$, where $[x - \mu] = [x_1 - \mu_2, \ldots, x_d - \mu_d]^T$. The joint CDF is

$$F_{\mu, \Sigma}(x_1, \ldots, x_d) = \frac{1}{\sqrt{(2\pi)^d \, \det \Sigma}}$$

$$\times \int_{-\infty}^{x_d} \cdots \int_{-\infty}^{x_1} \exp\left(-\frac{1}{2}(s - \mu)^T \Sigma^{-1}(s - \mu)\right) ds_1 \cdots ds_d, \tag{6.8}$$

for all $[x_1, \ldots, x_d]^T \in \mathbb{R}^d$.

---

[1] Σ is positive semi-definite if and only if $x^T \Sigma x \geq 0$ for all $x \in \mathbb{R}^d$. Σ is positive definite if and only if $x^T \Sigma x > 0$ for all $x \in \mathbb{R}^d$.

We write $X \sim \mathcal{N}(\mu, \Sigma)$, where $\mathcal{N}(\mu, \Sigma)$ denotes the distribution of a multivariate normal random variable in d-dimensions with d-dimensional mean vector μ and $d \times d$ covariance matrix Σ.

In the normal case specifically, since the distribution of that random variable X is decided, we switch notation for the joint PDF and CDF functions from f_X and F_X to $f_{\mu,\Sigma}$ and $F_{\mu,\Sigma}$ in order to emphasise the characteristic parameters μ and Σ. If $\mu = [0, \ldots, 0]^T$ is a vector of zeros, the distribution is referred to as centred, with joint PDF and CDF functions denoted f_Σ and F_Σ respectively.

For each $i = 1, \ldots, d$, set $\sigma_i^2 := \mathrm{Var}[X_i]$ and $\rho_{i,j} := \mathrm{Corr}(X_i, X_j) = \mathrm{Corr}(X_j, X_i)$. We can then rewrite Σ in the form

$$
\Sigma = \begin{pmatrix} \sigma_1 & & & 0 \\ & \sigma_2 & & \\ & & \ddots & \\ 0 & & & \sigma_d \end{pmatrix} \begin{pmatrix} 1 & \rho_{1,2} & \cdots & \rho_{1,d} \\ \rho_{1,2} & 1 & & \rho_{2,d} \\ \vdots & & \ddots & \vdots \\ \rho_{1,d} & \rho_{2,d} & \cdots & 1 \end{pmatrix} \begin{pmatrix} \sigma_1 & & & 0 \\ & \sigma_2 & & \\ & & \ddots & \\ 0 & & & \sigma_d \end{pmatrix}. \tag{6.9}
$$

Exercise 6.2 asks you to confirm (6.9).

6.3.1 Incorporating a Covariance Structure by Cholesky Factorisation

In order to simulate sample trajectories of the vector $S = [S_1, \ldots, S_d]$, we must know how to incorporate a covariance structure when sampling from the vector of perturbing standard Brownian motions $W^{\mathbb{Q}} = \left[W_1^{\mathbb{Q}}, \ldots, W_d^{\mathbb{Q}} \right]^T$.

Suppose that for any $t \in [0, T]$, $W^{\mathbb{Q}}(t)$ satisfies,

$$
W^{\mathbb{Q}}(t) \sim \mathcal{N}(0, \Sigma), \quad \text{where} \quad \Sigma = \begin{pmatrix} 1 & \rho_{1,2} & \cdots & \rho_{1,d} \\ \rho_{1,2} & 1 & & \rho_{2,d} \\ \vdots & & \ddots & \vdots \\ \rho_{1,d} & \rho_{2,d} & \cdots & 1 \end{pmatrix} t, \tag{6.10}
$$

for some correlations $\rho_{i,j} \in [-1, 1]$, $i, j = 1, \ldots, d$ and $i \neq j$ such that $\det \Sigma \neq 0$. Notice that we set $\sigma_i = \sqrt{t}$ for all $i = 1, \ldots, d$ in Eq. (6.9), which results in a scaling by $t > 0$ in (6.10).

We use a technique known as Cholesky factorisation to recast $W^{\mathbb{Q}}$ as a d-dimensional function of independent standard Brownian motions, which can then be sampled using a call to `rng.normal()`. This relies on the following property of multivariate normal random variables.

Proposition 6.2 (The Linear Transformation Property) *Let A be any real-valued $k \times d$ matrix, for some $k \in \mathbb{N}$. If $X \sim \mathcal{N}(\mu, \Sigma)$ then*

$$
AX \sim \mathcal{N}(A\mu, A\Sigma A^T).
$$

See Theorem 4.9.6 in Grimmett and Stirzaker [26] for a proof in the case where $\mu = 0$ and $k = d$. The authors of that text point us to Kingman and Taylor [42] for a proof of the general case.

Let us start with a vector of independent standard Brownian motions $B^{\mathbb{Q}} = [B_1^{\mathbb{Q}}, \ldots, B_d^{\mathbb{Q}}]^T$, so that $B^{\mathbb{Q}}(t) \sim \mathcal{N}(0, \mathbb{I}t)$, $t \in [0, T]$, where $\mathbb{I}$ is the $d \times d$ identity matrix. Now, by Proposition 6.2, for any $k \times d$ matrix A we have

$$AB^{\mathbb{Q}}(t) \sim \mathcal{N}(0, AA^T t). \tag{6.11}$$

By comparing (6.11) with (6.10), we see that it is enough to identify a matrix A such that

$$AA^T = \begin{pmatrix} 1 & \rho_{1,2} & \cdots & \rho_{1,d} \\ \rho_{1,2} & 1 & & \rho_{2,d} \\ \vdots & & \ddots & \vdots \\ \rho_{1,d} & \rho_{2,d} & \cdots & 1 \end{pmatrix}. \tag{6.12}$$

There is no unique way to choose A, but in the case where Σ is positive definite[2] it is computationally convenient to let A be lower triangular, in which case we can solve (6.12) for A by directly multiplying out the LHS of (6.12) and using back-substitution to solve for each entry in turn. The matrices A and A^T that result are known as the *Cholesky factors* of Σ.

6.3.2 A Python Implementation in Two Dimensions

We can illustrate this technique in the two-dimensional case as follows. Set $W^{\mathbb{Q}} = \left[W_1^{\mathbb{Q}}(t), W_2^{\mathbb{Q}}(t)\right]^T$, and $\rho := \rho_{1,2}$ in (6.10). We wish to sample from the change in $W^{\mathbb{Q}}$ over a single step of length t, so that

$$W^{\mathbb{Q}}(t) \sim \mathcal{N}\left(0, \begin{pmatrix} 1 & \rho \\ \rho & 1 \end{pmatrix} t\right).$$

Part 1 of Exercise 6.1 confirms that the matrix A in the Cholesky factorisation of Σ / t is

$$A = \begin{pmatrix} 1 & 0 \\ \rho & \sqrt{1 - \rho} \end{pmatrix}.$$

---

[2] A symmetric matrix is positive definite if and only if all its eigenvalues are positive.

Therefore by (6.10) and (6.11) with $d = 2$ we can write

$$\begin{pmatrix} W_1^{\mathbb{Q}}(t) \\ W_2^{\mathbb{Q}}(t) \end{pmatrix} = \begin{pmatrix} 1 & 0 \\ \rho & \sqrt{1-\rho} \end{pmatrix} \begin{pmatrix} B_1^{\mathbb{Q}}(t) \\ B_2^{\mathbb{Q}}(t) \end{pmatrix}. \tag{6.13}$$

We can now sample from $W^{\mathbb{Q}}(t)$ in Python. Set up the covariance matrix Σ as:

```
In []: T=1; rho=0.5
       sigma=T*np.array([[1,rho],[rho,1]])
```

Now compute the Cholesky factor using a built-in method from the `numpy.linalg` library:

```
In []: A=np.linalg.cholesky(sigma)
```

Confirm that $AA^T = \Sigma$ by matrix multiplication:

```
In []: np.dot(A,A.T)
```

We can produce 3000 samples of $W^{\mathbb{Q}}(1)$ as

```
In []: M=3000
       Z=rng.normal(0,1,(2,M))
       W=A.dot(Z)
```

The left column of Fig. 6.1 displays scatterplots of these samples for $\rho = 0.8, 0, -0.8$.

6.3.3　The Probability Integral Transform

Consider an $\mathbb{R}$-valued random variable X with continuous and strictly increasing CDF $F_X(x) = \mathbb{Q}[X \leq x]$, $x \in \mathbb{R}$. Define a random variable U by applying the CDF of X to the random variable X, so that $U := F_X(X)$. Then the CDF of U satisfies

$$F_U(u) = \mathbb{Q}[U \leq u] = \mathbb{Q}[F_X(X) \leq u]$$

$$= \mathbb{Q}[X \leq F_X^{-1}(u)] = F_X(F_X^{-1}(u)) = u, \quad 0 \leq u \leq 1. \tag{6.14}$$

The random variable $U = F_X(X)$ is uniformly distributed on $[0, 1]$. This is illustrated for a standard normal distribution in Fig. 6.2.

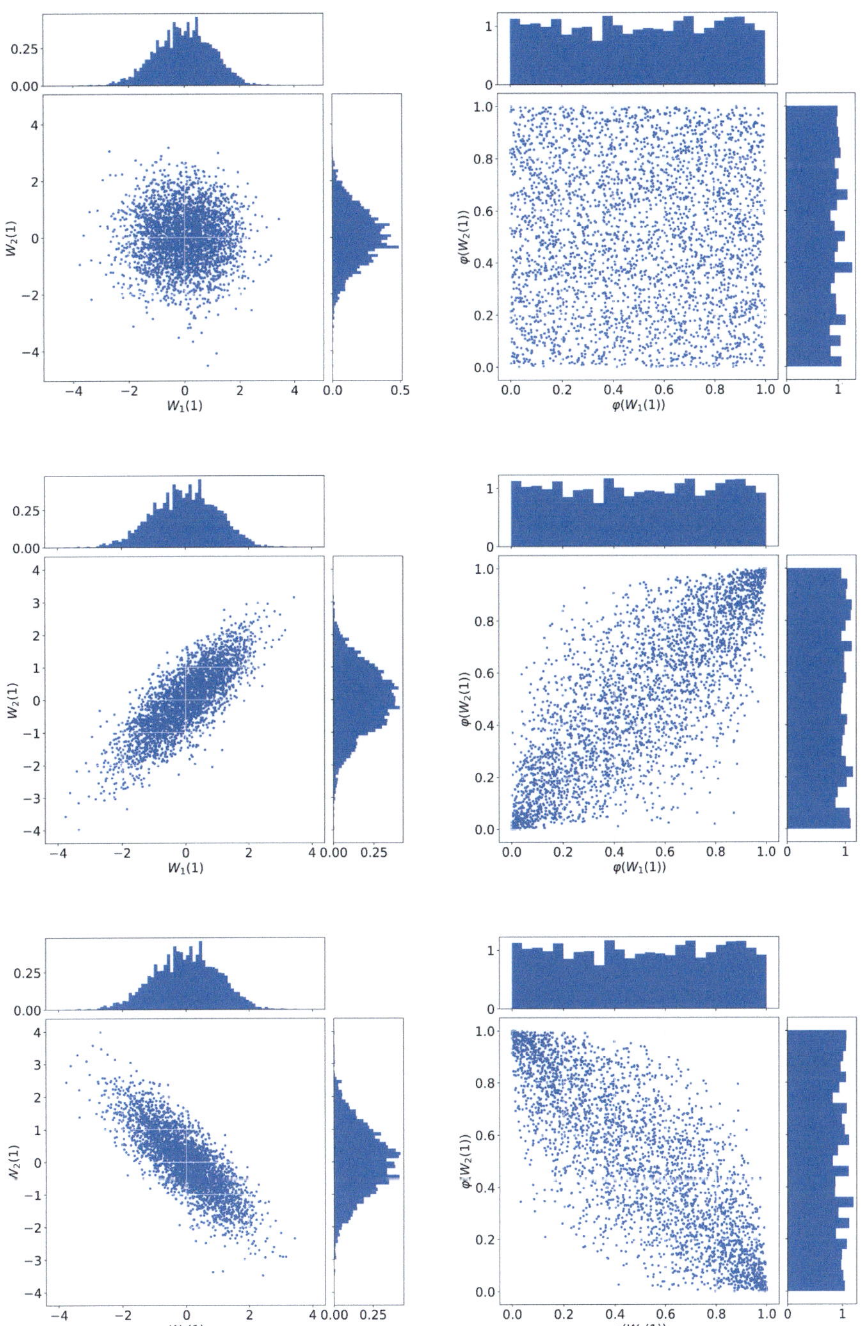

Fig. 6.1 Scatterplots showing $M = 3 \times 10^3$ samples at time $T = 1$ from a pair of correlated standard Brownian motion processes with $\rho = 0$ (top row), $\rho = 0.8$ (middle row), and $\rho = -0.8$ (bottom row). The values of Z have been reused in each case. Direct observations are on the left. On the right we have applied Φ to make the marginal data uniform, and observe that the covariance structure of the bivariate distribution is still visible

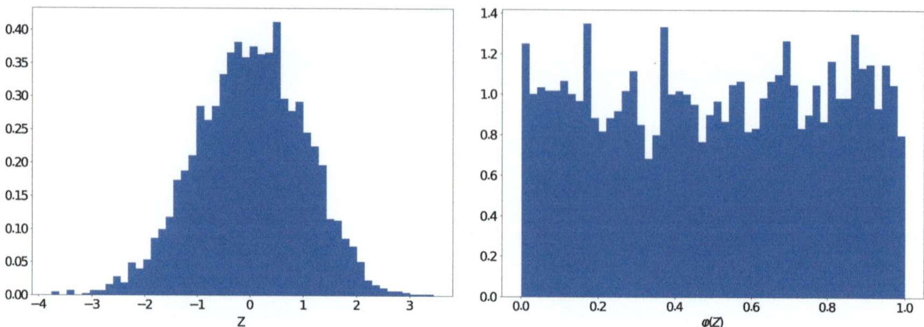

Fig. 6.2 Density histograms of 3×10^3 samples from a $\mathcal{N}(0, 1)$ distribution (left), and the same samples after a standard normal CDF Φ has been applied

Conversely, suppose that U is uniformly distributed on $[0, 1]$. Then

$$\mathbb{Q}[F_X^{-1}(U) \leq x] = \mathbb{Q}[U \leq F_X(x)] = F_X(x), \quad x \in \mathbb{R},$$

and therefore $F_X^{-1}(U)$ has CDF given by F_X.

The right column of Fig. 6.1 shows scatterplots of the samples in the left column, which were generated in Sect. 6.3.2, after the marginal data have been *uniformised* by applying the univariate standard normal CDF Φ given in (6.5). Notice that the covariance structure of the bivariate distribution is still visible in the scatterplot after this transformation. This method of representing the dependence structure of a multivariate distribution separately from the marginal densities leads to the notion of a *copula*, and will be treated later in the chapter.

6.3.4 Simulating Correlated Black-Scholes Asset Models

Suppose that, instead of sampling $W^{\mathbb{Q}}(t)$ at a single point in time, we wish to generate trajectories of $W^{\mathbb{Q}}$ over the interval $[0, T]$ using a mesh with N nodes and timestep $\delta t = T/N$. Then we can use the fact that the Brownian increment has distribution

$$W^{\mathbb{Q}}(t + \delta t) - W^{\mathbb{Q}}(t) \sim \mathcal{N}\left(0, \begin{pmatrix} 1 & \rho \\ \rho & 1 \end{pmatrix} \delta t\right). \tag{6.15}$$

The following code uses (6.15) to generate full trajectories of $W^{\mathbb{Q}}$:

```
In []: # Take 100 uniform steps in time over the interval [0,1]
        N=100; T=1; dt=T/N

        # Create a time set
        t=np.linspace(0,T,N+1)
```

```
# Sample independent Gaussian values
Z=rng.normal(0,1,(2,N))


# Construct covariance matrix and Cholesky factorisation
Sigma=np.array([[1,rho], [rho,1]])
A=np.sqrt(dt)*np.linalg.cholesky(Sigma)


# Generate correlated increments for W
X=A.dot(Z)


# Aggregate to construct the trajectory and
# insert the zero initial values
W=np.c_[np.zeros(2),np.cumsum(X, axis=1)]
```

We can now simulate trajectories for pricing under the risk-neutral measure $\mathbb{Q}$ of the pair of correlated asset models $S = [S_1, S_2]^T$ governed by the system of SDEs (6.2). We use $T = 1$, $r = 0.03$, $\sigma_1 = 0.2$ and $\sigma_2 = 0.3$ and initial data $S_0 = [S_1(0), S_2(0)]^T = [1, 1]^T$, retaining the covariance structure of $W^{\mathbb{Q}}$ from Sect. 6.3.2 by reusing the variable W.

Let's apply this technique to our asset model over the same timeset. By direct transformation of the Brownian values W, use the `np.tile()` call to construct a matrix of asset prices where the ith row represents a single trajectory of S_i, $i = 1, 2$ over the interval $[0, 1]$.

```
In []: # Model parameters
       S0=np.ones(2); r=0.03; sig=np.array([0.2, 0.3])


       # Simulation parameters
       N=100; dt=1/N; t=np.linspace(0,1,N+1)
       tt=np.tile(t,(2,1))


       # Asset price values
       S=S0*np.exp((r-0.5*sig**2)*tt.T+sig*W.T)
```

S will be an $(N + 1) \times 2$ array where the ith column contains a trajectory of S_i for $i = 1, 2$. Trajectories of both asset price processes are displayed in Fig. 6.3. Notice that the trajectory for S_1 (in blue) is identical in each case, whereas the trajectory for S_2 (in orange) changes with ρ.

It's worth examining in more detail the role of `np.tile()` in this procedure. Create an array containing a discrete time set over the interval $[0, 1]$ with 5 uniform steps:

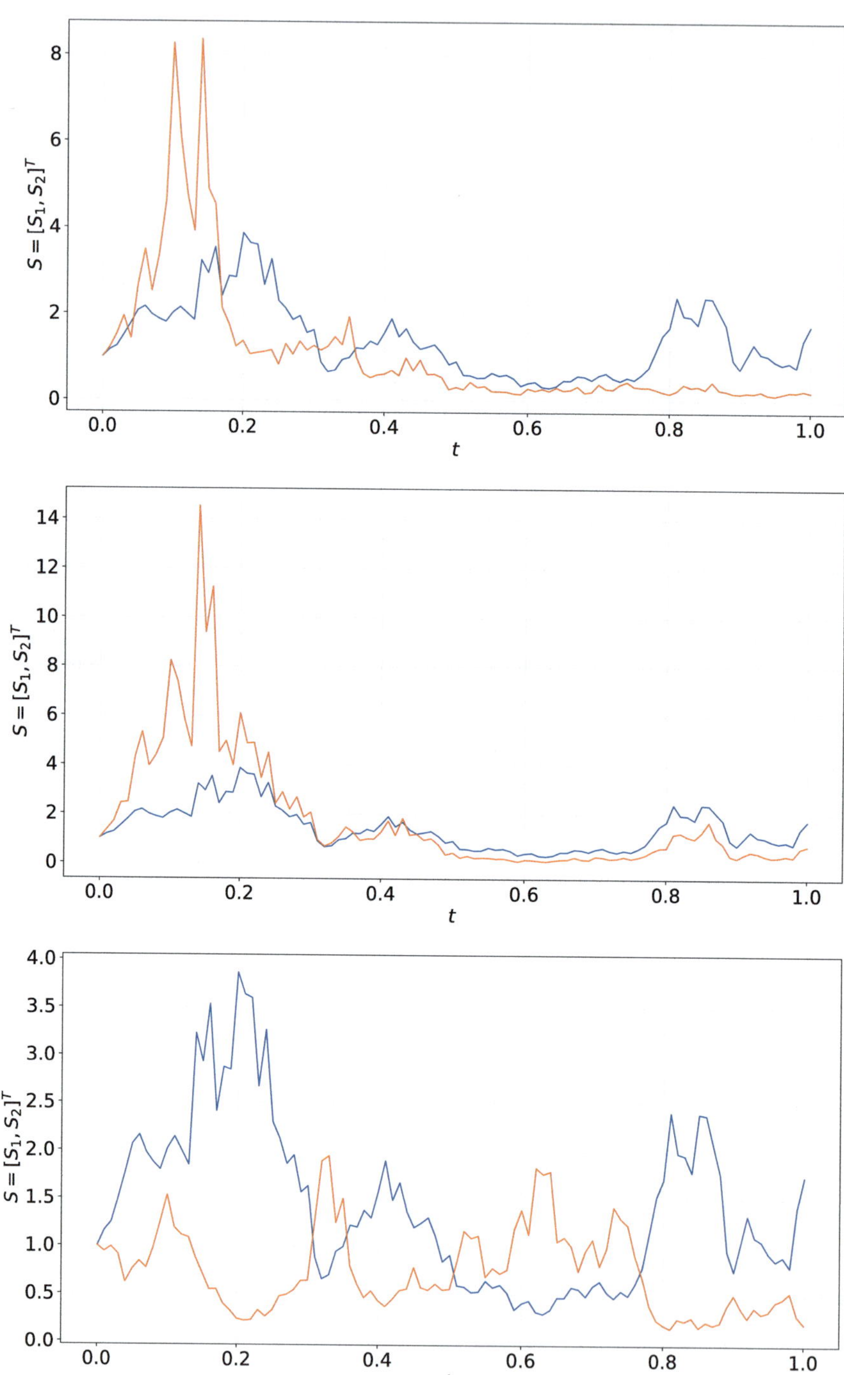

Fig. 6.3 Asset price trajectories for the 2-dimensional system of Black-Scholes models with $\rho = 0$ (top), $\rho = 0.8$ (middle), and $\rho = -0.8$ (bottom). The values of Z have been reused in each case

```
In []: tDemo=np.linspace(0,1,5); print(tDemo)
```

```
Out[]: [0.   0.25 0.5  0.75 1.  ]
```

Set up a pair of constants $a_1 = 2$ and $a_2 = 4$:

```
In []: a=np.array([2,4])
```

We wish to separately multiply each component of the 1×2 array a term-by-term into the 1×5 array t. To achieve this we can use the `np.tile()` call to create a 2×5 array where each row corresponds to t. Then we can use the vectorised multiplication operator to scale each row by the respective values of a_1 and a_2. In effect, we are creating a separate time set for each component of the 2-dimensional stochastic system that can be rescaled separately according to the parameters that hold for each component.

```
In []: tt=np.tile(t,(2,1))
       new=(a*tt.T).T; print(new)
```

```
Out[]: [[0.  0.5 1.  1.5 2. ]
        [0.  1.  2.  3.  4. ]]
```

Note the use of transposes to ensure that the array dimensions line up appropriately for the multiplication operation.

Exercise 6.5 asks you to use the Cholesky factorisation technique to produce point and interval estimates of the value of a swaption with value (6.3) using brute-force Monte Carlo with $M = 10^4$ samples, then modify your approach to implement variance reduction by antithetic variate.

Since we do not have to compute the Cholesky factor of Λ by hand, it is straightforward to apply this technique to larger systems of correlated assets. Exercise 6.6 asks you to modify your code to sample from a correlated system of 5 assets and produce a Monte Carlo valuation of a basket call option.

6.4   Dimension Reduction via Change-of-Measure

Consider again the swaption written on two correlated assets. The numeraire for the option value in (6.3) is the value of the risk-free investment at time $t \in [0, T]$, given by $\beta(t) = e^{rt}$. We can choose a different asset as numeraire, say S_2. This means switching to a probability measure $\mathbb{P}_{S_2}$ under which all of $S_1(t)/S_2(t)$, $S_2(t)/S_2(t) \equiv 1$, and $\beta(t)/S_2(t)$ are martingales.

The Radon-Nikodym derivative for changing from an expectation with respect to $\mathbb{P}_{S_2}$ to one with respect to $\mathbb{Q}$ is given a.s. by (see Glasserman [25])

$$\frac{d\mathbb{Q}}{d\mathbb{P}_{S_2}} = \frac{\beta(T)}{S_2(T)} \cdot \frac{S_2(0)}{\beta(0)} = \frac{\beta(T)S_2(0)}{S_2(T)},$$

since $\beta(0) = 1$. Using the fact that $\beta(T) = e^{rT}$, the value of the swaption at time $t = 0$ can be written as

$$e^{-rT}\mathbb{E}_{\mathbb{Q}}\left[(c_1 S_1(T) - c_2 S_2(T))^+\right]$$

$$= e^{-rT}\mathbb{E}_{\mathbb{P}_{S_2}}\left[(c_1 S_1(T) - c_2 S_2(T))^+ \left(\frac{d\mathbb{Q}}{d\mathbb{P}_{S_2}}\right)\right]$$

$$= \frac{1}{\beta(T)}\mathbb{E}_{\mathbb{P}_{S_2}}\left[(c_1 S_1(T) - c_2 S_2(T))^+ \left(\frac{\beta(T)S_2(0)}{S_2(T)}\right)\right]$$

$$= S_2(0)\mathbb{E}_{\mathbb{P}_{S_2}}\left[\left(c_1 \frac{S_1(T)}{S_2(T)} - c_2\right)^+\right].$$

To estimate the value of this option by Monte-Carlo, we only need to sample from the ratio $S_1(T)/S_2(T)$. In the case of a 2-dimensional Black-Scholes model with correlation ρ governed by (6.2) we can write

$$\frac{S_1(T)}{S_2(T)} = \frac{S_1(0)}{S_2(0)} e^{-(\sigma_1^2 - \sigma_2^2)T/2 + \sqrt{T}(Z_1(\sigma_1 - \sigma_2\rho) - \sigma_2\sqrt{1-\rho^2}Z_2)}$$

where Z_1 and Z_2 are i.i.d. $N(0, 1)$ random variables. Recall by the affine property of normal random variables that for any $a, b \in \mathbb{R}$ we have

$$aZ_1 + bZ_2 \sim N(0, a^2 + b^2).$$

Therefore

$$\tilde{Z} = Z_1(\sigma_1 - \sigma_2\rho) - \sigma_2\sqrt{1-\rho^2}Z_2 \sim N\left(0, (\sigma_1 - \sigma_2\rho)^2 + \sigma_2^2(1 - \rho^2)\right).$$

So by sampling instead from

$$\frac{S_1(0)}{S_2(0)} e^{-(\sigma_1^2 - \sigma_2^2)T/2 + \sqrt{T}\tilde{Z}},$$

we can estimate the value of the swaption using half as many calls to rng. This is a considerable computational advantage, gained as a result of reducing the dimension of the financial problem from 2 to 1.

In general, for any option on d correlated assets, we can reduce by one the number of stochastic variables by taking one of the risky assets as numeraire as long as the payoff $\Lambda : \mathbb{R}^d, \mathbb{R}$ satisfies

$$\Lambda(\alpha s_1, \ldots, \alpha s_d) = \alpha \Lambda(s_1, \ldots, s_d) \tag{6.16}$$

for all $\alpha \in \mathbb{R}$ and $(s_1, \ldots, s_d) \in \mathbb{R}^d$. Such functions are referred to as *homogeneous of degree 1*.

For example, the payoff of a spread call option with strike E, written as the function

$$\Lambda(s_1, s_2) = ((s_1 - s_2) - E)^+ ,$$

is not homogeneous of degree 1 (see Exercise 6.8), and multiplication by the same Radon-Nikodym derivative leads to an option value characterised by

$$S_2(0) \, \mathbb{E}_{\mathbb{P}_{S_2}} \left[\left(\left[\frac{S_1(T)}{S_2(T)} - 1 \right] - \frac{E}{S_2(T)} \right)^+ \right]$$

A Monte Carlo valuation of the expectation requires that we sample separately from $(S_1(T)/S_2(T)$ and $E/S_2(T))$. Clearly, the dimension of the problem has not been reduced and no efficiency will be gained by this approach.

6.5   Fitting a Bivariate Normal Distribution to Equity Log-Returns

Using Pandas datareader we can import pricing data for a pair of equity assets and examine their joint distribution. Let's examine Microsoft Corporation and Intel Corporation, both of which are listed on the NASDAQ stock exchange.

```
In []: asts=web.DataReader(['MSFT','INTC'],'stooq',
                    start='11-17-2020',end='11-16-2023')
```

`asts` is a Pandas DataFrame and indicative contents are displayed in Fig. 6.4. For the shares of each company we have acquired 3 years of opening and closing prices, as well as daily highs, lows and trading volume. The reader can adjust the `start` and `end` arguments as needed.

We will only consider closing share prices in this exercise, and remove the other values from the DataFrame using the `drop()` call.

```
In []: asts=asts.drop(['High','Low','Open','Volume'],axis=1)
```

Now compute the daily log-returns, defined to be $\ln(X_{i+1}/X_i)$ where X_i is the closing price on the ith day of trading, and remove the first (undefined) value from the resulting table using the Pandas `iloc[]` method, which allows us to select from parts of a DataFrame by specifying the appropriate index or index range.

```
In []: DLR=np.log(asts.Close.shift(1))-np.log(asts.Close)
       DLR=DLR.iloc[1:]
```

Attributes	Close		High		Low		Open		Volume	
Symbols	MSFT	INTC	MSFT	INTC	MSFT	INTC	MSFT	INTC	MSFT	INTC
Date										
2023-11-16	376.170	43.3500	376.350	43.3950	370.1800	40.8200	370.960	41.0000	2.718232e+07	8.640569e+07
2023-11-15	369.670	40.6100	373.130	40.8400	367.1100	39.5800	371.280	39.5850	2.686010e+07	4.739214e+07
2023-11-14	370.270	39.4100	371.950	39.4900	367.3454	38.0700	371.010	38.8000	2.768386e+07	4.545736e+07
2023-11-13	366.680	38.2300	368.470	38.5650	365.9000	38.1400	368.220	38.3500	1.998651e+07	2.455826e+07
2023-11-10	369.670	38.8600	370.100	38.9900	361.0700	38.2400	361.490	38.2400	2.806516e+07	3.887128e+07
...	...	...	...	...	...	...	...	...	...	...
2020-11-23	206.436	43.1908	208.570	43.3305	204.5000	42.6183	207.270	42.6476	2.614147e+07	3.784787e+07
2020-11-20	206.713	42.5617	209.533	43.0217	206.3270	42.5060	208.490	42.8146	2.325044e+07	2.580728e+07
2020-11-19	208.700	42.7785	209.285	42.8234	206.2570	41.9160	207.677	42.1201	2.523483e+07	3.142726e+07
2020-11-18	207.380	42.2520	211.409	42.8371	207.2500	42.2345	209.900	42.7111	2.887871e+07	3.499862e+07
2020-11-17	210.149	42.6926	213.316	43.2464	209.7710	42.5744	211.756	42.8997	2.464917e+07	3.313675e+07

755 rows × 10 columns

Fig. 6.4 Three years of daily asset price data for MSFT and INTC imported using Pandas DataReader

Notice that the `Close` method returns columns for both AAPL and INTC from the DataFrame `asts`. We can access individual share information from the DataFrame DLR by referencing directly the relevant column labels, and apply NumPy calls as we would to a NumPy array. For example, we can compute the covariance and correlation matrices for both sets of log-returns thus:

```
In []: sigma=np.cov(DLR.MSFT,DLR.INTC)
       rho=np.corrcoef(DLR.MSFT,DLR.INTC)
       print(sigma); print(rho)

Out[]: [[0.00030669 0.00019686]
        [0.00019686 0.00052629]]
       [[1.         0.48999256]
        [0.48999256 1.        ]]
```

The volatilities of MSFT and INTC, denoted σ_M and σ_I respectively, are computed as the square roots of the diagonal entries of `sigma`, and to 16 significant digits are:

$$\sigma_M = 0.0175126871815483; \quad \sigma_I = 0.02294111183686578.$$

Since they will be used later we store them without rounding as `sigMSFT` and `sigINTC` respectively. The correlation of the daily log-returns of this pair of assets

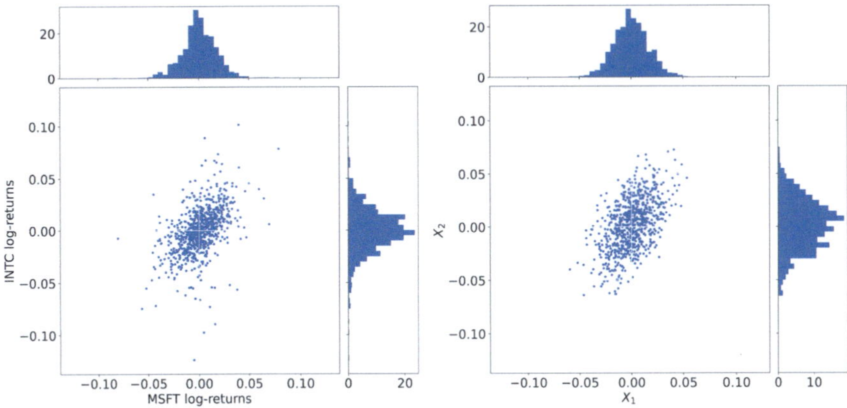

Fig. 6.5 Scatterplots and marginal histograms for log-returns of MSFT and INTC, and the same number of samples from a bivariate normal distribution with the same covariance structure and mean (right). Note the difference in the patterns of deviation from the mean, even though the covariance structure is the same

is $\rho = 0.48999256$. Mean values can be accessed as

```
In []: muMSFT=np.mean(DLR.MSFT); muINTC=np.mean(DLR.INTC)
       print(muMSFT); print(muINTC)

Out[]: 0.0007721808566798152
       2.026671103215652e-05
```

Using the techniques described earlier in the section, we now have sufficient information to simulate a pair of Black-Scholes asset models driven by a pair of Brownian motions with the same covariance structure as displayed by the data. This is the subject of Exercise 6.7. In Fig. 6.5, we compare scatterplots and marginal histograms of the real-world data, to their synthetic counterparts simulated under the real-world measure. The real-world equity data displays larger and more frequent deviations from the mean than the synthetic data simulated using a bivariate normal distribution. It is visually clear that the normal samples concentrate in a much smaller region of $\mathbb{R}^2$, despite having the same covariance matrix. This is our first indication that a bivariate normal distribution may not be a good choice to model these log-returns.

We can examine this further by generating a Q-Q plot for each of the marginal distributions. These are shown in Fig. 6.6, and show strong evidence of being drawn from a distribution with tails that are heavier than those of the normal distribution.

This empirical evidence demonstrates that the multivariate normal distribution is insufficient to model the log-returns of correlated assets in a portfolio. In fact, as confirmed by the evidence for heavy tails observed in the $Q-Q$ plots of the marginal distributions, this is an issue that also arises in the single asset case. In order to have

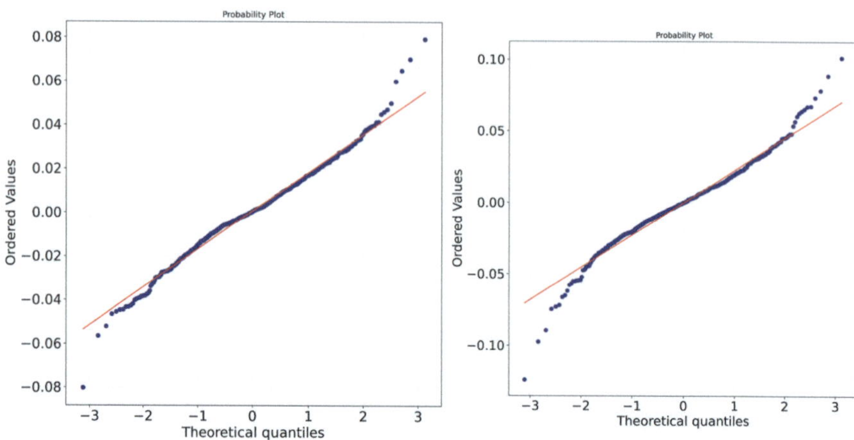

Fig. 6.6 Q-Q plots for the MSFT daily log-returns (left) and INTC daily log-returns (right). Both are characteristic of distributions with heavy tails

a chance at properly generating samples from a distribution that better-characterises these log-returns we next examine how to construct samples from an alternative multivariate distribution with heavier tails.

6.6   Heavy-Tails: The Multivariate Student's t_ν Distribution

In this section we illustrate one possible way to simulate a portfolio of correlated assets with log-returns displaying a heavy tailed distribution, by sampling from a multivariate Student's t_ν distribution.

6.6.1   Mathematical Characterisation of the Distribution

First, let us construct the random variable in d-dimensions and characterise its joint PDF, the derivation of which may be found in Chapter 4 of [26].

Definition 6.3 Let $Z \sim \mathcal{N}(0, \Sigma)$ be a d-dimensional centred normal random variable with covariance matrix Σ and joint probability density function given by (6.7) with the mean vector set to zero. Let $S \sim \chi_\nu^2$ be independent of Z and, for any $\nu > 2$ and $\mu = [\mu_1, \ldots, \mu_d]^T \in \mathbb{R}^d$, construct a d-dimensional random variable $X = [X_1, \ldots, X_d]^T$ as

$$X := \mu + \frac{\sqrt{\nu}}{\sqrt{S}} Z. \tag{6.17}$$

X is said to have a *joint Student's t_v distribution with $v > 2$ degrees of freedom,* mean vector $\mu \in \mathbb{R}^d$, *shape matrix* $\Sigma \in \mathbb{R}^{d \times d}$, and covariance matrix $\frac{v}{v-2}\Sigma \in \mathbb{R}^{d \times d}$. We denote

$$R_{ij} := \frac{\Sigma_{ij}}{\sqrt{\Sigma_{ii}\Sigma_{jj}}}, \quad i, j = 1, \ldots, d.$$

The distinction between the shape matrix Σ and the covariance matrix $\frac{v}{v-2}\Sigma$ is important to highlight. In the bivariate case, suppose that $Z \sim \mathcal{N}(0, \Sigma)$, where

$$\Sigma = \begin{pmatrix} \sigma_1 & 0 \\ 0 & \sigma_2 \end{pmatrix}\begin{pmatrix} 1 & \rho \\ \rho & 1 \end{pmatrix}\begin{pmatrix} \sigma_1 & 0 \\ 0 & \sigma_2 \end{pmatrix} = \begin{pmatrix} \sigma_1^2 & \rho\sigma_1\sigma_2 \\ \rho\sigma_1\sigma_2 & \sigma_2^2 \end{pmatrix}$$

is the covariance matrix of the bivariate centred normal random variable Z. Then the covariance matrix of the Student's t_v random variable X constructed according to (6.17) in Definition 6.3 is written

$$\frac{v}{v-2}\Sigma = \begin{pmatrix} \frac{v\sigma_1^2}{v-2} & \frac{v\rho\sigma_1\sigma_2}{v-2} \\ \frac{v\rho\sigma_1\sigma_2}{v-2} & \frac{v\sigma_2^2}{v-2} \end{pmatrix} = \begin{pmatrix} \sqrt{\frac{v}{v-2}}\sigma_1 & 0 \\ 0 & \sqrt{\frac{v}{v-2}}\sigma_2 \end{pmatrix}\begin{pmatrix} 1 & \rho \\ \rho & 1 \end{pmatrix}\begin{pmatrix} \sqrt{\frac{v}{v-2}}\sigma_1 & 0 \\ 0 & \sqrt{\frac{v}{v-2}}\sigma_2 \end{pmatrix}.$$

It follows that a bivariate Student's t_v random variable constructed in this way will have linear correlation coefficient $R_{12} = R_{21} = \rho$ and standard deviations $\sqrt{\frac{v}{v-2}}\sigma_i$, $i = 1, 2$.

The PDF of X for $d \geq 2$ is given by

$$f_{t_v}(x) = \frac{\Gamma((v+d)/2)}{\Gamma(v/2)v^{d/2}\pi^{d/2}}\left(1 + \frac{1}{v}[x - \mu]^T \Sigma^{-1}[x - \mu]\right)^{-(v+d)/2}, \quad x \in \mathbb{R}^d,$$

where the Gamma function Γ is given by $\Gamma(z) := \int_0^\infty t^{z-1}e^{-t}dt$.

6.6.1.1 The Univariate Student's t_v Distribution

To further build intuition about this distribution, let us consider the special (centred, univariate) case where $d = 1$ and $\mu = 0$, in which the density may be written

$$f_{t_v}(x) = \frac{\Gamma((v+1)/2)}{\Gamma(v/2)\sqrt{v\pi}}\left(1 + \frac{x^2}{v}\right)^{-(v+1)/2}, \quad x \in \mathbb{R}. \tag{6.18}$$

The parameter v represents the degrees of freedom of the distribution, and as $v \to \infty$ the density given by (6.18) converges to that of a standard normal random variable. The first 3 moments of a random variable $T \sim t_v$, along with the range of

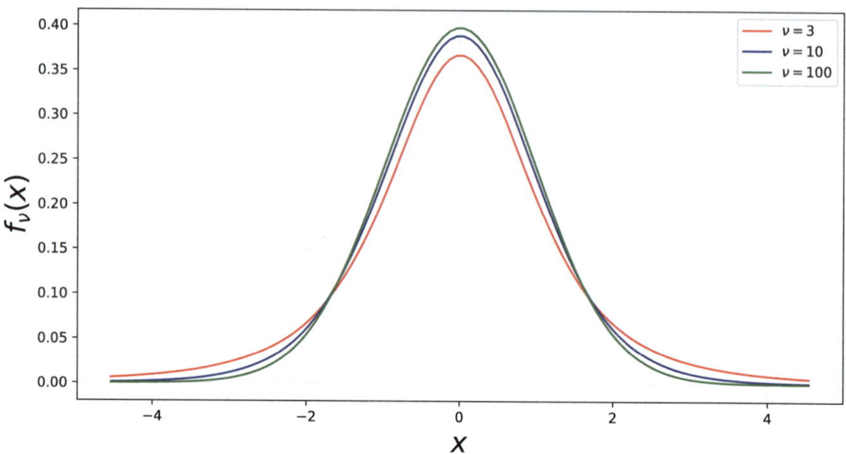

Fig. 6.7 Univariate student's t_ν density plots with increasing degrees of freedom

values of ν for which they exist, are given by

$$\mathbb{E}[T] = 0, \ \nu > 1; \quad \mathrm{Var}[T] = \frac{\nu}{\nu - 2}, \ \nu > 2; \quad \mathbb{E}[T^3] = 0, \ \nu > 3.$$

We can characterise the heaviness of the tails via the 4th moment, or kurtosis, of the distribution. This is given by

$$\mathbb{E}\left[T^4\right] = 3 + \frac{6}{\nu - 4}, \quad \nu > 4,$$

which, as $\nu \to \infty$, converges from above upon the value 3, the kurtosis of a standard normal random variable. So we can control the heaviness of the tails by varying ν: smaller values lead to heavier tails: see Fig. 6.7. This distribution generalises to the so-called *location-scale-t_ν* distribution. Let $T \sim t_\nu$, and let $\mu, \tau \in \mathbb{R}^+$ denote location and scale parameters respectively. Then

$$X = \mu + \tau T$$

has a location-scale-t_ν distribution ($X \sim \mathrm{lst}(\mu, \tau, \nu)$), with

$$\mathbb{E}[X] = \mu; \quad \mathrm{Var}[X] = \frac{\nu}{\nu - 2}\tau^2.$$

In the multivariate case, a d-dimensional random variable X with the Student's t_ν distribution displays a property known as *tail dependence*, where individual elements of the random vector X_i, $i = 1, \ldots, d$ tend to display extreme co-movements, even when the shape matrix is diagonal, so that the individual elements are pairwise mutually uncorrelated. This property is characterised mathematically in terms of the Student copula later in the chapter: see Sect. 6.7.3.3.

6.6.2 Sampling from a Bivariate Student's t_ν Distribution Using Python

Definition 6.3 gives us a method for sampling from this distribution. Here is a 2-dimensional example with the same mean and covariance structure computed for MSFT and INTC data in Sect. 6.5. Note that we multiply the covariance matrix from the data (stored in the array variable `sigma`) by $(\nu - 2)/\nu$ in order to ensure the correct scaling according to the construction in Definition 6.3.

```
In []: # Generate 3000 observations on a bivariate
       # centred Normal distribution
       nu=25; M=3000
       tsigma=((nu-2)/nu)*sigma
       mu=np.array([[muMSFT,muINTC],]*M).T

       A=np.linalg.cholesky(tsigma)
       Y=rng.normal(0,1,(2,M))
       Z=A.dot(Y)

       # Generate 3000 observations on a chi-square
       # distribution
       S=rng.chisquare(nu,M)

       # Combine to generate 3000 observations on a bivariate
       # Student's t distribution
       X=mu+(np.sqrt(nu)/np.sqrt(S))*Z
```

Note the syntax used to construct `mu` as a $2 \times M$ array created by cloning the 2×1 mean vector M times and transposing the result. Output for $\nu = 3$ and $\nu = 25$ is represented in Fig. 6.8.

In Fig. 6.9 we show scatterplots of $M = 3 \times 10^3$ samples from a bivariate Student's t_ν distribution with $\nu = 2 + 10^{-6}$ (so that Σ is still an invertible matrix) with $\rho = 0.8, 0, -0.8$. If we compare the output to that for the bivariate normal distribution with the same values of ρ in Fig. 6.1, we observe more extreme values in the empirical density plots in the left column, and evidence of tail dependence (see Sect. 6.7.3.3) in the samples from the bivariate Student's t_ν distribution, manifesting in the increased density of samples appearing in the extreme corners of the empirical distribution plots in the right column, even when $\rho = 0$.

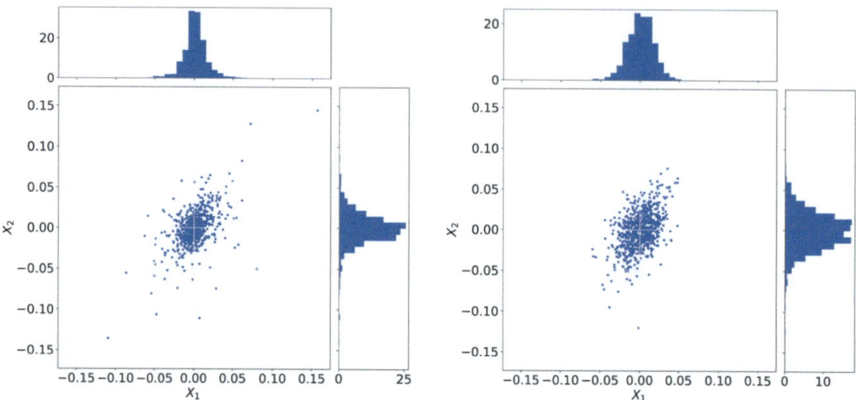

Fig. 6.8 Scatterplots and marginal histograms for $M = 754$ samples from a bivariate Student's t_ν distribution with covariance structures and mean taken from the daily log-returns of the MSFT-INTC equity pair. The degrees of freedom are set as $\nu = 3$ (left), and $\nu = 25$ (right)

6.6.2.1 Modelling the Equity Log-Returns with a Bivariate Student's t_ν Distribution

Finally, let us return to the log-returns of MSFT and INTC. By generating a probability plot for each of the marginal distributions where the percentiles of the data for each equity asset are compared to the theoretical percentiles for two univariate location-scale-t_ν distribution with means muMSFT and nuINTC, and scale parameters computed as follows:

```
In []:  sig1=np.cov(DLR.MSFT,DLR.INTC)[0,0]
        sig2=np.cov(DLR.MSFT,DLR.INTC)[1,1]

        scaleMSFT=np.sqrt((nu-2)/nu)*sig1)
        scaleINTC=np.sqrt((nu-2)/nu)*sig2)
```

We can see in Fig. 6.10 that there is better agreement between the marginal data and a location-scale-t_ν distribution with $\nu = 5$ in both cases, a value that has been arrived at for this demonstration by visually comparing probability plots for a range of values of ν.

In Fig. 6.11 we use the probability integral transform from Sect. 6.3.3 to isolate the dependence structure of MSFT and INTC daily log-returns by applying a univariate Student's t_ν CDF with $\nu = 5$ to each.

We could now simulate synthetic values for the daily log-returns of the MSFT-INTC equity asset price pair by sampling from a bivariate Student's t_5 distribution. Notice that this approach requires that the marginal data sets both display a good fit to two univariate location-scale-t_ν distributions with the same value of ν. There is also no guarantee that the dependence structure in the data is well-characterised by that of a bivariate Student's t_5 distribution. For greater flexibility, we should ideally like to apply a target dependence structure to our choice of marginal distributions, even if they differ.

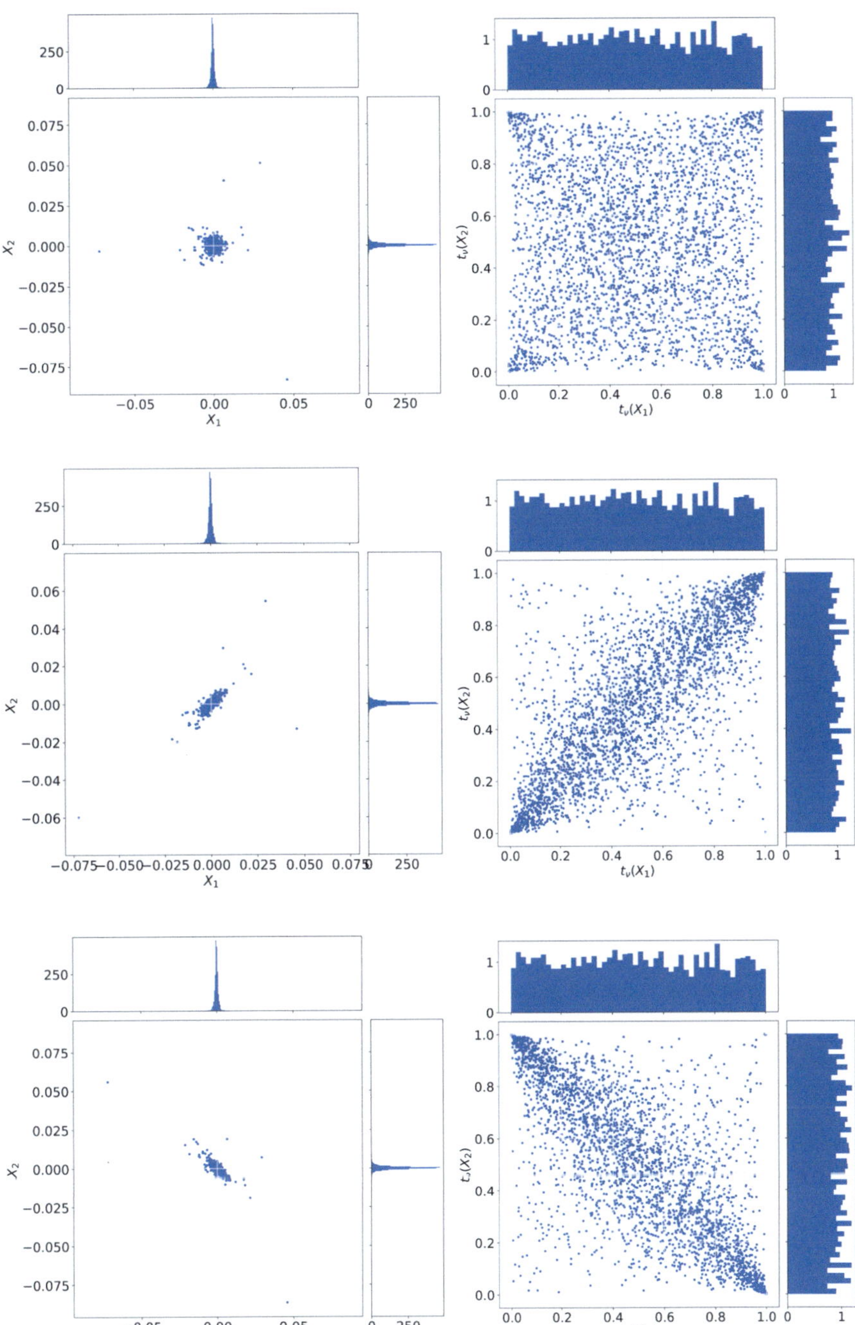

Fig. 6.9 Scatterplots showing observations (left column) and empirical density (right column) of $M = 3 \times 10^3$ samples from a pair of correlated Student's t_ν random variables with $\nu = 2 + 10^{-6}$ degrees of freedom, $\mu = 0$, $\rho = 0$ (top row), $\rho = 0.8$ (middle row), and $\rho = -0.8$ (bottom row). The values of Z and S have been reused in each case

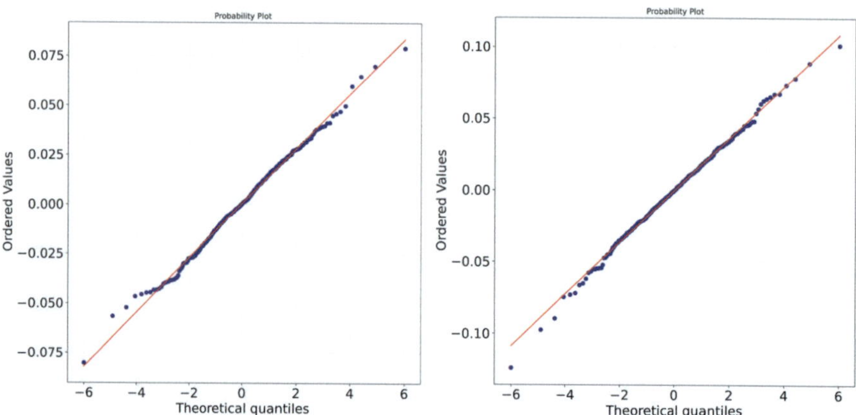

Fig. 6.10 Probability plots for the MSFT daily log-returns (top) and INTC daily log-returns (bottom) against theoretical quantiles of the Student's t_ν distribution with $\nu = 5$

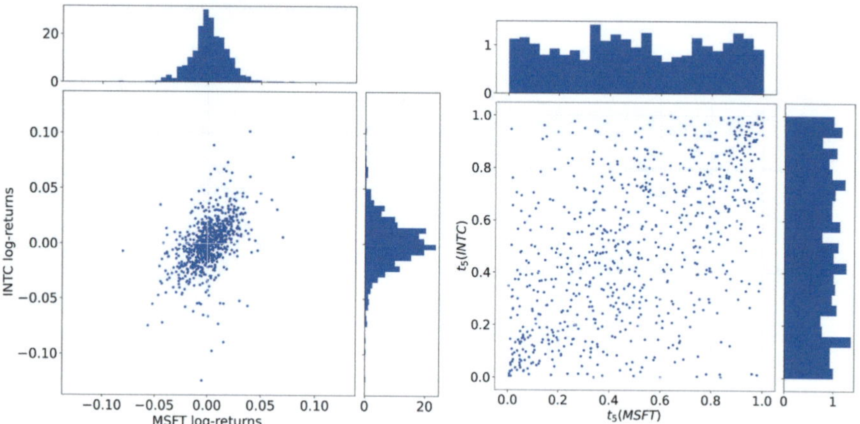

Fig. 6.11 Scatterplots of daily log-returns for the MSFT-INTC equity pair with associated marginal histograms (left), and of the transformed values where we have attempted to *uniformise* the marginal data using a univariate Student's t_ν distribution with $\nu = 5$ in each case

6.7   Copulas for d-Dimensional Random Variables

Copulas allow us to characterise (and therefore sample from) a target joint distribution with specified dependence structure while maintaining great flexibility over the choice of marginal distributions. In this section we will present the general definition of a d-dimensional copula, along with the most important associated theoretical results. We then provide examples based upon the bivariate Gaussian and Student's t_ν distributions.

6.7.1 The Definition and Role of a Copula

As before, let $X = [X_1, \ldots X_d]^T$ be a d-dimensional random vector on the probability space $(\Omega, \mathcal{F}, \mathbb{P})$.

Definition 6.4 A *d-dimensional copula* is any joint cumulative distribution function (CDF), $C : [0, 1]^d \mapsto [0, 1]$, which can be written as

$$C(u_1, \ldots, u_d) = \mathbb{P}[U_1 \le u_1, \ldots, U_d \le u_d], \quad 0 \le u_1, \ldots, u_d \le 1,$$

where $U_1, \ldots, U_d$ are $\mathbb{P}$-uniformly distributed random variables on $[0, 1]$.

The following important theorem confirms that the joint CDF of X can be expressed in terms of the marginal CDFs through the copula.

Proposition 6.5 (Sklar's Theorem) *Given F_X, the joint CDF of X, there exists a copula $C : [0, 1]^d \to [0, 1]$ such that*

$$F_X(x_1, \ldots, x_d) = C(F_{X_1}(x_1), \ldots, F_{X_d}(x_d)),$$

for $x_1, \ldots, x_d \in \mathbb{R}$.

Since the copula C encodes the dependence structure of the multivariate distribution, it is instructive to examine its form in simple special cases.

1. Suppose $U_1, \ldots, U_d$ are mutually independent random variables, each with a uniform distribution on the interval $[0, 1]$. Then the joint CDF is the product of the marginal CDFs, and

$$\begin{aligned}
F_{(U_1, \ldots, U_d)}(u_1, \ldots, u_d) &= \mathbb{P}[U_1 \le u_1, \ldots, U_d \le u_d] \\
&= \mathbb{P}[U_1 \le u_1] \cdots \mathbb{P}[U_d \le u_d] \\
&= u_1 \cdots u_d.
\end{aligned}$$

   In this case the copula takes the form $C(u_1, \ldots, u_d) = u_1 \cdots u_d$, where $u_i \in [0, 1]$ for each $i = 1 \ldots, d$.
2. Suppose $U_1, \ldots, U_d$ are fully correlated random variables, each with a uniform distribution on the interval $[0, 1]$. Then

$$\begin{aligned}
F_{(U_1, \ldots, U_d)}(u_1, \ldots, u_d) &= \mathbb{P}[U_1 \le u_1, \ldots, U_d \le u_d] \\
&= \mathbb{P}[U_1 \le \min\{u_1, \ldots, u_d\}] \\
&= \min\{u_1, \ldots, u_d\}.
\end{aligned}$$

In this case $C(u_1, \ldots, u_d) = \min\{u_1, \ldots, u_d\}$, where $u_i \in [0, 1]$ for each $i = 1 \ldots, d$.

3. In Exercise 6.9 you are asked to show that the copula corresponding to the case of two fully anticorrelated random variables with uniform distribution on the interval $[0, 1]$, so that $U_1 = 1 - U_2$, is given by

$$C(u_1, u_2) = (u_1 + u_2 - 1)^+, \quad 0 \le u_1, u_2 \le 1.$$

6.7.2 Defining a Copula from a Joint CDF, and Vice Versa

We can use the dependence structure in a known joint CDF to induce a copula as follows. Suppose that $X_1, \ldots, X_d$ are $\mathbb{R}$-valued random variables with CDFs F_{X_i}, $i = 1, \ldots d$ that are continuous and strictly increasing. It follows from Proposition 6.5 that the joint CDF $F_{(X_1, \ldots, X_d)}$ defines a d-dimensional copula

$$C(u_1, \ldots, u_d) := F_X \left(F_{X_1}^{-1}(u_1), \ldots, F_{X_d}^{-1}(u_d) \right), \tag{6.19}$$

where $u_1, \ldots, u_d \in [0, 1]$.

Conversely, we can apply the dependence structure characterised by a copula to a set of marginal distributions to create a joint CDF as follows. Suppose that $Y_1, \ldots, Y_d$ are $\mathbb{R}$-valued random variables with CDFs $F_{Y_1}, \ldots, F_{Y_d}$. Given a copula $C : [0, 1]^d \to [0, 1]$, the function

$$F_{(Y_1, \ldots, Y_d)}^C(y_1, \ldots, y_d) = C \left(F_{Y_1}(y_1), \ldots, F_{Y_d}(y_d) \right), \tag{6.20}$$

for $y_1, \ldots, y_d \in \mathbb{R}$, defines a joint CDF with marginal CDFs $F_{Y_1}, \ldots, F_{Y_d}$.

6.7.3 Examples of Bivariate Copulas

Here, we demonstrate how to construct and sample from two types of copula, which capture the dependence structure of the bivariate Gaussian and Student's t_ν distributions respectively. These copulas are classified as *elliptical*, after the class of probability distribution from which they arise, and are widely used in practice.

Another commonly used class of copula is that of the *Archimedean copulas*. We do not treat these directly here, but a definition is provided in Exercise 6.11. This, along with Exercise 6.12, gives readers an opportunity to examine the properties of several bivariate examples.

6.7.3.1 The Gaussian Copula

The multivariate centred normal distribution induces dependence structure between the normally distributed marginal random variables that can be expressed as a copula. We describe this in the bivariate case.

Set $d = 2$ and $\sigma_1 = \sigma_2 = 1$ in (6.9) and consider the bivariate centred standard normal random pair $X = [X_1, X_2]^T$ such that, with respect to a probability measure $\mathbb{P}$,

$$X \sim \mathcal{N}(0, \Sigma); \quad \Sigma = \begin{pmatrix} 1 & \rho \\ \rho & 1 \end{pmatrix}.$$

When $\rho \in (-1, 1)$, the matrix Σ is invertible and by integrating the joint probability density function (6.7), the joint CDF is given by

$$\Phi_\Sigma(x_1, x_2) = \mathbb{P}[X_1 \le x_1, \ X_2 \le x_2]$$

$$= \frac{1}{2\pi \sqrt{\det \Sigma}} \int_{-\infty}^{x_1} \int_{-\infty}^{x_2} \exp\left(-\frac{1}{2}(s, w)\Sigma^{-1}\begin{pmatrix} s \\ w \end{pmatrix}\right) ds\,dw$$

$$= \int_{-\infty}^{x_1} \int_{-\infty}^{x_2} \frac{1}{2\pi \sqrt{1 - \rho^2}} \exp\left(-\frac{s^2 - 2\rho s w + w^2}{2(1 - \rho^2)}\right) ds\,dw.$$

The marginal distributions of X_1 and X_2 are both given by:

$$\Phi(x) := \frac{1}{\sqrt{2\pi}} \int_{-\infty}^{x} e^{-s^2/2} ds. \tag{6.21}$$

By (6.14) in Sect. 6.3.3, $\Phi(X_1)$ and $\Phi(X_2)$ are each uniformly distributed on $[0, 1]$. Therefore, by (6.19) we can construct the *Gaussian copula* as

$$C_\Sigma(u_1, u_2)$$

$$= \mathbb{P}\left[\Phi(X_1) \le u_1, \Phi(X_2) \le u_2\right]$$

$$= \mathbb{P}\left[X_1 \le \Phi^{-1}(u_1), X_2 \le \Phi^{-1}(u_2)\right]$$

$$= \Phi_\Sigma\left(\Phi^{-1}(u_1), \Phi^{-1}(u_2)\right)$$

$$= \int_{-\infty}^{\Phi^{-1}(u_1)} \int_{-\infty}^{\Phi^{-1}(u_2)} \frac{1}{2\pi \sqrt{1 - \rho^2}} \exp\left(-\frac{s^2 - 2\rho s w + w^2}{2(1 - \rho^2)}\right) ds\,dw.$$

Contour plots of C_Σ for $\rho = 0.8$ and $\rho = -0.8$ are shown in Fig. 6.12. The Gaussian copula with correlation coefficient $\rho = R$, is constructed as follows:

```
In []: def gaussCopula(u1,u2,R):

           sig=np.array([[1,R],[R,1]])

           dist=stats.multivariate_normal(mean=None,cov=sig)
           phi1=stats.norm.ppf(u1)
           phi2=stats.norm.ppf(u2)

           return dist.cdf([phi1,phi2])
```

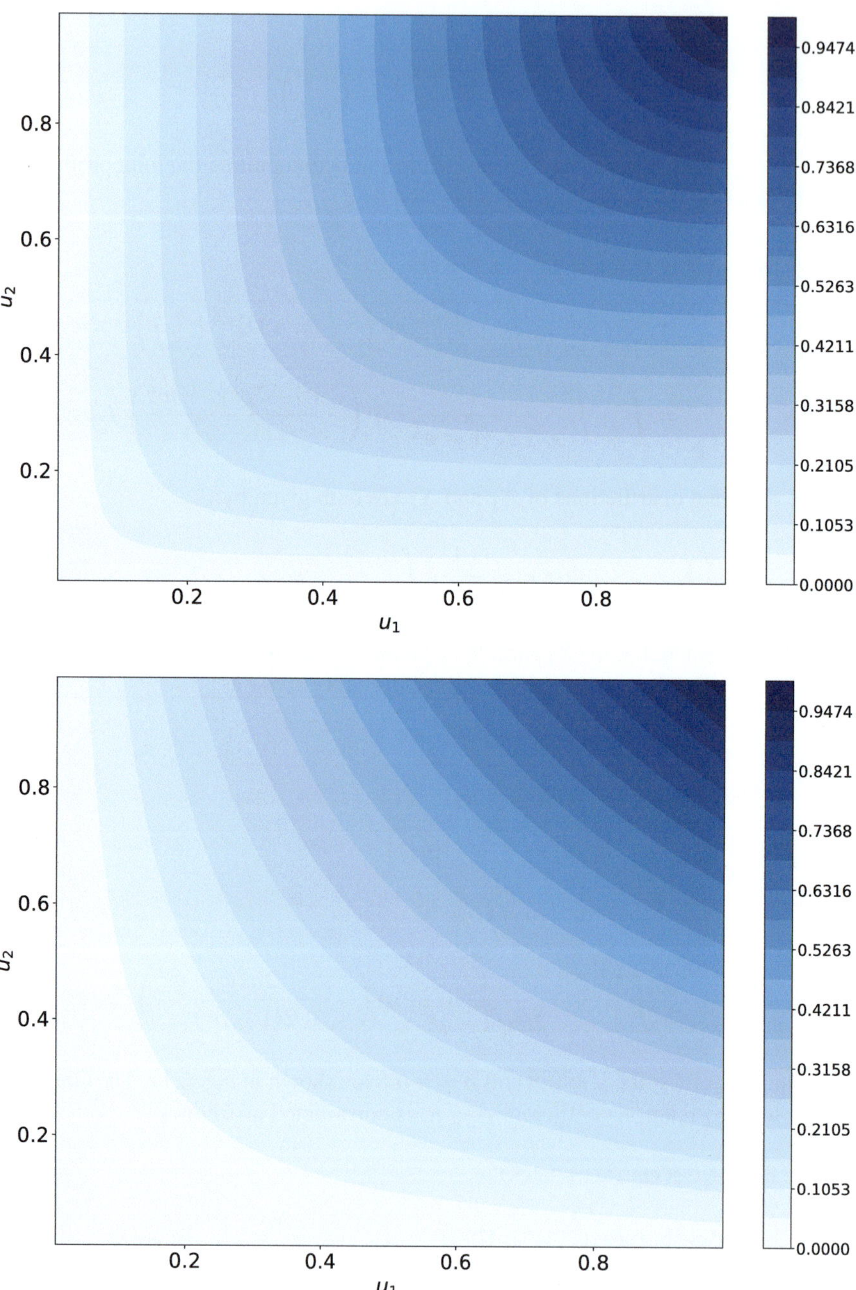

Fig. 6.12 Contour plots showing the bivariate Gaussian copula $C_\Sigma(u_1, u_2)$ as a joint CDF over the unit square with $\rho = 0.8$ (top) and $\rho = -0.8$ (bottom)

We see from (6.20) in Sect. 6.7.2 that the Gaussian copula can be used to construct other joint CDFs with the same dependence structure based upon non-Gaussian marginal distributions F_{X_1} and F_{X_2}. As we saw in Sect. 6.3.3, the random variables $F_{X_1}(X_1)$ and $F_{X_2}(X_2)$ are also uniformly distributed on $[0, 1]$. Joint CDFs based upon the Gaussian copula with these marginal CDFs then take the form

$$F_{\Sigma}^C(x_1, x_2) := C_{\Sigma}\left(F_{X_1}(x_1), F_{X_2}(x_2)\right)$$

$$= \int_{-\infty}^{F_{X_1}^{-1}(u_1)} \int_{-\infty}^{F_{X_2}^{-1}(u_2)} \frac{1}{2\pi\sqrt{1-\rho^2}} \exp\left(-\frac{s^2 - 2\rho s w + w^2}{2(1-\rho^2)}\right) ds\,dw.$$

How do we sample from a distribution thus constructed via a copula? There is no general method that works efficiently in all cases, but for the Gaussian copula we proceed as follows.

Suppose first that the $2 \times M$ array variable W has been constructed according to the method in Sect. 6.3.2, so that it contains M observations on a bivariate normal distribution with correlation coefficient ρ. If we apply the CDF of a standard normal distribution Φ given by (6.21) to each row:

```
In []: gCopula1=stats.norm.cdf(W[0,:])
       gCopula2=stats.norm.cdf(W[1,:])
```

then the observations contained in each of `gCopula1` and `gCopula2` will be drawn from a uniform distribution on $[0, 1]$. The `stats.norm.cdf()` call has not implemented an `axis` argument at the time of writing and so each row of W must be transformed separately.

If we then apply the inverse CDF of our choice to each set of marginal observations we can produce a sample from a new bivariate distribution with our target marginal distributions and dependence structure determined by the copula. For example, we can apply the inverse of a univariate Student's t_5 distribution to each set of uniform marginals using the `ppf()` call:

```
In []: gCopTMarg1=stats.t.ppf(gCopula1,df=5)
       gCopTMarg2=stats.t.ppf(gCopula2,df=5)
```

The result will be a sample from a bivariate distribution with dependence structure characterised by the Gaussian copula and Student's t_5 distributed margins.

In Fig. 6.13, we illustrate the output of a Gaussian copula applied to pseudo-random samples from three sets of marginal distributions: uniform over $[0, 1]$, Student's t_5, and exponential with rate parameter $\lambda = 1$.

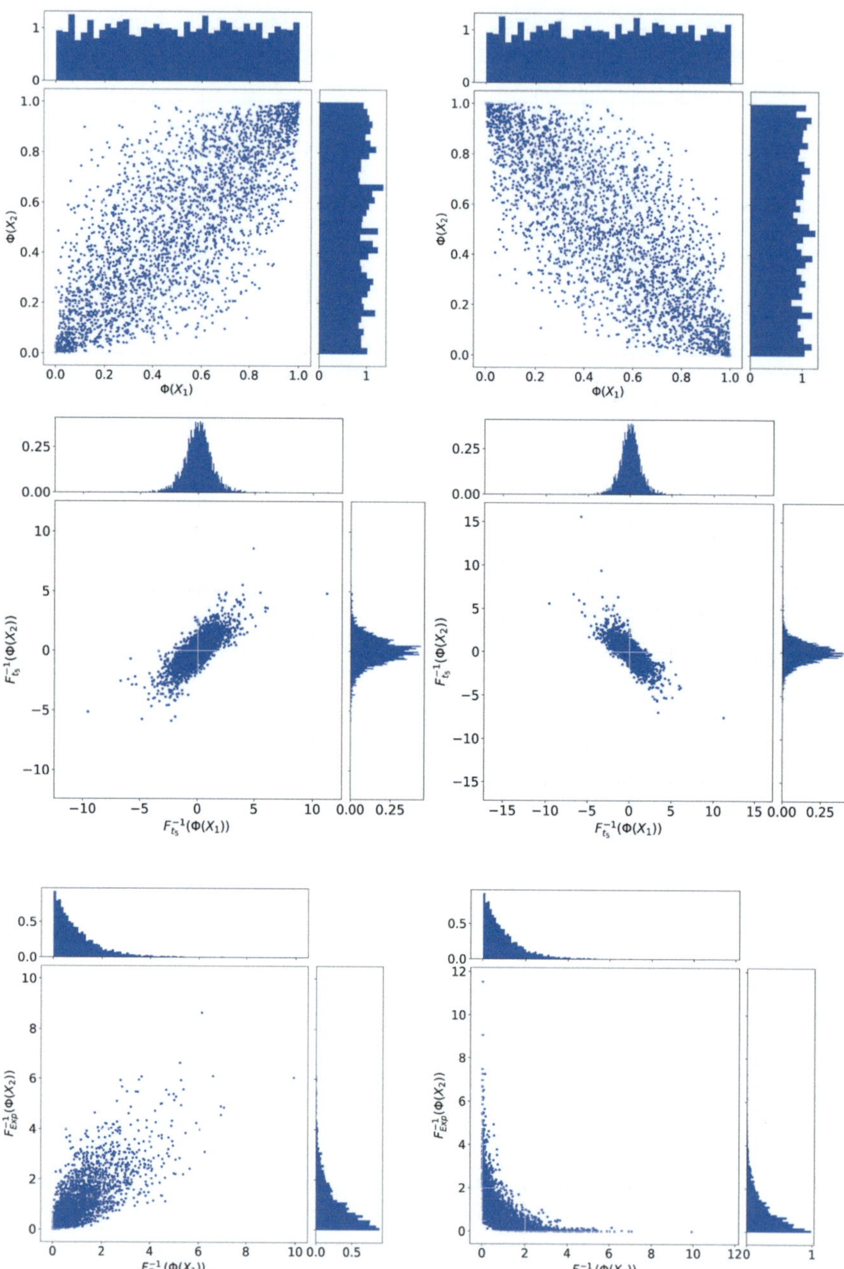

Fig. 6.13 Samples from bivariate distributions created via a Gaussian copula with $\rho = 0.8$ (left) and $\rho = -0.8$ (right). Marginal distributions are uniform on [0, 1] (top), Student's t_5 (middle) and exponential with $\lambda = 1$ (bottom)

6.7.3.2 The Student Copula

The *Student copula* is generated by the multivariate Student's t_ν distribution. In the bivariate case with t_ν-distributed marginal CDFs it has the form

$$C_{\nu,R}(u_1, u_2)$$

$$= \int_{-\infty}^{t_\nu^{-1}(u_1)} \int_{-\infty}^{t_\nu^{-1}(u_2)} \frac{1}{2\pi\sqrt{1-R^2}} \exp\left(1 + \frac{s^2 - 2Rsw + w^2}{\nu(1-R^2)}\right)^{-(\nu+2)/2} ds\,dw,$$

where $R = R_{12}$, using the notation given in Definition 6.3. Contour plots of $C_{\nu,R}$ with $\nu = 3$, $R = 0.8$ and $R = -0.8$ are shown in Fig. 6.14. The Student copula with nu degrees of freedom and linear correlation coefficient $R_{12} = R$ is constructed as

```
In []: def studentCopula(u1,u2,nu,R):

           shapeM=np.array([[1,R],[R,1]])

           dist=stats.multivariate_t(shape=shapeM)
           phi1=stats.t.ppf(u1,df=nu)
           phi2=stats.t.ppf(u2,df=nu)

           return dist.cdf([phi1,phi2])
```

Just as for the Gaussian copula, we can generate a joint CDF from alternative marginal distributions F_{X_1} and F_{X_2} as the function

$$F_{\nu,R}^C(x_1, x_2) := C_{\nu,R}\left(F_{X_1}(x_1), F_{X_2}(x_2)\right), \quad x_1, x_2 \in \mathbb{R}.$$

Sampling from a bivariate distribution with a Student copula is similar to the case of the Gaussian copula. Suppose that X is a $2 \times M$ array containing samples from a bivariate Student's t_5 distribution constructed according to Sect. 6.6.2. We first make the marginal data uniform:

```
In []: tCopula1=stats.t.cdf(X[0,:],df=5)
       tCopula2=stats.t.cdf(X[1,:],df=5)
```

Then apply the univariate inverse CDF of the target distribution for each margin using the `ppf()` call. For example we could make the marginal data exponentially distributed with rate parameter $\lambda = 1$:

```
In []: tCopExpMarg1=stats.expon.ppf(tCopula1)
       tCopExpMarg2=stats.expon.ppf(tCopula2)
```

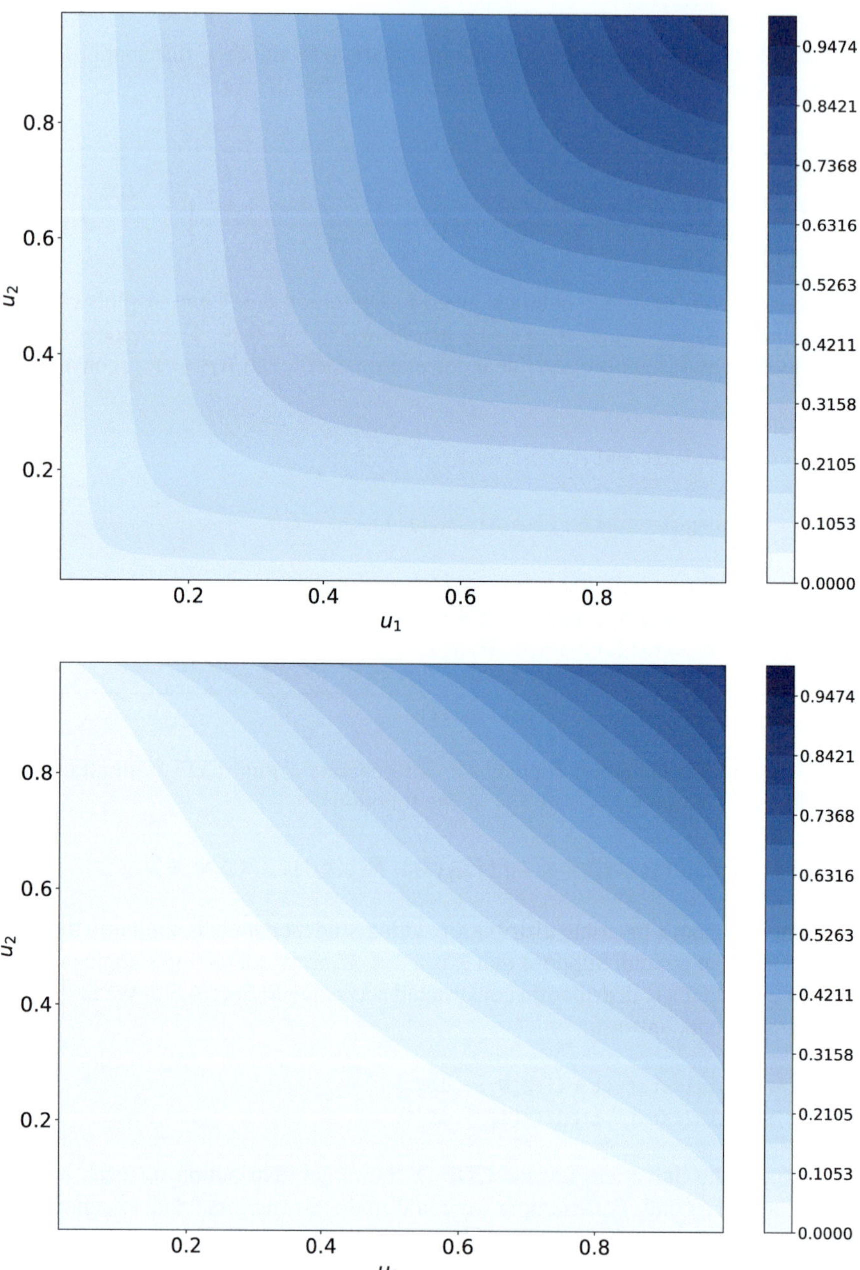

Fig. 6.14 Contour plots showing the bivariate Student copula $C_{\nu,\rho}(u_1, u_2)$ as a joint CDF over the unit square with $\rho = 0.8$ (top) and $\rho = -0.8$ (bottom). In both cases $\nu = 3$

In Fig. 6.15, we illustrate the output of a Student copula applied to pseudo-random samples from three sets of marginal distributions: uniform on $[0, 1]$, standard normal, and exponential with $\lambda = 1$.

6.7.3.3 Tail Dependence

Tail dependence characterises the tendency of a set of random variables to exhibit extreme co-movements. We consider here only *upper tail dependence*, though the notion of *lower tail dependence* can be defined analogously, and we refer the reader to Nelson [55] for more detail on the material presented in this section.

Definition 6.6 Let X_1 and X_2 be random variables on a probability space $(\Omega, \mathcal{F}, \mathbb{P})$ with continuous and strictly increasing CDFs F_{X_1} and F_{X_2} respectively.

1. The *upper tail dependence coefficient* of X_1 and X_2 is

$$\lambda_U = \lim_{u \to 1^-} \mathbb{P}\left[X_2 > F_{X_2}^{-1}(u) | X_1 > F_{X_1}^{-1}(u) \right],$$

   provided that the limit λ_U exists.
2. If $\lambda_U > 0$, X_1 and X_2 are *asymptotically dependent in the upper tail*.
3. If $\lambda_U = 0$, X_1 and X_2 are *asymptotically independent in the upper tail*.

The upper tail dependence for a bivariate pair with dependence structure characterised by the Gaussian copula with parameter R is

$$\lambda_U = 2 \lim_{x \to \infty} \left(1 - \Phi\left(x \frac{\sqrt{1 - R}}{\sqrt{1 + R}} \right) \right).$$

Notice that for any $R \in [0, 1)$, $\lambda_U = 0$ and there is zero tail dependence. In the fully correlated case (when $R = 1$) we have $\lambda_U = 1$.

Contrast this with the upper tail dependence for a bivariate pair with dependence structure characterised by the Student copula with parameters v, R, which is given by

$$\lambda_U = 2 \left(1 - t_{v+1} \left(\frac{\sqrt{v + 1}\sqrt{1 - R}}{\sqrt{1 + R}} \right) \right)$$

Note that if we set $R = 0$, upper tail dependence becomes

$$\lambda_U = 2 \left(1 - t_{v+1} \left(\sqrt{v + 1} \right) \right) > 0, \quad 2 < v < \infty.$$

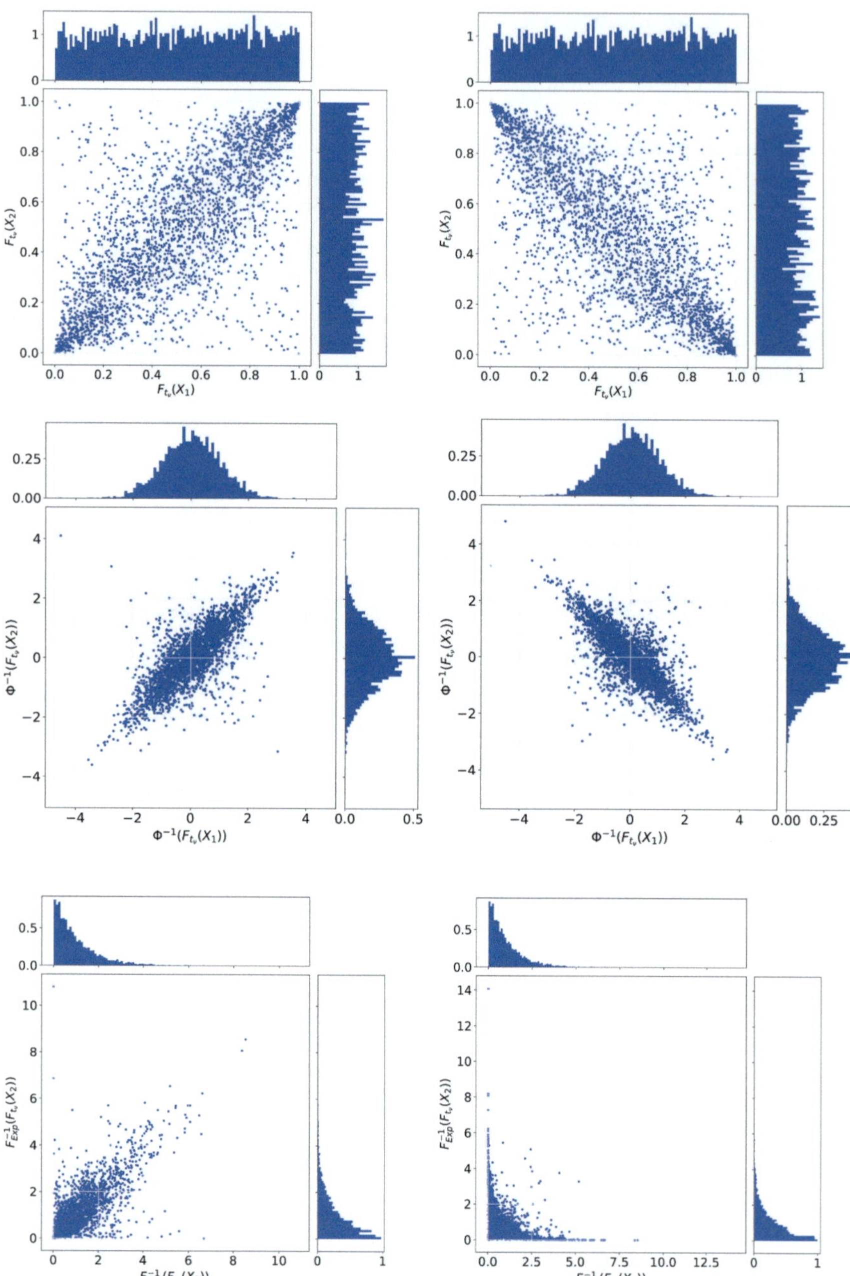

Fig. 6.15 Samples from bivariate distributions created via a Student's t_ν copula with $\nu = 2 + 10^{-6}$, $\rho = 0.8$ (left) and $\rho = -0.8$ (right). Marginal distributions are uniform on $[0, 1]$ (top), standard normal (middle) and exponential with $\lambda = 1$ (bottom)

We see that, with a Student copula, upper tail dependence is present even if $R = 0$, and in this case is controlled by the size of the parameter v. For example, setting $v = 5$ we can compute

```
In []: 2*(1-stats.t.cdf(np.sqrt(6),df=6))
```

```
Out[]: 0.04982526278057664
```

6.7.4 Fitting a Copula to a Data Set

6.7.4.1 The Empirical Copula of a Sample

Suppose we have a set of M observations of d financial asset log returns. The *empirical copula* is constructed from the sample data via its rank statistics $\{r_1^m, \ldots, r_d^m\}$ for $m = 1, \ldots, M$.

Definition 6.7 Any function

$$\widehat{C}\left(\frac{m_1}{M}, \ldots, \frac{m_d}{M}\right) = \frac{1}{M} \sum_{m=1}^{M} \prod_{i=1}^{d} \mathcal{I}_{\{r_i^m \leq m_i\}}$$

defined on a lattice

$$l = \left\{ \left(\frac{m_1}{M}, \ldots, \frac{m_d}{M}\right) : m_i = 0, \ldots, M; \ 1 \leq i \leq d \right\}$$

is an empirical copula.

In the case of the daily log-returns of the equity asset price pair MSFT-INTC, contained in the 754×2 array DLR, $d = 2$, $M = 754$, and we can construct an array of the same dimensions containing the rank statistics as

```
In []: DLRranks = stats.rankdata(DLR,axis=0)
```

Making use of conditional arrays via the `np.where()` and `np.all()` calls, we can then construct an empirical copula function in the following way.

```
in []: def empiricalCopula(u1,u2,M,ranks):

           m1=M*u1; m2=M*u2

           # Identify the row numbers of 'ranks' where
           # inequality is satisfied for both columns.
           # Returns an M by 1 array of boolean expressions
           bothLessThan=np.all(ranks<=[m1,m2],axis=1)
```

```
# Count the number of rows with value 'True'.
rowsKeep=len(np.where(bothLessThan)[0])


# Return value from CDF
return (rowsKeep/M)
```

6.7.4.2 Fitting a Copula to the MSFT-INTC Daily Log-Returns

We proceed in three stages. First, make the marginal data uniform by applying a location-scale-t_5 CDF function as determined in Sect. 6.6.2.1, then produce the ranks of each in turn:

```
In []: nu = 5
       x=stats.t.cdf(DLR.MSFT,df=nu,
                     loc=muMSFT,scale=np.sqrt((nu-2)/nu)*sigMSFT)
       y=stats.t.cdf(DLR.INTC,df=nu,
                     loc=muINTC,scale=np.sqrt((nu-2)/nu)*sigINTC)


       xRank=stats.rankdata(x,axis=0)
       yRank=stats.rankdata(y,axis=0)
```

Second, we estimate the the correlation parameter R for both the Gaussian and Student copulas using a transformation of Kendall's τ, which is a statistical estimator of linear correlation for ranked data and satisfies the relation $R = \sin(\pi\tau/2)$ for elliptical copulas (see Lindskog et al. [47]):

```
In []: tau,p_val=stats.kendalltau(xRank,yRank)
       R=np.sin(0.5*np.pi*tau)
       print(tau); print(R)

Out[]: 0.38388930293963386
       0.5671257426153531
```

Finally, we wish to choose an elliptical copula to fit our data from a set of candidates $\{C_k\}_{1\leq k\leq K}$. One way to do this is to select the copula that minimises the Pythagorean distance

$$d_2(\widehat{C}, C_k) = \left(\sum_{m_1,m_2=0}^{M} \left(\widehat{C}\left(\frac{m_1}{M}, \frac{m_2}{M}\right) - C_k\left(\frac{m_1}{M}, \frac{m_2}{M}\right) \right)^2 \right)^{1/2}. \tag{6.22}$$

Table 6.1 computes $d_2(\widehat{C}, C)$ for three candidate copulas.

Table 6.1 Estimated parameters and values of the distance d_2 defined in (6.22) for three copulas

Copula	$d_2(\widehat{C}, C)$
Gaussian	3.704904194406361
Student's t_5	38.145398730931866
Student's t_{10}	43.39931413149358

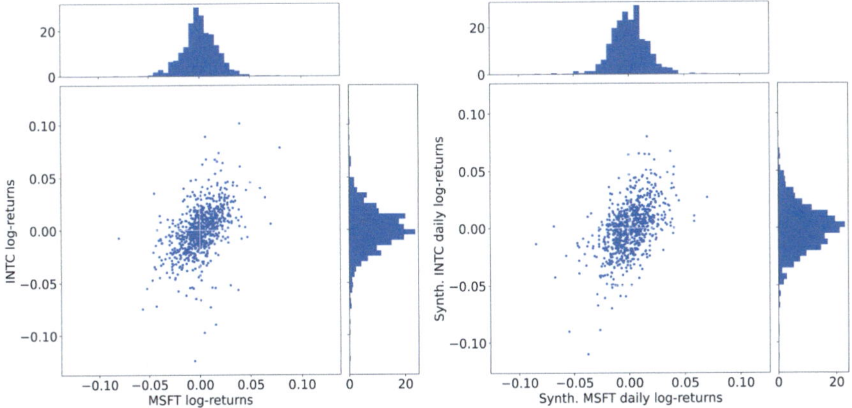

Fig. 6.16 Scatterplots and marginal histograms of observed daily log-returns for the MSFT-INTC equity pair (left), and a sample of $M = 754$ synthetic values (right) from a distribution constructed using a Gaussian copula with $R = 0.5671$ applied to marginal samples from location-scale-t_ν distributions with $\nu = 5$ and empirically determined location and scale values

Based upon this output we would select the Gaussian copula with $R \approx 0.5671$ as the best fit to the empirical copula generated by our data. We can then produce synthetic daily log-returns distributed approximately like those of the MSFT-INTC pair by applying it to two sets of marginal samples from the univariate location-scale-t_5 distributions identified in Sect. 6.6.2.1 with location parameters muMSFT and muINTC respectively, and scale parameters scaleMSFT and scaleINTC. Output is displayed in Fig. 6.16.

6.8 Further Reading

A general treatment of the Black-Scholes framework that extends the argument contained in Chap. 2 to value derivative securities in a complete market that includes a risk-free bond and d risky assets, may be found in Bjork [7], and a more advanced treatment of some of the topics here and in the next chapter is given in Glasserman [25].

Copulas are a large topic and an active area of research, and we have barely scratched the surface in this chapter. An in-depth treatment available in the book by Nelson [55], and next steps for applications of copulas in finance may be found in [14]. A short and approachable demonstration of the use of copulas for modelling financial data is contained in the article by Di Clemente and Romano [21], which treats elliptical, Archimedean, and empirical copulas, as well as providing more detail on parameter estimation.

The inappropriate use of Gaussian copulas has been suggested as a contributing factor in the 2008 financial crisis, due to their inability to adequately model tail dependence. An interesting commentary on the role that they played is available in MacKenzie and Spears [51].

Machine learning techniques are increasingly being deployed as an alternative way to identify hidden structure in financial time series, including asset price log-returns. See [68] for an example that uses a machine-learning framework known as Generative Adversarial Networks. An introduction to machine learning techniques for finance may be found in [22].

6.8.1 Exercises

6.1 Suppose that $W = [W_1, W_2]^T$ is a pair of correlated standard Brownian motions such that

$$W(t) \sim \mathcal{N}(0, \Sigma), \quad \text{where} \quad \Sigma = \begin{pmatrix} 1 & \rho \\ \rho & 1 \end{pmatrix} t,$$

and where $\rho \in (-1, 1)$ and $t \in [0, T]$.

1. Compute A, the Cholesky factor of Σ;
2. Hence, show that

$$\mathbb{E}[W_1(t) W_2(t)] = \rho t.$$

6.2 By direct computation, show that (6.9) is a valid decomposition of Σ as defined in (6.6).

6.3 Let $f_{\mu, \Sigma}$ be the joint PDF of a bivariate normal random variable. Explain why $f_{\mu, \Sigma}$ is undefined if $\rho = \pm 1$.

6.4 Compute by hand the Cholesky factors of

$$\Sigma = \begin{pmatrix} 1 & \rho_{1,2} & \rho_{1,3} \\ \rho_{1,2} & 1 & \rho_{2,3} \\ \rho_{1,3} & \rho_{2,3} & 1 \end{pmatrix}.$$

6.5 Consider a swaption written at time $t = 0$ on holdings of two assets S_1 and S_2 with payoff given by (6.1), where the size of the asset holding in each case is

$c_1 = 106$, $c_2 = 92$, the expiry is $T = 0.5$, and the price processes satisfy the 2-dimensional system of SDEs (6.2) with

- initial values $[S_1(0), S_2(0)]^T = [1.0, 1.2]^T$,
- risk free rate $r = 0.03$,
- individual asset price volatilities $[\sigma_1, \sigma_2]^T = [0.3, 0.2]^T$,
- correlation of the standard Brownian motions W_1^Q and W_2^Q set as $\rho = 0.3$.

1. Write Python code to value this swaption by brute-force Monte Carlo.
2. Modify your code from Part 1 to implement dimension reduction by change-of-measure, as described in Sect. 6.4.
3. Use Python's `timeit` module to compare the execution-time of the brute-force implementation in Part 1 to the dimensionally reduced implementation in Part 2, and comment.
4. Modify this code further to implement variance reduction by antithetic variate, as described in Sect. 4.4.2 of Chap. 4.

6.6 Consider a portfolio of 5 correlated Black-Scholes assets $S = [S_1, \ldots, S_5]^T$ held in the amounts $[c_1, \ldots, c_5]^T = [100, 102, 67, 163, 78]$ respectively, and

- initial values

$$[S_1(0), \ldots, S_5(0)]^T = [1.0, 1.10, 1.50, 0.82, 1.46]^T,$$

- risk free rate $r = 0.03$,
- individual asset price volatilities

$$[\sigma_1, \ldots, \sigma_5]^T = [0.3, 0.2, 0.25, 0.18, 0.32]^T,$$

- covariance matrix for the standard Brownian motions W_i^Q, $i = 1, \ldots 5$ at time $t \in [0, T]$ given by

$$\Sigma = t \begin{pmatrix} 1 & 0.1 & 0.42 & -0.3 & 0.6 \\ 0.1 & 1 & -0.34 & 0.55 & 0.2 \\ 0.42 & -0.34 & 1 & -0.4 & 0.35 \\ -0.3 & 0.55 & -0.4 & 1 & 0.45 \\ 0.6 & 0.2 & 0.35 & 0.45 & 1 \end{pmatrix}.$$

1. Confirm that Σ is a positive definite matrix and hence compute the Cholesky factor A.
2. Write the vector of asset price processes S as a 5-dimensional system of SDEs with independent noise terms.

3. Write Python code to value by Monte Carlo a basket call option on this portfolio struck at $E = 550$ with time to expiry $T = 0.25$, using $M = 10^5$ samples. Use variance reduction by antithetic variate. Your output should include both a point and 95% confidence interval.

6.7 Generate a plot of asset price trajectories, as in Fig. 6.3, for the 5 correlated Black-Scholes asset models detailed in Exercise 6.6, using $N = 25$ steps over the interval $[0, 0.25]$.

6.8 Which (if any) of the rainbow options listed at the start of Sect. 6.1 yield to successful use of dimension reduction by change-of-measure when valued by Monte Carlo methods? Justify your answer.

6.9 Let (U, V) be a pair of uniformly distributed random variables such that $U = 1 - V$. Show that the copula takes the form

$$C(u, v) = (u + v - 1)^+, \quad 0 \le u, v \le 1.$$

6.10 Generate a 500×2 NumPy array consisting of random draws from a Bernoulli distribution on $\{0, 1\}$:

```
In []: test=rng.binomial(size=(500,2),n=1,p=0.5)
```

Use the calls `np.where()`, `np.all()`, and `len()` in combination to count the number of rows in the array variable `test` where both entries have value 1.

6.11 In d-dimensions, an Archimedean copula may be written in the form

$$C(u_1, \ldots, u_d) = \Psi^{-1} \left(\Psi(u_1) + \cdots + \Psi(u_d) \right), \quad u_1, \ldots, u_d \in [0, 1],$$

where Ψ satisfies the following conditions:

- $\Psi(1) = 0$;
- $\Psi'(t) < 0, t \in (0, 1)$;
- $\Psi''(t) \ge 0, t \in (0, 1)$.

The function Ψ is convex and decreasing on the interval $[0, 1]$ and is called the generator. Confirm each of the following:

1. $\Psi(t) = -\ln(t)$ is the generator for the *product copula*:

$$C(u_1, u_2) = u_1 u_2, \quad 0 < u_1, u_2 \le 1.$$

2. $\Psi(t) = t^{-\alpha} - 1, \alpha > 1$, is the generator for the *Clayton copula*:

$$C(u_1, u_2) = \left(u_1^{-\alpha} + u_2^{-\alpha} - 1\right)^{-1/\alpha}, \quad 0 < u_1, u_2 \leq 1.$$

3. $\Psi(t) = (-\ln(t))^{\alpha}, \alpha \geq 1$, is the generator for the *Gumbel copula*

$$C(u_1, u_2) = e^{-((-\ln(u_1))^{\alpha} + (-\ln(u_2))^{\alpha})^{1/\alpha}}, \quad 0 < u_1, u_2 \leq 1.$$

6.12 Plot the joint CDF, as in Figs. 6.12 and 6.14, produced by each of the Clayton and Gumbel copulas for your choice of α in each case. Use the `plt.contourf()` call from the PyPlot package.

6.13 The example of the IMSF-INTC equity log-returns used throughout the chapter to illustrate empirical bivariate distributions with heavy tailed marginals and the selection of an appropriate copula may be used as basis for a group or individual project. Readers should choose a pair of equity assets trading on a publicly listed market and, using Pandas Datareader, download 3 years of daily price data. Then follow the procedure demonstrated over the course of this chapter to identify and fit an appropriate bivariate distribution for the daily log-returns, from which synthetic data can be generated.

Stochastic Models for Interest Rates 7

In practice, the assumption within the Black-Scholes framework that there is a single constant risk-free annualised rate of interest is a simplification that ignores empirical observations such as term-structures and variation in interest rate data over time. As an assumption when valuing derivative securities it is difficult to justify when the time-to-expiry is greater than approximately one year.

Our first task in this chapter will be to show how a term structure reflected in published bond prices may be used to calibrate the annualised risk-free rate of interest for a Black-Scholes asset model, and how sampled trajectories can be interpolated using a Brownian bridge to incorporate new information.

Next, we review the theory of risk-neutral pricing of bonds and interest rate derivatives with expiry date S, which have a value taking the form

$$V(t) = \mathbb{E}_{\mathbb{Q}}\left[e^{-\int_t^S r(s)ds} V_{r,S} \middle| \mathcal{F}_t\right], \quad t \in [0, S],$$

where stochastic variation in an interest rate $r(t)$ is modelled as an SDE, and the payoff $V_{r,S}$ itself depends on the same interest rate through the underlying asset. Bonds themselves fall into this category of derivatives, as do options on interest rates, and options written on a bond. To this end we study specific SDE models for the short rate, exploring their properties both theoretically and computationally. Specifically we examine the Vasicek, Ho Lee, and Cox-Ingersoll-Ross SDE interest rate models, giving examples of calibration methods, illustrations of their stochastic dynamics, and demonstrating the term structures that emerge from the use of different SDE models in bond pricing.

C. Kelly, *Computation and Simulation for Finance*,
Springer Undergraduate Texts in Mathematics and Technology,
https://doi.org/10.1007/978-3-031-60575-8_7

Table 7.1 User-defined function required in Chap. 7

Section	Function name	Function output	Parameter	Parameter meaning
1.2.4	dFac()	$e^{-r(T-t)}$	r	r
			dt	$T-t$

Computational approaches for the valuation of interest rate derivatives that use Monte Carlo sampling from a specific SDE model of the short rate introduce the necessity to sample from the distribution of the SDE model. After motivating our approach via the Fundamental Theorem of Bond Pricing and the discounted Feynman-Kac theorem for interest rate derivatives, we apply techniques from Chap. 6 to the problem of exact sampling of the bivariate joint distribution of an interest rate model with its associated numeraire asset, and highlight how a carefully chosen change-of-measure can sidestep the need to sample from a stochastic discount factor.

Though our emphasis in this chapter is on exact sampling, we identify a role for the numerical approximation of SDEs, which will then be treated in Chap. 8.

Python Setup To follow the Python code in this chapter, you will need to import the NumPy and PyPlot packages, as well as the `fsolve` module from the `scipy.optimize` package.

```
In []: import numpy as np
        import matplotlib.pyplot as plt
        from scipy.optimize import fsolve
```

As usual, instantiate a local RNG:

```
In []: rng=np.random.default_rng()
```

We will also make use of the user-defined function from Chap. 1 listed in Table 7.1.

7.1   Calibrating a Black-Scholes Asset Model to Bond Prices

Recall from Sect. 1.2.3 in Chap. 1 that a *bond* is a fixed-income instrument that represents a loan made by an investor (the bondholder) to a borrower (the issuer). We review first some terminology relevant for discussing bonds and their properties.

The *issue price* is the price at which the bond issuer originally sells the bond. The *maturity date* is the date when the bond issuer will pay the bondholder the face value of the bond, the *face value* being the amount that the bond will be worth at maturity. *Coupon dates* are the dates on which the bond issuer makes interest payments, and the *coupon rate* is the rate of interest per annum the bond issuer will pay on the face value. The *par yield* for a given maturity date is the coupon rate that causes the bond price to equal its face value.

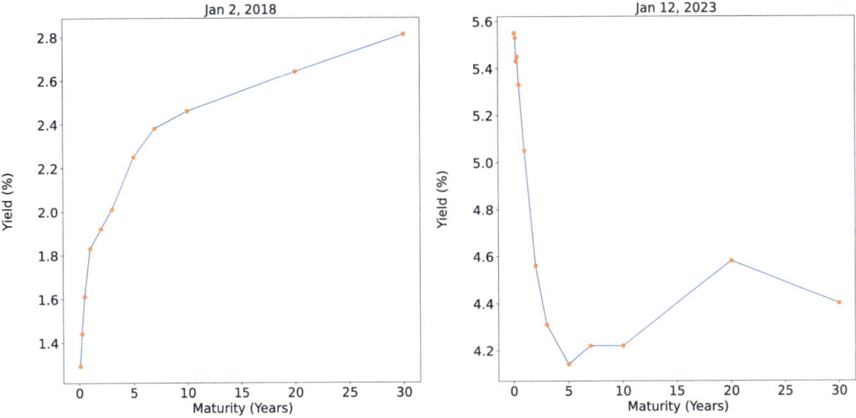

Fig. 7.1 US Treasury par yield curves showing the term structure of interest rates on US government bonds prevailing in January 2018 (left) and January 2023 (right). Source of data: https://home.treasury.gov/policy-issues/financing-the-government/interest-rate-statistics

If the annualised rate of interest r paid on the bond is constant, then the price $P(t, T)$ of a zero-coupon bond at time $t \in [0, T]$ maturing at time T with face value $P(T, T) = 1$ is

$$P(t, T) = e^{-r(T-t)}, \quad t \in [0, T].$$

Interest rates based on current bond prices and yields are observable and therefore deterministic. Realistically, at the time of issue we will observe prices for bonds with different maturities T that reflect different annualised rates of interest for each maturity T: a term structure of interest rates. Figure 7.1 shows the par yields for US Treasuries over a range of maturity dates, from 1 month to 30 years, that prevailed in January 2018 and again in January 2023. For the 2018 data notice that longer term bonds have a higher par yield, reflecting a market view that lending for longer periods carries greater risk for the bond issuer. This is the typical situation. An inverted yield curve is one where yields for longer duration bonds are lower than for shorter term bonds, and it is often taken as a market indicator of imminent economic recession in the US. The 2023 data shows a partially inverted yield curve.

We can use observed term structures like these to help calibrate the risk-neutral form of a Black-Scholes asset model.

7.1.1   A Risk Free Rate that Varies in Time

Suppose now that r is deterministic but time varying. The value of a bond with face value $P(T, T) = 1$ at time $t \in [0, T]$ is then

$$P(t, T) = e^{-\int_t^T r(u)du}, \tag{7.1}$$

or, equivalently,

$$r(u) = -\left.\frac{\partial}{\partial T}\ln P(0, T)\right|_{T=u}.$$

The risk-neutral form of the Black-Scholes asset price model becomes

$$dS(t) = r(t)S(t)dt + \sigma S(t)dW^{\mathbb{Q}}(t), \quad t \in [0, T],$$

with solution

$$S(t) = S_0 e^{\int_0^t r(u)du - \frac{1}{2}\sigma^2 t + \sigma W^{\mathbb{Q}}(t)}, \quad t \in [0, T]. \tag{7.2}$$

Suppose that we have already estimated σ from market data (see Sect. 2.3 in Chap. 2). It remains to estimate $\int_0^t r(s)ds$, or equivalently, $e^{\int_0^t r(s)ds}$ for $t \in [0, T]$. If we observe bond prices with N distinct maturities $0 = T_0 < T_1 < \ldots < T_N = T$ as

$$P(0, T_0), \ P(0, T_1), \ldots, P(0, T_N),$$

noting that $P(0, T_0) = P(0, 0) = 1$, then by (7.1) the following relation holds:

$$\frac{P(0, T_i)}{P(0, T_{i+1})} = e^{\int_{T_i}^{T_{i+1}} r(u)du}, \quad i = 0, \ldots, N - 1.$$

We can then simulate the asset price over a grid of times corresponding to the bond maturities $T_0, \ldots T_N$ as

$$S(T_{i+1}) = S(T_i)e^{\int_{T_i}^{T_{i+1}} r(u)du - \frac{1}{2}(T_{i+1} - T_i)\sigma^2 + \sigma \Delta W^{\mathbb{Q}}_{i+1}}$$

$$= S(T_i)\frac{P(0, T_i)}{P(0, T_{i+1})}e^{-\frac{1}{2}(T_{i+1} - T_i)\sigma^2 + \sigma \Delta W^{\mathbb{Q}}_{i+1}}, \tag{7.3}$$

for $i = 0, \ldots N - 1$, where each $\Delta W^{\mathbb{Q}}_{i+1} = W^{\mathbb{Q}}(T_{i+1}) - W^{\mathbb{Q}}(T_i)$.

7.1.2 An Example of Risk-Neutral Model Calibration

Consider the Black-Scholes asset model (7.2) with $\sigma = 0.3$. Suppose we wish to simulate values of $S(T)$, where $T = 2$, from this model while ensuring that the time-varying risk-free annualised rate of interest is consistent with an observed set of bond prices.

In the first two columns of Table 1.1, we reproduce the term structure of interest rates encountered in Table 1.1 of Chap. 1, and will use this to create a set of bond prices, observable at time $t = 0$, to which we calibrate the asset model (7.2). These prices are displayed in the third column of the table. For example, the price of a bond maturing in 18 months is computed to be $P(0, 1.5) = 100 \times e^{-0.025 \times 1.5} = 96.32$. Exercise 7.2 asks you to repeat the process using real yield curve data from the US Treasury website.

Set up array variables containing the maturities and the corresponding zero rates:

```
In []: mat=np.array([0.5,1.0,1.5,2.0])
       rates=np.array([0.02,0.022,0.025,0.03])
```

The time-t value of a zero coupon bond with face value $P(T, T) = 100$ maturing at time T is $100 \times e^{-r_T(T-t)}$, so we can create an array of bond values as follows:

```
In []: bonds=100*dFac(rates,mat)
```

We will calibrate our model (7.2) to these bond prices. The time set

$$[T_0, T_1, T_2, T_3, T_4] = [0, 0.5, 1.0, 1.5, 2.0],$$

and array of bond values

$$[P(0, 0), P(0, 0.5), P(0, 1.0), P(0, 1.5), P(0, 2.0)],$$

are constructed by prepending $T_0 = 0$ and $P(0, 0) = 100$ to the beginning of each of mat and bonds:

```
In []: t=np.insert(mat,0,0)
       bonds=np.insert(bonds,0,100)
```

It is clear that we should set N=4 and dt=0.5 to represent N and δt in (7.3).

We can now simulate our asset at 6-monthly intervals to be consistent with these bond values using the iteration (7.3) with $N = 4$ and $\delta t = 0.5$.

```
In []: T=2.0; sig=0.3
       N=4; dt=T/N

       inc=rng.normal(0,np.sqrt(dt),N)
       S=np.ones(N+1)
       factor=np.exp(-0.5*(sig**2)*dt+sig*inc)
       bondRatio=bonds[:-1]/bonds[1:]
       update=bondRatio*factor

       for i in range(N):
           S[i+1]=S[i]*update[i]
```

Finally, let's consider how we would generate quarterly values for S calibrated to the same set of observed bond values. Since these observed values are for maturities separated by 6 months each, we should interpolate the values of r_T using the `np.interp()` function, and then compute the additional bond values.

First, create the new time set containing values every 3 months:

```
In []: matNew=np.arange(0.25,2.25,0.25)
```

Now interpolate the observed interest rates and compute the new list of bond prices:

```
In []: ratesNew=np.interp(matNew,mat,rates)
       bondsNew=100*dFac(ratesNew,matNew)


       tNew=np.insert(matNew,0,0)
       bondsNew=np.insert(bondsNew,0,100)
```

Notice that in order to produce the bond value $P(0, 0.25)$ we should interpolate the yield curve between r_0 and $r_{0.5}$. Since there is no available value for r_0 the `np.interp()` call sets $r_{0.25} = r_{0.5}$. This is not ideal, and a suitable proxy for r_0 (the short rate at time 0) is suggested later in the chapter: see in Sect. 7.2.4.

We can now generate trajectories of S over the expanded time set `tNew` using the following script, adapted from above.

```
In []: N=8; dt=0.25;


       S=np.ones(N+1)
       inc=rng.normal(0,np.sqrt(dt),N)
       factor=np.exp(-0.5*(sig**2)*dt+sig*inc)
       bondRatio=bondsNew[:-1]/bondsNew[1:]
       update=bondRatio*factor


       for i in range(N):
           S[i+1]=S[i]*update[i]
```

However once we have added intermediate time points and interpolated the observed bond prices, we are resampling the increments of the Brownian motion $W^{\mathbb{Q}}$ over the new finer mesh and the output will show a new trajectory. We cannot return to the same trajectory as before unless we also interpolate the trajectory of $W^{\mathbb{Q}}$. We will look at how this can be done next.

7.1.3   Interpolating Individual Asset Price Trajectories

In our example from the previous section, we have constructed an array S containing values from a particular trajectory of the asset price model $S(i\delta t)$, $i = 0, \ldots, 4$, and wish to include intermediate values from the same path. In particular let us consider the question of interpolating the trajectory from $S(0.5)$ to $S(1)$, to produce an observation $S(0.75)$ from the same trajectory, i.e. without changing the observed values of $S(0.5)$ and $S(1)$. Since the randomness of S is coming from the standard Brownian motion $W^{\mathbb{Q}}$, the problem is equivalent to sampling a value for $W^{\mathbb{Q}}(0.75)$ without changing the observed values of $W^{\mathbb{Q}}(0.5)$ and $W^{\mathbb{Q}}(1)$.

Let $W_3 = [W(t_1), W(t_2), W(t_3)]^T$ be a vector containing values of a standard Brownian motion at three successive times $0 < t_1 < t_2 < t_3 < T$. We have omitted the probability measure $\mathbb{Q}$ from our notation since the specific choice of measure is not relevant to what follows. W_3 has a multivariate centred normal distribution with $d = 3$, and we write $W_3 \sim \mathcal{N}(0, \Sigma)$: see Definition 6.1 in Chap. 6.

First, let us compute the covariance matrix Σ. For $i, j = 1, 2, 3$ and $i \leq j$,

$$\mathrm{Cov}\left[W(t_i), W(t_j)\right] = \mathrm{Cov}\left[W(t_i), W(t_i) + \left(W(t_j) - W(t_i)\right)\right]$$

$$= \mathrm{Cov}\left[W(t_i), W(t_i)\right] + \mathrm{Cov}\left[W(t_i), W(t_j) - W(t_i)\right]$$

$$= t_i + 0$$

$$= t_i.$$

So the covariance matrix is given by

$$\Sigma = \begin{pmatrix} t_1 & t_1 & t_1 \\ t_1 & t_2 & t_2 \\ t_1 & t_2 & t_3 \end{pmatrix}$$

and

$$W = \begin{pmatrix} W(t_1) \\ W(t_2) \\ W(t_3) \end{pmatrix} \sim \mathcal{N}\left(\begin{pmatrix} 0 \\ 0 \\ 0 \end{pmatrix}, \begin{pmatrix} t_1 & t_1 & t_1 \\ t_1 & t_2 & t_2 \\ t_1 & t_2 & t_3 \end{pmatrix} \right)$$

or, by rearranging the order of times,

$$\begin{pmatrix} W(t_2) \\ W(t_1) \\ W(t_3) \end{pmatrix} \sim \mathcal{N}\left(\begin{pmatrix} 0 \\ 0 \\ 0 \end{pmatrix}, \begin{pmatrix} t_2 & t_1 & t_2 \\ t_1 & t_1 & t_1 \\ t_2 & t_1 & t_3 \end{pmatrix} \right). \tag{7.4}$$

Partition the vector W into two parts: a scalar and a 2×1 vector, and denote each in turn

$$X_{[1]} := W(t_2); \quad X_{[2]} := [W(t_1), W(t_3)]^T.$$

Then the mean vectors associated with the partition are denoted

$$\mu_{[1]} := 0 \quad \text{and} \quad \mu_{[2]} := [0, 0]^T.$$

The covariance matrices associated with the partition are denoted:

$$\Sigma_{[1,1]} := \mathrm{Cov}[W(t_2), W(t_2)] = \mathrm{Var}[W(t_2)] = t_2;$$

$$\Sigma_{[2,2]} := \begin{pmatrix} \mathrm{Cov}[W(t_1), W(t_1)] & \mathrm{Cov}[W(t_1), W(t_3)] \\ \mathrm{Cov}[W(t_3), W(t_1)] & \mathrm{Cov}[W(t_3), W(t_3)] \end{pmatrix} = \begin{pmatrix} t_1 & t_1 \\ t_1 & t_3 \end{pmatrix};$$

$$\Sigma_{[1,2]} := [\mathrm{Cov}[W(t_2), W(t_1)], \mathrm{Cov}[W(t_2), W(t_3)]] = [t_1, t_2];$$

$$\Sigma_{[2,1]} := \begin{pmatrix} \mathrm{Cov}[W(t_1), W(t_2)] \\ \mathrm{Cov}[W(t_2), W(t_3)] \end{pmatrix} = \begin{pmatrix} t_1 \\ t_2 \end{pmatrix}.$$

Assembling these components into a block matrix allows us to confirm by comparison with (7.4) that

$$\begin{pmatrix} \Sigma_{[1,1]} & \Sigma_{[1,2]} \\ \Sigma_{[2,1]} & \Sigma_{[2,2]} \end{pmatrix} = \begin{pmatrix} t_2 & t_1 & t_2 \\ t_1 & t_1 & t_1 \\ t_2 & t_1 & t_3 \end{pmatrix} = \Sigma.$$

We state the following general result on the conditional distribution of multivariate normal random variables.

Proposition 7.1 *Suppose that the partitioned vector $(X_{[1]}, X_{[2]})^T$ is a multivariate normal random variable with specified distribution*

$$\begin{pmatrix} X_{[1]} \\ X_{[2]} \end{pmatrix} \sim \mathcal{N} \left(\begin{pmatrix} \mu_{[1]} \\ \mu_{[2]} \end{pmatrix}, \begin{pmatrix} \Sigma_{[11]} & \Sigma_{[12]} \\ \Sigma_{[21]} & \Sigma_{[22]} \end{pmatrix} \right),$$

where $\Sigma_{[22]}$ is an invertible square matrix. Then the random vector $X_{[1]}$ has the conditional distribution

$$(X_{[1]}|X_{[2]} = x) \sim N \left(\mu_{[1]} + \Sigma_{[12]} \Sigma_{[22]}^{-1} (x - \mu_{[2]}), \ \Sigma_{[11]} - \Sigma_{[12]} \Sigma_{[22]}^{-1} \Sigma_{[21]} \right).$$

We do not prove Proposition 7.1 here, but it can be immediately applied to give us a technique for sampling from $W(t_2)$ when the values of $W(t_1)$ and $W(t_3)$ are already known and fixed.

Theorem 7.2 (The Brownian Bridge) *Let $(W(t_1) = x_1, \ldots, W(t_n) = x_n)$ be observed values on a standard Brownian motion trajectory. Suppose that $t_i \le t \le t_{i+1}$ for any $i = 1, \ldots, n$. Then the value $W(t)$ has conditional distribution*

$$(W(t)|W(t_i) = x_i, W(t_{i+1}) = x_{i+1})$$

$$\sim \frac{(t_{i+1} - t)x_i + (t - t_i)x_{i+1}}{(t_{i+1} - t_i)} + \sqrt{\frac{(t_{i+1} - t)(t - t_i)}{(t_{i+1} - t_i)}} Z, \qquad (7.5)$$

where $Z \sim N(0, 1)$ is independent of all of $W(t_1), \ldots, W(t_n)$.

Proof If we apply Proposition 7.1 with

$$X_{[1]} = W(t_2); \quad X_{[2]} = [W(t_1), W(t_3)]^T; \quad x = [x_1, x_3]^T,$$

so that

$$\left(X_{[1]}|X_{[2]} = x\right) = \left(W(t_2)|W(t_1) = x_1, W(t_3) = x_3\right),$$

we get

$$\mathbb{E}\,[W(t_2)|W(t_1) = x_1, W(t_3) = x_3] = \mu_{[1]} + \Sigma_{[1,2]}\Sigma_{[2,2]}^{-1}\left(x - \mu_{[2]}\right)$$

$$= 0 + [t_1, t_2]\begin{pmatrix} t_1 & t_1 \\ t_1 & t_3 \end{pmatrix}^{-1}\left(\begin{pmatrix} x_1 \\ x_3 \end{pmatrix} - \begin{pmatrix} 0 \\ 0 \end{pmatrix}\right)$$

$$= \frac{(t_3 - t_2)x_1 + (t_2 - t_1)x_3}{(t_3 - t_1)}.$$

We see that the conditional expected value of the interpolated value $W(t_2)$ is determined as the vertical component of an intermediate point (t, x) along the interpolated line between (t_1, x_1) and (t_3, x_3). Moreover,

$$\mathrm{Var}[W(t_2)|W(t_1) = x_1, W(t_3) = x_3] = \Sigma_{[1,1]} - \Sigma_{[1,2]}\Sigma_{[2,2]}^{-1}\Sigma_{[2,1]}$$

$$= t_2 - [t_1, t_2]\left(\frac{1}{t_1 - t_3}\right)\begin{pmatrix} -t_3/t_1 & 1 \\ 1 & -1 \end{pmatrix}\begin{pmatrix} t_1 \\ t_2 \end{pmatrix}$$

$$= \frac{(t_3 - t_2)(t_2 - t_1)}{(t_3 - t_1)}.$$

We can conclude that

$$(W(t_2)|W(t_1) = x_1, W(t_3) = x_3)$$

$$\sim \frac{(t_3 - t_2)x_1 + (t_2 - t_1)x_3}{(t_3 - t_1)} + \sqrt{\frac{(t_3 - t_2)(t_2 - t_1)}{(t_3 - t_1)}}\,Z,$$

where $Z \sim \mathcal{N}(0, 1)$ is a standard normal random variable that is independent of all of $W(t_1), \ldots, W(t_n)$.

7.1.4   Brownian Bridging in Python

We seek to take a Brownian path over the interval $[0, T]$ generated with N steps, each of length dt, and apply a Brownian bridge to add an additional observation at the midpoint of each step, thus doubling the resolution of the trajectory.

First, generate the Brownian trajectory at low resolution and store in the array variable `W`:

```
In []: T=2.0; N=4; dt=T/N

       inc=rng.normal(0,np.sqrt(dt),N)
       W=np.insert(np.cumsum(inc),0,0)

Out[]: array([0.        , 0.79987423, 1.15305443, 0.48146785,
           1.02112108])
```

Next, expand `W` to accommodate the extra values by inserting a placeholder zero at every second position in the array:

```
In []: Wbr=np.insert(np.zeros(N+1),np.arange(N+1),W)[:-1]

Out[]: array([0.        , 0., 0.79987423, 0.,
           1.15305443, 0., 0.48146785, 0., 1.02112108])
```

Notice that we removed a superfluous zero at the end by appending `[:-1]` to the array produced by the `np.insert()` call. Next set up a new time set `t` with the correct time points:

```
In []: N1=2*N; dt=T/N1; t=np.arange(0,T+dt,dt)
```

We want to replace all those new zero values at indices $1, 3, 5, \ldots, 2N - 1$ with values produced by a Brownian bridge. Define a function based upon (7.5) in the statement of Theorem 7.2 that we can call in order to generate the bridge observation:

```
In []: def bbridge(t1,t,t2,x1,x2,rng):
           Z=rng.normal(0,np.sqrt((t2-t)*(t-t1)/(t2-t1)),1)
           return ((t2-t)*x1+(t-t1)*x2)/(t2-t1)+Z
```

Now use a `for` loop to insert bridge observations at each odd indexed entry of `Wbr`:

```
In []: for i in range(1,len(Wbr),2):
           Wbr[i]=bbridge(t[i-1],t[i],t[i+1],
                          Wbr[i-1],Wbr[i+1],rng)
```

Successive applications of a Brownian bridge to increase the resolution of a single Brownian trajectory are illustrated in Fig. 7.2. Exercise 7.2 asks you to apply this technique, combined with the method described in Sect. 7.1.2, to the interpolation of trajectories of the Black-Scholes model calibrated to the (interpolated) bond prices in Table 7.2. Output is shown in Fig. 7.3 using the same trajectory of $W^{\mathbb{Q}}$ as displayed in Fig. 7.2.

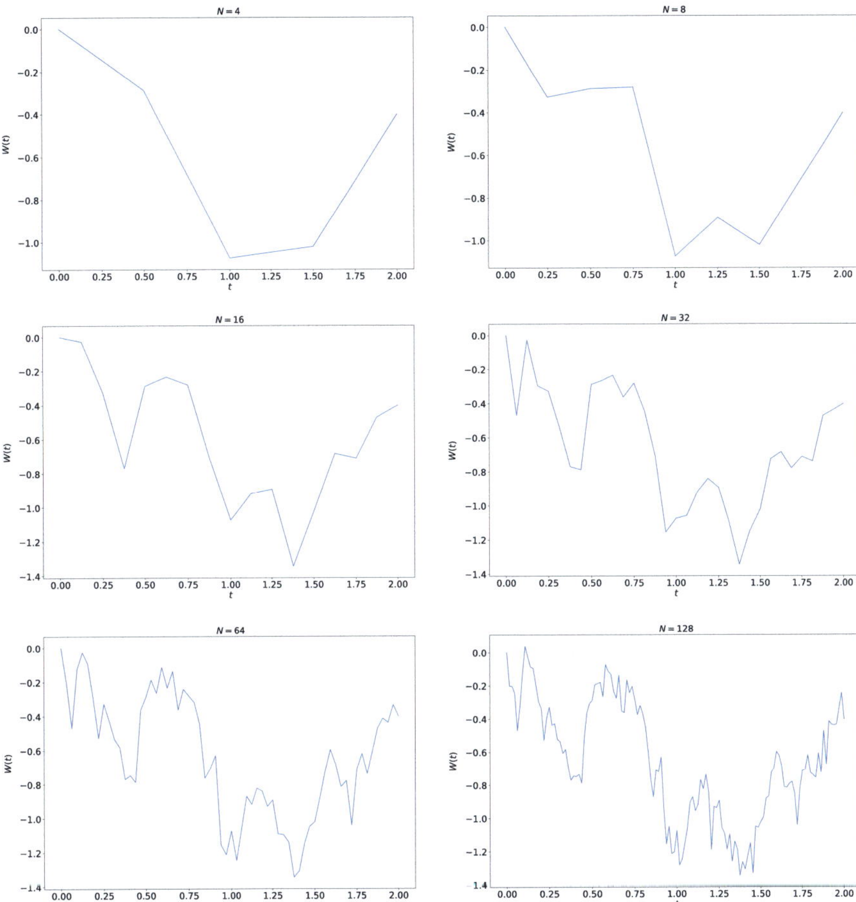

Fig. 7.2 Brownian bridging has been used to increase the resolution of a single Brownian trajectory over the interval $[0, 2]$ with N steps, where $N = 2^k$ for $k = 2, 3, 4, 5, 6, 7$

Table 7.2 Maturity dates, interest rates, bond prices. Computations have been made at double precision, with output reported to two decimal places

T (years)	r_T (%)	$P(0, T)$ ($)
0	–	100
0.5	2.0	99.00
1.0	2.2	97.82
1.5	2.5	96.32
2.0	3.0	94.18

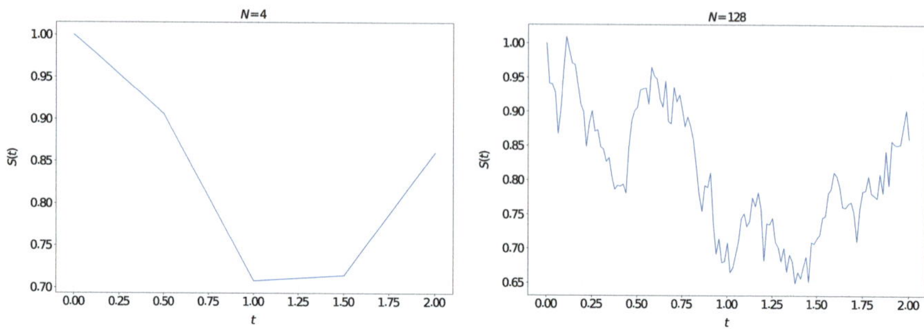

Fig. 7.3 Using the same Brownian trajectory from Fig. 7.2, we display a single trajectory of a Black-Scholes asset model evolving over the interval $[0, 2]$ on two meshes with $N = 4$ and $N = 128$ steps respectively. The risk-free rate for the model has been calibrated directly to bond prices from Table 7.2 on the left, and to interpolations of those prices on the right

7.2 A Review of Bonds and Interest Rates

7.2.1 The Spot Rate and the Term Structure

Let $P(t, T)$ denote the value of a zero-coupon bond at time t maturing at time T with face-value $P(T, T) = 1$.

Definition 7.3 The *spot rate* $R(t, T)$ at time t for maturity T is defined as the yield-to-maturity of the T-maturity bond:

$$R(t, T) = -\frac{\log P(t, T)}{T - t}, \quad t \in [0, T]. \tag{7.6}$$

The spot rate can be interpreted as the effective rate of interest applicable from time t to time T implied by the market price of the bond. It is also referred to as the *zero rate* or *long rate*.

We can see from Definition 7.3 that the price of a zero coupon bond maturing at time T with face value $P(T, T) = 1$ may be expressed in terms of the spot rate as

$$P(t, T) = e^{-R(t,T)(T-t)}, \quad t \in [0, T].$$

So we can alternatively view the spot rate as the discount rate that equates the face value of a bond to its market value at time $t \in [0, T]$. This notion extends to coupon-bearing bonds: see Sect. 7.2.3.

In fact we can view the term structure of interest rates presented in Table 7.2 as representing the time $t = 0$ values of spot rates, so that each $r_{T_i} = R(0, T_i)$ where $T_i = i\delta t$ for $i = 1, 2, 3, 4$ and $\delta t = 0.5$. Definition 7.3 generalises each of these rates so that they may vary continuously in time for each fixed maturity date T. In this more general setting the prevailing term structure at a fixed time t can be understood as $R(t, T)$, expressed as a function of the maturities $T > t \geq 0$.

7.2.2   The Short Rate

Interest rates that will hold in the future cannot be observed in advance and must be modelled as stochastic processes. An example of such a rate is the short rate which, from the point of view of an observer at time $s \geq 0$, will be random at future times $t > s$.

Definition 7.4 The *short rate r* is defined by

$$r(t) = \lim_{T \downarrow t} R(t, T), \quad t \in [0, T].$$

The short rate is a continuous stochastic process that, for each $t \in [0, T]$, describes the dynamics of the interest rate earned on a riskless investment over the interval $[t, t + \delta t]$ in the limit as $\delta t \to 0$. It is often used as the rate of interest paid on the numeraire asset for risk-neutral pricing of derivatives. The properties and sampling of stochastic models for these rates will be a significant focus later in this chapter.

7.2.3   Yield-to-Maturity for Coupon-Bearing Bonds

The notion of a single yield-to-maturity also applies to bonds that provide coupon payments to the holder. The *present value of a coupon bearing bond* is the present value of all future cash flows received by the holder using the appropriate spot rates as discount rates. The *yield-to-maturity* on a coupon-bearing bond is the single discount rate R that equates the cash flows on the bond to its present value.

For example, consider a 2-year risk-free bond with face value $P(2, 2) = 100$ paying coupons at 6% semi-annually, where the spot rates $R(0, T_i)$ have the term structure provided in Table 7.2 for $T_i = i\delta t, i = 1, \ldots 4$ and $\delta t = 0.5$. The value of each coupon payment is \$3, computed explicitly as $C = 0.06 \times P(2, 2) \times \delta t = 3$, and the present value of the bond at time $t = 0$ may be computed as the sum of all four coupon payments, each discounted back to the present time $t = 0$ from its respective payment date T_i, and the face value of the bond discounted back to $t = 0$ from the maturity date:

$$P(0, T_4)$$
$$= Ce^{-R(0, T_1) \times T_1} + Ce^{-R(0, T_2) \times T_2} + Ce^{-R(0, T_3) \times T_3} + (P(T_4, T_4) + C)e^{-R(0, T_4) \times T_4}$$
$$= 3e^{-0.02 \times 0.5} + 3e^{-0.022 \times 1} + 3e^{-0.025 \times 1.5} + 103e^{-0.03 \times 2}$$
$$= 105.80.$$

In fact Python returns the value `105.79620041874122` which we assign to a variable PV. In this example, the yield-to-maturity is the solution of the nonlinear equation

$$3e^{-0.5 \times R} + 3e^{-R \times 1} + 3e^{-1.5 \times R} + 103e^{-2 \times R} = P(0, T_4).$$

We can use `fsolve()` to find R using iterative root-finding, just as we did to compute implied volatility in Sect. 2.3.2 of Chap. 2:

```
In []: def bondyield(R):
           return (3*np.exp(-R*0.5)+3*np.exp(-R*1)
               +3*np.exp(-R*1.5)+103*np.exp(-R*2)-PV)


       R0=0.02
       ytm=fsolve(bondyield,R0)
       print(ytm[0])


Out[]: 0.0297051163897335
```

So the yield-to-maturity of this bond, expressed as a percentage to two decimal places, is 2.97%.

7.2.4 Forward Rates

Let $P(t, T')$ denote the value of a zero-coupon bond at time t maturing at time T' with with face-value $P(T', T') = 1$.

Interest rates that are agreed upon at time t for the purchase of a bond at some future time $T > t$ and with maturity date $S > T$ are known as forward rates. Suppose that the bond has face value $P(S, S) = 1$. Then the bond price agreed at time t for this trade is given by

$$e^{-F(t,T,S)(S-T)}, \quad 0 \leq t < T < S,$$

where $F(t, T, S)$ is the (implied) forward rate at time t for delivery of an S-bond at time T. The forward rate can be computed from current bond prices and an arbitrage argument. Consider the following pair of strategies for delivering \$1 at time S. At time t:

1. Purchase a bond that matures at time S with price $P(t, S)$.
2. Enter into a forward rate agreement starting at time T that delivers \$1 at time S. This commits us to paying $e^{-F(t,T,S)(S-T)}$ at time T. We can fund this commitment by buying a bond that delivers $e^{-F(t,T,S)(S-T)}$ at time T. The price of this bond is $e^{-F(t,T,S)(S-T)}P(t, T)$.

Both strategies deliver exactly \$1 at time S. To avoid an arbitrage opportunity, both should require the same initial outlay at time t. Therefore we must have

$$P(t, S) = e^{-F(t,T,S)(S-T)} P(t, T), \quad 0 \le t < T < S. \tag{7.7}$$

Making $F(t, T, S)$ the subject of (7.7) leads to the following definition.

Definition 7.5 The *forward rate* $F(t, T, S)$ at t for $T < S$ is defined by

$$F(t, T, S) = \frac{1}{S - T} \ln \left(\frac{P(t, T)}{P(t, S)} \right), \quad 0 \le t < T < S. \tag{7.8}$$

The forward rate can be interpreted as the future spot rate that will hold at time T that is implied by observed bond prices at time $t \in [0, T]$.

It is also possible to express the forward rate in terms of the spot rates $R(t, T)$ and $R(t, S)$ instead of the bond prices $P(t, T)$ and $P(t, S)$. By (7.6), $-\ln(P(t, T')) = R(t, T')(T' - t)$ for both $T' = T$ and $T' = S$. Thus

$$
\begin{aligned}
F(t, T, S) &= \frac{-\ln(P(t, S)) + \ln(P(t, T))}{S - T} \\
&= \frac{(S - t)R(t, S) - (T - t)R(t, T)}{S - T} \\
&= \frac{(S - t)(R(t, S) - R(t, T)) + (S - T)R(t, T)}{S - T}, \quad 0 \le t < T < S.
\end{aligned}
$$

It follows that

$$F(t, T, S) = R(t, T) + \frac{(S - t)(R(t, S) - R(t, T))}{S - T}, \quad 0 \le t < T < S. \tag{7.9}$$

From the forward rate we can define an interest rate that describes the instantaneous interest rate agreeable at time t that will be paid on a bond to be traded at a future time $T > t$.

Definition 7.6 The *instantaneous forward rate* $f(t, T)$ is defined by

$$f(t, T) = \lim_{S \downarrow T} F(t, T, S).$$

where $F(t, T, S)$ is as given in Definition 7.5. For any fixed t, the values of $f(t, T)$ expressed as a function of maturity times $T > t$ make up the *forward curve*. If $t = 0$ this is called the *initial forward curve*.

Table 7.3 Maturity dates, spot rates, bond prices, initial instantaneous forward rates. Computations have been made at double precision, values are reported to two decimal places

T (years)	$R(0, T)$ (%)	$P(0, T)$ ($)	$\ln(P(0, T))$	$f(0, T)$ (%)
0	–	100	4.61	–
0.5	2.0	99.00	4.60	2.0
1.0	2.2	97.82	4.58	2.4
1.5	2.5	96.32	4.57	3.1
2.0	3.0	94.18	4.55	4.5

Instantaneous forward rates can be used to calibrate stochastic models of the short rate (we will demonstrate this in Sect. 7.4.2), and so we must know how to compute them from observable data. We can derive formulae for the instantaneous forward rate separately in terms of bond prices and spot rates.

First, by (7.8),

$$f(t, T) = \lim_{S \downarrow T} F(t, T, S) = -\lim_{S \downarrow T} \frac{\ln(P(t, S)) - \ln(P(t, T))}{S - T}$$

$$= -\frac{\partial}{\partial T} \ln(P(t, T)), \tag{7.10}$$

giving a means to construct a forward curve from a set of observed bond prices $P(t, T)$. We can approximate the derivative on the RHS of (7.10) using the bond price data provided in Table 7.2 using backward differencing (see Sect. 5.4 in Chap. 5), so that for each $i = 1, \ldots, 4$,

$$f(0, T_i) = -\frac{\partial}{\partial T} \ln(P(0, T)) \bigg|_{T=T_i} \approx -\left[\frac{\ln(P(0, T_i)) - \ln(P(0, T_{i-1}))}{T_i - T_{i-1}} \right],$$

where $T_i = i \times 0.5$ for $i = 0, \ldots, 4$. The results are given in Table 7.3. This approach uses the slopes of a piecewise linear interpolation of the values of $\ln(P(0, T))$ to approximate the derivative in (7.10), and those who enjoy smoothness may prefer a more sophisticated interpolation method: see Further Reading. Notice that at the first maturity date $T_1 = 0.5$ the initial instantaneous forward rate $f(0, T_1)$ and the spot rate $R(0, T_1)$ coincide, a relation that holds in general.

To construct a forward curve $f(t, T)$ from a known term structure $R(t, T)$, by (7.9) we have, for $0 \le t \le T < S$

$$f(t, T) = \lim_{S \downarrow T} F(t, T, S) = R(t, T) + \lim_{S \downarrow T} (S - t) \frac{R(t, S) - R(t, T)}{S - T}$$

$$= R(t, T) + (T - t) \frac{\partial}{\partial T} R(t, T)$$

$$= \frac{\partial}{\partial T} ((T - t) R(t, T)), \quad t \in [0, T]. \tag{7.11}$$

Notice in Table 7.3 there is no value for $R(0, 0)$, and recall that we had a similar issue in Sect. 7.1.2. In order to apply backward differencing to approximate the derivative on the RHS of (7.11) at time T_1 we could instead use an observation on the short rate $r(0) = \lim_{T \downarrow 0} R(0, T)$ as a proxy for $R(0, 0)$. If we wish to move in the other direction, and construct a term structure from a known forward curve, we can use the inverse relation that expresses the spot rate as an average over observations on the curve.

$$R(t, T) = \frac{1}{T - t} \int_t^T f(t, u) du, \quad t \in [0, T].\tag{7.12}$$

Moreover, the instantaneous forward rate is related to the short rate by

$$r(t) = \lim_{T \downarrow t} R(t, T) = f(t, t).$$

7.3 Risk-Neutral Valuation of Bonds and Interest Rate Derivatives

7.3.1 The Fundamental Equation for Bond Pricing

Suppose that the short rate r has real-world dynamics that are described by an SDE of the form

$$dr(t) = \mu(r(t), t)dt + \sigma(r(t), t)dW^{\mathbb{P}}(t)\tag{7.13}$$

for some drift coefficient μ and diffusion coefficient σ such that $\sigma(r, t) \neq 0$. Let $V(r, t)$ denote the price at time t of an interest rate derivative that depends on an observed value r of the interest rate. A simple example of such a derivative would be a zero-coupon bond maturing at time T with face-value $P(T, T) = 1$. To value this bond from r we cannot apply the Black-Scholes framework developed Chap. 2 because there is no underlying asset that may be used to construct a risk-free portfolio.

Instead, Vasicek [66] proposed a construction whereby one bond is hedged by another bond with a different maturity date, leading to the Fundamental Equation for Bond Pricing, stated as follows:

Proposition 7.7 (Fundamental Equation for Bond Pricing) *Let r be the short rate of interest with real-world dynamics governed by the SDE (7.13), and let $V(r, t)$ denote the price at time t of an interest rate derivative with payoff $g(r(T))$ at time T. Then V satisfies the PDE*

$$V_t(r, t) + \frac{1}{2}\sigma^2 V_{rr}(r, t) + m(r, t)V_r(r, t) - rV(r, t) = 0,\tag{7.14}$$

for some function $m(r, t)$, and with final condition $V(r, T) = g(r(T))$.

Common examples of interest rate derivatives and their payoffs are:

1. A zero-coupon bond with face value $P(T, T) = 1$ has payoff $V(r, T) = 1$.
2. Let $P(r, t, S)$ denote the time-t price of a bond maturing at time S for a given value of the short rate r. A *European option on the bond*, with strike price E and expiring at time $T \leq S$, has payoff $V_{r,T} = (P(r, T, S) - E)^+$ for the call and $V_{r,T} = (E - P(r, T, S))^+$ for the put.

If V denotes the price of a coupon-bearing bond that pays continuously at a rate of $K(r, t)$, the PDE (7.14) becomes

$$V_t(r, t) + \frac{1}{2}\sigma^2 V_{rr}(r, t) + m(r, t)V_r(r, t) - rV(r, t) + K(r, t) = 0.$$

7.3.2 Interest Rates Derivatives with Payoff Depending on a Forward Curve

Interest rate derivatives may also have a payoff that is expressed in terms of an instantaneous forward curve f. For example a *caplet* with tenor δ for the interval $[T, T + \delta]$ is a European call option struck at E on the interest paid on a floating rate liability with principle value c. It has payoff

$$V(f, \delta) = c \left(\left[e^{\int_0^\delta f(T, T+u)du} - 1 \right] - \delta E \right)^+. \tag{7.15}$$

A *cap* is a portfolio of caplets, each active over a period $[T_i, T_{i+1}]$, where $0 = T_0 < T_1 < \cdots < T_N = T$ is a set of *tenor dates*. A cap can be valued as the sum of the values of all its constituent caplets, each discounted from time T_i back to the present. The European put variant of a caplet is called a *floorlet*, and the corresponding portfolio of floorlets active between a set of tenor dates is known as a *floor*. Valuation of such derivatives is beyond the scope of this textbook: see Further Reading.

7.3.3 The Discounted Feynman-Kac Theorem for Interest Rate Derivatives

Let $m(r, t)$ be the function in (7.14), and define the *market price of interest rate risk* to be

$$\theta(r, t) := \frac{\mu(r, t) - m(r, t)}{\sigma(r, t)}.$$

Suppose that Novikov's condition (see Theorem 2.6 in Sect. 2.1.5 of Chap. 2) holds for θ. Then by Girsanov's Theorem (see Theorem 2.7 in Sect. 2.1.5 of Chap. 2) there is a probability measure $\mathbb{Q}$ equivalent to $\mathbb{P}$ such that

$$W^{\mathbb{Q}}(t) = W^{\mathbb{P}}(t) + \int_0^t \theta(u)du$$

is a standard Brownian motion under the measure $\mathbb{Q}$ (see Definition 1.3 in Chap. 1). The dynamics of the interest rate may now be written as

$$dr(t) = m(r(t), t)dt + \sigma(r(t), t)dW^{\mathbb{Q}}(t), \quad t \in [0, T]. \tag{7.16}$$

and in this form is referred to as the *risk-neutral interest rate*.

Next consider a derivative written on an interest rate r with payoff $V_{r,S}$ at expiry S. Let $V(r, t, S)$ be the price of the derivative at time $0 \leq t \leq S$. An initial investment with value 1 in an account earning interest at a variable and stochastic rate of $r(u)$ at time $u \in [0, S]$ has value given by

$$\beta(t) = e^{\int_0^t r(u)du}, \quad t \in [0, S].$$

We will use this as our numeraire for risk-neutral pricing (recall that we also used a stochastic numeraire in Sect. 6.4 of Chap. 6 to improve computational efficiency for Monte Carlo pricing of certain rainbow options).

Define the discounted derivative price process as

$$\widetilde{V}(r, t, S) := \beta(t)^{-1}V(r, t, S) = e^{-\int_0^t r(u)du}V(r, t, S), \quad t \in [0, S].$$

We arrive at the following result, a proof of which may be found in Choe [15].

Proposition 7.8 (The Discounted Feynman-Kac Theorem for Interest Rate Derivatives) *Assume that there exists a nonnegative random variable Y such that $\mathbb{E}_{\mathbb{Q}}[Y] < \infty$ and $|\widetilde{V}(r, t)| \leq Y$ for all $t \in [0, T]$. Then*

$$V(r, t, S) = \mathbb{E}_{\mathbb{Q}}\left[e^{-\int_t^S r(u)du}V_{r,S}\Big|\mathcal{F}_t\right], \quad t \in [0, S], \tag{7.17}$$

where r is the risk-neutral interest rate with dynamics governed by the SDE (7.16), and the first argument in the function V on the LHS of (7.17) represents the observed value of the short rate r at time t.

In particular, at time $t \in [0, S]$ the price of a zero-coupon bond maturing at time S with face value $P(S, S) = 1$ is given by

$$P(t, S) = \mathbb{E}_{\mathbb{Q}}\left[e^{-\int_t^S r(u)du}\Big|\mathcal{F}_t\right], \quad t \in [0, S].$$

Proposition 7.8 motivates us to examine suitable stochastic models for risk-neutral interest rates, so that both they and the corresponding numeraire β can be sampled either exactly or approximately. This will allow us to use Monte Carlo techniques to value bonds and interest rate derivatives.

7.4   SDEs for Risk-Neutral Models of the Short Rate

7.4.1   Affine Term Structures Arising from Linear SDE Models

A short rate model satisfying an SDE of the form (7.16) has an affine term structure if

$$P(r, t, T) = e^{A(t,T)-B(t,T)r} \tag{7.18}$$

for some sufficiently smooth deterministic functions A and B, where $P(r, t, T)$ denotes the price at time $t \in [0, T]$ of a zero-coupon bond with maturity T, unit face value, and where r is the current observed value of the short rate at time t. Given the dependence on r we will, from now on, include it as an argument of the function P.

Consider a risk-neutral model of the short rate satisfying (7.16) with drift and diffusion coefficients satisfying

$$m(r, t) = a(t)+b(t)r; \quad \sigma^2(r, t) = c(t)+d(t)r, \quad r \in \mathbb{R}, \quad t \in [0, T], \tag{7.19}$$

for some deterministic functions a, b, c, d. We can substitute these choices of m and σ^2 back into the PDE (7.14) with $V(r, t) = P(r, t, T)$ to get, for any $r \in \mathbb{R}$,

$$P_t(r, t, T) + \frac{1}{2}[c(t) + d(t)r]P_{rr}(r, t, T)$$

$$+ [a(t) + b(t)r]P_r(r, t, T) - rP(r, t, T) = 0, \quad t \in [0, T]. \tag{7.20}$$

If we proceed according to the assumption that (7.18) holds, we have

$$P_t(r, t, T) = [A_t(t, T) - B_t(t, T)r]P(r, t, T);$$

$$P_r(r, t, T) = -B(t, T)P(r, t, T);$$

$$P_{rr}(r, t, T) = B^2(t, T)P(r, t, T).$$

Substituting back into (7.20) and dividing through by $P(r, t, T) > 0$ leads to the equation

$$[A_t(t, T) - B_t(t, T)r] + \frac{1}{2}[c(t) + d(t)r]B^2(t, T)$$

$$- [a(t) + b(t)r]B(t, T) - r = 0, \quad t \in [0, T]. \tag{7.21}$$

which must hold for all $r \in \mathbb{R}$. Therefore the sum of all terms depending on r must be zero, as must be the sum of all terms that do not depend on r. Gathering the terms that depend on r and equating to zero gives

$$-r B_t(t, T) + \frac{1}{2} d(t) r B^2(t, T) - b(t) r B(t, T) - r = 0. \tag{7.22}$$

The equation holds trivially when $r = 0$. Suppose that $r \neq 0$. Dividing though (7.22) by $-r \neq 0$ leads to the following, known as a *Riccati equation*:

$$B_t(t, T) - \frac{1}{2} d(t) B^2(t, T) + b(t) B(t, T) + 1 = 0. \tag{7.23}$$

Gathering the terms that do not depend on r leads to a second equation

$$A_t(t, T) + \frac{1}{2} c(t) B^2(t, T) - a(t) B(t, T) = 0. \tag{7.24}$$

The final-time condition associated with the bond is

$$1 = P(r, T, T) = e^{A(T,T) + B(T,T)r}, \quad r > 0,$$

or equivalently,

$$A(T, T) + B(T, T)r = 0, \quad r > 0.$$

Therefore we can associate the final-time conditions $B(T, T) = 0$ to (7.23) and $A(T, T) = 0$ to (7.24).

If the functions a, b, c, d that characterise a linear risk-neutral model of the short rate satisfying (7.16) are known, then we can determine the bond value $P(r, t, T)$ according to (7.18) by solving (7.23) first for B, followed by (7.24) for A.

We can then construct the term structure produced by the short rate as follows. Define the time-to-maturity $\tau := T - t$. Then making use of (7.18) and Definition 7.3, the spot rate is given by

$$R(t, t + \tau) = -\frac{\ln(P(r(t), t + \tau))}{\tau} = -\frac{A(t, t + \tau)}{\tau} + \frac{B(t, t + \tau)}{\tau} r(t). \tag{7.25}$$

The term structure associated with a specific model of r may then be constructed by setting $t = 0$ in (7.25) and computing $R(0, \tau)$ as a function of τ for varying $r(0) = r_0$. We do this explicitly for the Vasicek model in the next section.

7.4.2   Linear SDE Models of the Short Rate

There are several commonly used SDE models for the interest rate r evolving in time over the interval $[0, T]$. Let $(\Omega, \mathcal{F}, (\mathcal{F}_t)_{t \in [0,T]}, \mathbb{Q})$ be a filtered probability space where $\mathbb{Q}$ is the risk neutral measure constructed in Sect. 7.3.3, and $W^{\mathbb{Q}}$ is a standard Brownian motion under the measure $\mathbb{Q}$ adapted to the filtration $(\mathcal{F}_t)_{t \in [0,T]}$.

7.4.2.1 The Class of General Gaussian Markov Processes

Stochastic processes that satisfy an SDE of the form

$$dr(t) = [g(t) + h(t)r(t)]\, dt + \sigma(t)dW^{\mathbb{Q}}(t), \quad t \in [0, T], \tag{7.26}$$

where g, h, σ are deterministic functions, are known as *general Gaussian Markov processes*, and solutions of (7.26) may be written

$$r(t) = e^{\bar{h}(0,t)}r(0) + \int_0^t e^{\bar{h}(s,t)}g(s)ds$$

$$+ \int_0^t e^{\bar{h}(s,t)}\sigma(s)dW^{\mathbb{Q}}(s), \quad t \in [0, T], \tag{7.27}$$

where $\bar{h}(s, t) = \int_s^t h(v)dv$ for all $0 \leq s \leq t \leq T$.

Examples of commonly-used models that fall into this class are

1. The model of *Ho-Lee*, characterised within this class of models by a state-independent drift coefficient, describes the short rate r as the SDE

$$dr(t) = g(t)dt + \sigma dW^{\mathbb{Q}}(t), \tag{7.28}$$

   where g is a deterministic function of time selected to produce a term structure consistent with an observed set of bond prices.
2. The model of *Vasicek*, characterised within this class of models by a mean-reverting drift coefficient, describes the short rate as the SDE

$$dr(t) = \alpha(b - r(t))dt + \sigma dW^{\mathbb{Q}}(t), \quad t \in [0, T], \tag{7.29}$$

   where $\alpha, b, \sigma \in \mathbb{R}^+$.
3. The model of *Hull-White*, which is a generalisation of the Vasicek model with time-inhomogeneous linear coefficients. It describes the short rate as the SDE

$$dr(t) = \alpha(t)(b(t) - r(t))dt + \sigma(t)dW^{\mathbb{Q}}(t), \quad t \in [0, T], \tag{7.30}$$

   where $\alpha, b, \sigma : [0, T] \to \mathbb{R}^+$ are deterministic functions.

7.4.2.2 The Hull-White Model: Solutions and Conditional Distributions

Suppose that the short rate r satisfies the SDE (7.30). Applying (7.27) with $g(t) = \alpha(t)b(t)$, $h(t) = -\alpha(t)$ yields, for any $u \in [0, t)$,

$$r(t) = e^{-\int_u^t \alpha(v)dv} r(u)$$

$$+ \int_u^t e^{-\int_s^t \alpha(v)dv} \alpha(s)b(s)ds + \int_u^t e^{-\int_s^t \alpha(v)dv} \sigma(s)dW^{\mathbb{Q}}(s).$$

Let $0 \leq u < t \leq T$. Since the Itô integral of a deterministic function is normally distribution with mean and variance given by (1.9) in Chap. 1, the distribution of $r(t)$ conditional upon an observed value of $r(u)$ is

$$(r(t)|r(u) = r_u) \sim \mathcal{N}\left(\mu_r(u, t), \sigma_r^2(u, t)\right), \quad 0 \leq u < t \leq T, \tag{7.31}$$

where the mean is given by

$$\mu_r(u, t) = e^{-\int_u^t \alpha(v)dv} r(u) + \int_u^t e^{-\int_s^t \alpha(v)dv} \alpha(s)b(s)ds, \quad 0 \leq u < t \leq T, \tag{7.32}$$

and the variance by

$$\sigma_r^2(u, t) = \int_u^t e^{-2\int_s^t \alpha(v)dv} \sigma^2(s)ds, \quad 0 \leq u < t \leq T. \tag{7.33}$$

7.4.2.3 The Vasicek Model: Long-Run Behaviour and Sampling

Suppose now that the short rate satisfies the SDE (7.29), so that $\alpha(t) \equiv \alpha \in \mathbb{R}^+$, $b(t) = b \in \mathbb{R}^+$, and $\sigma(t) = \sigma \in \mathbb{R}^+$. Then, for $t \in [0, T]$,

$$r(t) = e^{-\alpha(t-u)} r(u) + \alpha \int_u^t e^{-\alpha(t-s)} b ds + \sigma \int_u^t e^{-\alpha(t-s)} dW^{\mathbb{Q}}(s)$$

$$= e^{-\alpha(t-u)} r(u) + b(1 - e^{-\alpha(t-u)}) + \sigma \int_u^t e^{-\alpha(t-s)} dW^{\mathbb{Q}}(s). \tag{7.34}$$

The distribution of $r(t)$, conditional upon an observed value of $r(u)$, denoted r_u, is again normal and characterised by (7.31) with

$$\mu_r(u, t) = e^{-\alpha(t-u)} r_u + b(1 - e^{-\alpha(t-u)});$$

$$\sigma_r^2(u, t) = \frac{\sigma^2}{2\alpha}\left(1 - e^{-2\alpha(t-u)}\right), \tag{7.35}$$

for $0 \leq u < t \leq T$. Exercise 7.5 asks you to confirm this. Since the mean and variance can be expressed in closed form for the Vasicek model and the coefficients are specified, we can determine the long-run statistical dynamics directly. Allowing for $T = \infty$, the constant $b > 0$ is the long-run mean of the process:

$$\lim_{t \to \infty} \mathbb{E}_{\mathbb{Q}}[r(t)] = \lim_{t \to \infty} \mu_r(0, t) = \lim_{t \to \infty} \left[e^{-\alpha t} r(0) + (1 - e^{-\alpha t}) b \right] = b.$$

In the same way, the variance converges to a limit in the long-run:

$$\lim_{t \to \infty} \mathrm{Var}_{\mathbb{Q}}[r(t)] = \lim_{t \to \infty} \sigma^2(0, t) = \lim_{t \to \infty} \frac{\sigma^2}{2\alpha}(1 - e^{-2\alpha t}) = \frac{\sigma^2}{2\alpha}.$$

Indeed, since a continuous limit of normal distributions is also a normal distribution, r converges in distribution to a normal random variable with mean b and variance $\sigma^2/2\alpha$.

Over an arbitrary mesh $0 = t_0 < t_1 < \cdots < t_N = T$ we can also use the exact conditional distribution of the Vasicek model to produce an ensemble of trajectories and examine its long-term behaviour using the relation

$$r(t_{i+1}) = e^{-\alpha(t_{i+1}-t_i)} r(t_i) + b\alpha(1 - e^{-\alpha(t_{i+1}-t_i)})$$

$$+ \sigma_r(t_i, t_{i+1}) Z_{i+1}, \quad i = 0, \ldots, N - 1,$$

where each Z_i is i.i.d. $\mathcal{N}(0, 1)$.

Figure 7.4 shows an ensemble of simulated trajectories of a Vasicek interest rate model generated using the following code:

```
In []: # Model and simulation parameters
       alpha=0.4; b=0.05; sigma=0.015; r0=0.01
       T=20; N=2000; dt=T/N; M=5

       # Function definitions
       def m(u,t):
           return b*(1-np.exp(-alpha*(t-u)))

       def sig(u,t):
           sigAlpha=(sigma**2)/(2*alpha)
           return np.sqrt(sigAlpha*(1-np.exp(-2*alpha*(t-u))))

       # Instantiate array for model values and set up time set
       r=r0*np.ones((M,len(t)))
       t=np.linspace(0,T,N+1)

       # Produce a trajectory
       Z=rng.normal(0,1,(M,len(t)-1))
```

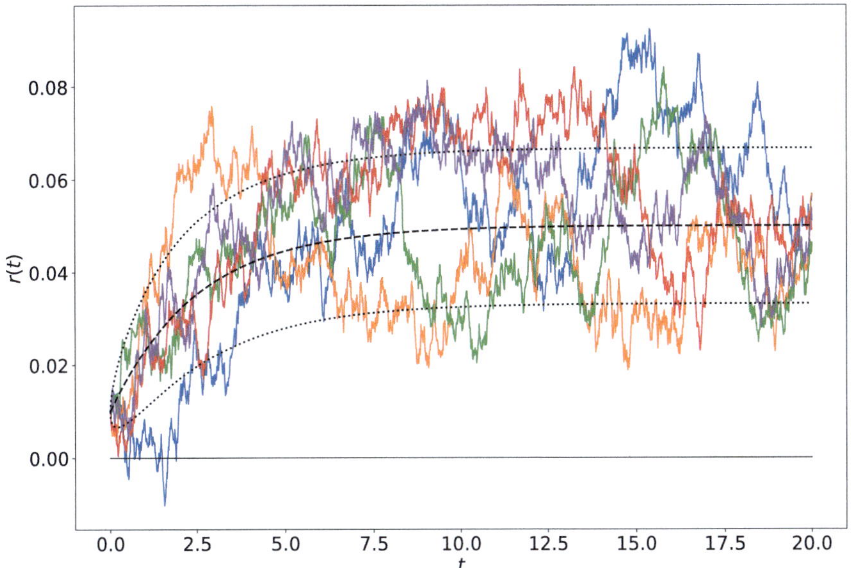

Fig. 7.4 An ensemble of $M = 5$ simulated trajectories of a Vasicek interest rate model with $\alpha = 0.4$, $b = 0.05$, $\sigma = 0.015$, $r_0 = 0.01$. Each trajectory is sampled over the interval $[0, 20]$ using a uniform mesh with $N = 2000$ steps. A dashed black line tracks the theoretical mean of the process and dotted lines mark upper and lower values that are one theoretical standard deviation from the mean in each direction

```
for i in range(len(t)-1):
    r[:,i+1]=np.exp(-alpha*(t[i+1]-t[i]))*r[:,i]
                +m(t[i],t[i+1])+sig(t[i],t[i+1])*Z[:,i]
```

Notice the loss of non-negativity in one of the trajectories, and convergence to the long term limits of both mean and standard deviation.

7.4.2.4 The Vasicek Model: Bond Pricing and Term Structure

When r satisfies (7.29), at time $t \in [0, T]$ the price of a zero-coupon bond maturing at time T with face value $P(T, T) = 1$ is given by (7.18) with

$$B(t, T) = \frac{1}{\alpha}\left(1 - e^{-\alpha(T-t)}\right);$$

$$A(t, T) = (B(t, T) - (T - t))\left(b - \frac{\sigma^2}{2\alpha^2}\right) - \frac{\sigma^2}{4\alpha}B(t, T)^2.$$

$$(7.36)$$

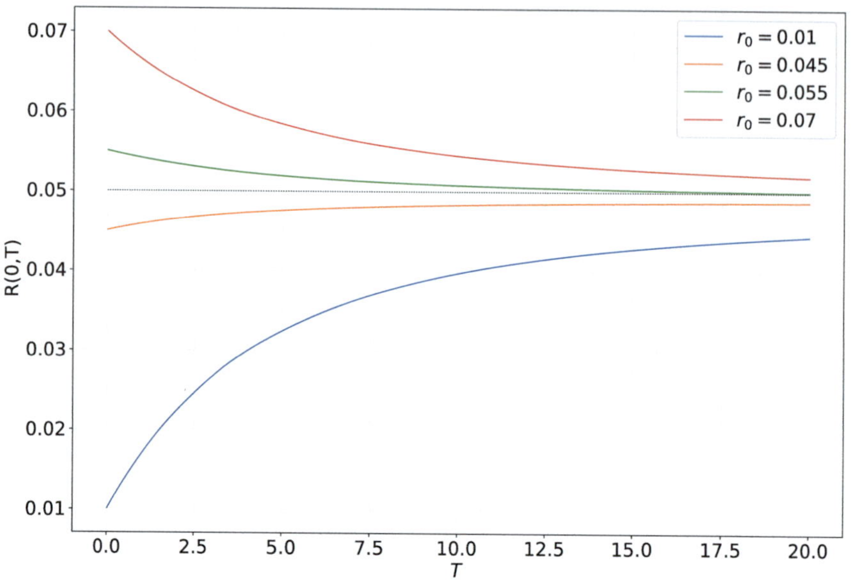

Fig. 7.5 Yield curves, parameterised in r_0, describing a term-structure of interest rates for bonds valued under the Vasicek model. Parameters are as in Fig. 7.4. A dotted line marks the value of b

The term structure generated by the Vasicek model is illustrated in Fig. 7.5, using the same model parameters as those used in Fig. 7.4. Part 2 of Exercise 7.6 asks you to confirm the form of A and B above. Exercise 7.9 asks you to produce a similar illustration using the technique described in Sect. 7.4.1.

7.4.2.5 The Ho-Lee Model: Calibration to an Initial Forward Curve

Suppose that the stochastic dynamics of r are governed by the Ho Lee model (7.28). Exercise 7.7 asks you to show that the price of a zero-coupon bond under this model satisfies

$$P(0, T) = \mathbb{E}\left[e^{-\int_0^T r(u)du}\right] = \exp\left(-r(0)T - \int_0^T \int_0^u g(s)dsdu + \frac{\sigma^2 T^3}{6}\right). \tag{7.37}$$

We are interested in choosing g so that this relation is satisfied for a set of observed values of $P(0, T)$. By (7.12) we can write

$$P(0, T) = e^{-\int_0^T f(0,t)dt}, \tag{7.38}$$

where $f(0, t)$ is the instantaneous forward rate for time t at time 0. Equate (7.37) and (7.38) and take logarithms on both sides to get

$$\int_0^T f(0, t)dt = r(0)T + \int_0^T \int_0^u g(s)dsdu - \frac{\sigma^2 T^3}{6}.$$

Differentiate both sides twice with respect to T:

$$\frac{\partial}{\partial T} f(0, T) = g(T) - \sigma^2 T.$$

Substitute T with t and make $g(t)$ the subject to get

$$g(t) = \left.\frac{\partial}{\partial T} f(0, T)\right|_{T=t} + \sigma^2 t.$$

Therefore the Ho-Lee model calibrated to the initial forward curve is given by

$$dr(t) = \left[\left.\frac{\partial}{\partial T} f(0, T)\right|_{T=t} + \sigma^2 t\right] dt + \sigma dW^{\mathbb{Q}}(t).$$

To simulate Ho-Lee on the grid $0 = T_0 < T_1 < \cdots < T_N = T$ made up of dates for which we can determine each value of $f(0, T_i)$, we write the SDE (7.28) in integral form over the step $[T_i, T_{i+1}]$:

$$r(T_{i+1}) = r(T_i) + \int_{T_i}^{T_{i+1}} \left(\left.\frac{\partial}{\partial T} f(0, T)\right|_{T=s} + \sigma^2 s\right) ds + \sigma (W^{\mathbb{Q}}(T_{i+1}) - W^{\mathbb{Q}}(T_i))$$

$$= r(T_i) + [f(0, T_{i+1}) - f(0, T_i)] + \frac{\sigma^2}{2}(T_{i+1}^2 - T_i^2)$$

$$+ \sigma (W^{\mathbb{Q}}(T_{i+1}) - W^{\mathbb{Q}}(T_i)),$$

for $i = 0, \ldots, N-1$, making use of the Fundamental Theorem of Calculus to ensure that no integration (exact or approximate) of the drift coefficient g is necessary.

Figure 7.6 shows an ensemble of $M = 50$ simulated trajectories of the Ho Lee interest rate model calibrated to the initial forward curve with values represented in Table 7.3, and with a typical choice of σ. To avoid specifying g we have chosen not to interpolate over a finer mesh than is induced by the observed instantaneous forward values. Notice again the loss of non-negativity in several trajectories. Since this model has a drift coefficient which is usually positive (except possibly in the case of an inverted yield curve), and which is not mean-reverting, trajectories may grow quite large over long periods of time. This is unrealistic for an interest rate model.

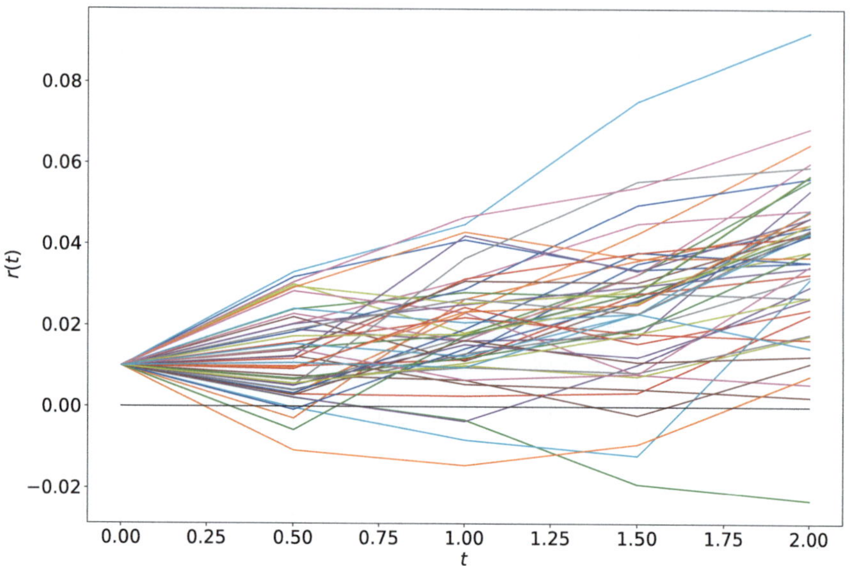

Fig. 7.6 An ensemble of $M = 50$ simulated trajectories of the Ho Lee interest rate model with $r_0 = 0.01$ and $\sigma = 0.015$ with drift coefficient g calibrated to values of the initial instantaneous forward curve displayed in Table 7.3

7.4.3 A Nonlinear SDE Model of the Short Rate: CIR

All general Gaussian Markov processes suffer from a common disadvantage as models of the short rate: they allow for solutions to become negative with positive probability. For example, consider the Vasicek model with stochastic dynamics under the risk-neutral measure $\mathbb{Q}$ governed by (7.29). Conditional upon the observation $r(u) = r_u$,

$$(r(t)|r(u) = r_u) \sim \mathcal{N}\left(\mu_r(u, t), \sigma_r^2(u, t)\right)$$

at times $0 \leq u < t \leq T$, where μ_r and σ_r are given by (7.35). Figure 7.7 shows the conditional probability $\mathbb{Q}\left[r(t) < 0|r(u) = r_u\right]$ as a function of r_u and the time interval $t - u$. We observe that the probability of the model returning a negative value is highest when r_u is close to zero and $t - u$ is large. The effect is more pronounced, and the probability of a negative model value higher, when σ is increased from $\sigma = 0.015$ (left) to $\sigma = 0.03$ (right). The other model parameters α and b are the same as those used to generate the trajectories in Fig. 7.4.

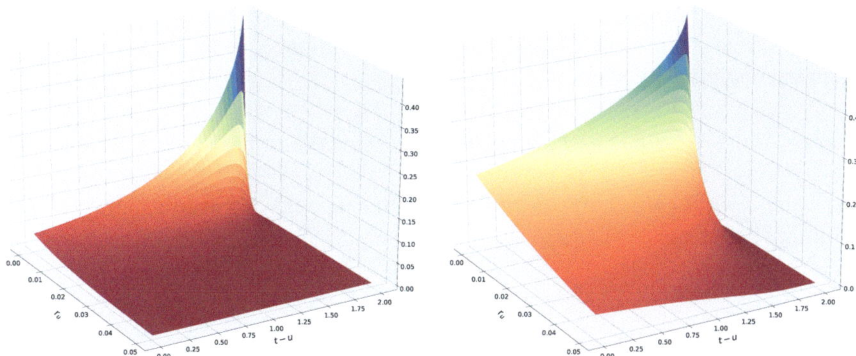

Fig. 7.7 A surface plot showing the conditional probability $\mathbb{Q}\,[r(t) < 0 | r(u) = r_u]$ (vertical axis) as a function of $r_u \in (0, 0.05]$ and $t - u \in (0, 2]$ for a Vasicek interest rate model with $\alpha = 0.4$, $b = 0.05$, $\sigma = 0.015$ (left) and $\sigma = 0.03$ (right)

An alternative SDE model that preserves the non-negativity of interest rates is the *Cox-Ingersoll-Ross* model:

$$dr(t) = \lambda(\mu - r(t))dt + \sigma\sqrt{r(t)}dW^{\mathbb{Q}}(t), \quad t \in [0, T], \quad r(0) = r_0 > 0, \tag{7.39}$$

where $W^{\mathbb{Q}}$ is a standard Brownian motion under the risk neutral measure, and $\lambda, \mu, \sigma > 0$. The CIR model (7.39) has a mean-reverting drift coefficient, just as the Vasicek model does, but it has a diffusion coefficient that depends in a nonlinear way on the current state $r(t)$. The standard theory for existence and uniqueness of solutions to SDEs does not apply to the SDE (7.39), since the diffusion coefficient $g(r) = \sigma\sqrt{r}$ does not satisfy even a local Lipschitz condition. However unique solutions of (7.39) exist and are a.s. non-negative.

In general, trajectories of (7.39) can achieve a value of zero but are reflected back into the positive half of the real line immediately. Further, when the *Feller condition*

$$2\lambda\mu > \sigma^2 \tag{7.40}$$

is satisfied, solutions are a.s. positive. The original article proposing (7.39) as a short rate model is [18].

7.4.3.1 The Non-central Chi-Squared Distribution

It is not possible to express the solutions of (7.39) in closed form in terms of the Brownian motion $W^{\mathbb{Q}}$. However, the distribution of $r(t)$ is known to be closely related to the following class of random variables.

Definition 7.9 For $\delta \in \mathbb{N}$, let $X_1, X_2, \ldots, X_\delta$ be independent normal random variables, each with mean $\mathbb{E}[X_i] = \mu_i$ and unit variance, so that $X_i \sim \mathcal{N}(\mu_i, 1)$. Then the random variable

$$Y := X_1^2 + X_2^2 + \cdots + X_\delta^2$$

has a *non-central chi-squared distribution* with δ degrees of freedom and with non-centrality parameter

$$\kappa = \mu_1^2 + \mu_2^2 + \cdots + \mu_\delta^2.$$

We write

$$Y \sim \chi^2(\delta, \kappa).$$

The CDF of Y is given by

$$F_{\chi^2(\delta, \kappa)}(y) = e^{-\kappa/2} \sum_{j=1}^{\infty} \left(\frac{(\kappa/2)^j}{j!} \right) \left(\frac{\gamma(j + \delta/2, y/2)}{\Gamma(j + \delta/2)} \right), \tag{7.41}$$

with lower incomplete Gamma function γ, and Gamma function Γ, defined respectively as

$$\gamma(a, z) := \int_0^z t^{a-1} e^{-t} dt; \quad \Gamma(z) := \int_0^\infty t^{z-1} e^{-t} dt.$$

The PDF of Y is given by

$$f_{\chi^2(\delta, \kappa)}(y) = \frac{1}{2} e^{-(y+\kappa)/2} \left(\frac{y}{\kappa} \right)^{(\delta/2-1)/2} \mathcal{B}_{\delta/2-1}(\sqrt{\kappa y}), \tag{7.42}$$

where

$$\mathcal{B}_a(z) := \left(\frac{z}{2} \right)^a \sum_{j=0}^{\infty} \frac{(z^2/4)^j}{j! \Gamma(a + j + 1)},$$

is a modified Bessel function of the first kind.

7.4.3.2 The Lamperti Transform Applied to the CIR Model

We will motivate the role of the non-central chi-squared distribution for a special case of the CIR model by applying a special transformation that converts the SDE (7.39), which has a diffusion coefficient $g(r) = \sigma \sqrt{r}$ that depends in a nonlinear way on the state value, to one which is independent of the state value.

Definition 7.10 The *Lamperti transform* for a general SDE of the form

$$dr(t) = \mu(r(t), t)dt + \sigma(r(t), t)dW^{\mathbb{Q}}(t); \quad r(0) = r_0,$$

is given by the indefinite integral

$$L(r) = \int^r \frac{1}{\sigma(x, t)} dx. \tag{7.43}$$

Exercise 7.10 asks you to verify, in the time-homogeneous case where $\sigma(x, t) = \sigma(x)$, that the process produced by the transform is governed by an SDE with state-independent diffusion coefficient.

In the case of the CIR model given by (7.39), we have $\sigma(r, t) = \sigma\sqrt{r}$ and therefore, for some constant of integration $c \in \mathbb{R}$,

$$L(r) := \frac{1}{\sigma} \int^r \frac{1}{\sqrt{x}} dx = \frac{2}{\sigma}\sqrt{r} + c.$$

To achieve the desired transformation it is sufficient to set $c = 0$, disregard the constant factor $2/\sigma$, and define

$$s(t) := \sqrt{r(t)}, \quad t \in [0, T].$$

Now set

$$\alpha := \frac{4\lambda\mu - \sigma^2}{8}, \quad \beta := \frac{\lambda}{2}, \quad \text{and} \quad \gamma := \frac{\sigma}{2}. \tag{7.44}$$

An application of Itô's formula yields the SDE

$$ds(t) = \left(\frac{\alpha}{s(t)} - \beta s(t)\right) dt + \gamma dW^{\mathbb{Q}}(t), \quad t \in [0, T], \tag{7.45}$$

with initial value $s(0) = s_0 = \sqrt{r_0}$. Notice that, when $\alpha \neq 0$, the SDE (7.45) is only well-defined if $s(t)$ is strictly positive, i.e. if the Feller condition (7.40) also holds, due to the inverse term in the drift. Eq. (7.45) in its turn may be written in variation of constants form as

$$s(t) = e^{-\beta t} s_0 + \int_0^t e^{-\beta t} \frac{\alpha}{s(u)} du + \gamma \int_0^t e^{-\beta(t-u)} dW^{\mathbb{Q}}(u).$$

The Feller condition as given in (7.40) may be expressed in terms of the transformed parameters as $2\alpha > \gamma^2$.

Suppose instead that $\alpha = 0$, so that $\lambda\mu = \sigma^2/4$ (equivalently stated as $4\lambda\mu/\sigma^2 = 1$). In this case $s(t)$ is governed by the Ornstein-Uhlenbeck process

$$ds(t) = -\beta s(t)dt + \gamma dW^{\mathbb{Q}}(t)$$

$$= -\frac{\lambda}{2}s(t)dt + \frac{\sigma}{2}dB(t), \quad t \in [0, T].$$

For all $\lambda, \sigma > 0$, $s(t)$ is a well-defined Gaussian Markov process of the form (7.26) with $g(t) \equiv 0$, $h(t) \equiv -\lambda/2$, $\sigma(t) = \sigma$ and solution via (7.27) satisfying

$$s(t) = s(u)e^{-\lambda(t-u)/2} + \frac{\sigma}{2}\int_u^t e^{-\lambda(t-s)/2}dW^{\mathbb{Q}}(s), \quad 0 \le u < t \le T. \qquad (7.46)$$

Since $r(t) = s^2(t)$, $t \in [0, T]$, it follows that, in this special case,

$$r(t) = \left[r_u^{1/2}e^{-\lambda(t-u)/2} + \frac{\sigma}{2}\int_u^t e^{-\lambda(t-s)/2}dW^{\mathbb{Q}}(s)\right]^2, \quad 0 \le u < t \le T.$$

From (7.46) and (1.9) in Chap. 1, we see that, conditional upon an observation $s(u) = s_u$, for $0 \le u < t \le T$,

$$s(t) \sim \mathcal{N}\left(\mu_s(u, t), \sigma_s^2(u, t)\right),$$

where the mean and variance satisfy

$$\mu_s(u, t) = s_u e^{-\lambda(t-u)/2};$$

$$\sigma_s^2(u, t) = \frac{\sigma^2}{4}\int_u^t e^{-\lambda(t-s)}ds = \frac{\sigma^2}{4\lambda}\left(1 - e^{-\lambda(t-u)}\right).$$

Now divide by the standard deviation to create the new random variable

$$X_u(t) := \frac{1}{\sigma_s(u, t)}s(t) \sim \mathcal{N}\left(\frac{\mu_s(u, t)}{\sigma_s(u, t)}, 1\right), \quad 0 \le u < t \le T.$$

By Definition 7.9 the square of this random variable has a non-central chi-squared distribution with $\delta = 1$ degrees of freedom and non-centrality parameter $\kappa = \mu_s^2(u, t)/\sigma_s^2(u, t)$, written

$$X_u^2(t) \sim \chi^2\left(1; \frac{\mu_s^2(u, t)}{\sigma_s^2(u, t)}\right).$$

Therefore in the case where $4\lambda\mu/\sigma^2 = 1$, the distribution of the CIR process $r(t)$ at any $t \in [0, T]$ may be expressed as a scaling of $X_u^2(t)$ as follows:

$$r(t) = \sigma_s^2(u, t)X_u^2(t) = \frac{\sigma^2}{4\lambda}\left(1 - e^{-\lambda(t-u)}\right)X_u^2(t)$$

$$= \mu\left(1 - e^{-\lambda(t-u)}\right)X_u^2(t), \quad 0 \le u < t \le T.$$

In fact it is usual to define a normalising term

$$c(u, t) := \frac{4\lambda}{\sigma^2(1 - e^{-\lambda(t-u)})}$$

and write, when $r(u) = r_u = s_u^2$ is known,

$$r(t) = \frac{X_u^2(t)}{c(u, t)}, \quad 0 \le u < t \le T.$$

7.4.3.3 The Conditional Distribution of the CIR Model

The characterisation of the conditional distribution of $r(t)$ provided in the last subsection in the special case where $4\lambda\mu/\sigma^2 = 1$ generalises in a natural way to all other parameter combinations. The main change is in the degrees of freedom parameter δ.

Proposition 7.11 *Let $r(t)$ satisfy the SDE (7.39) for any $\lambda, \mu, \sigma > 0$ and initial value $r(0) = r_0 > 0$. If we have observed the value of $r(u) = r_u$ for some time $u \in [0, T]$, then for $u < t \le T$, we may write*

$$(r(t)|r(u) = r_u) \sim \frac{X_u^2(t)}{c(u, t)}, \quad \text{where} \quad c(u, t) = \frac{4\lambda}{\sigma^2(1 - e^{-\lambda(t-u)})}, \tag{7.47}$$

and $X_u^2(t) \sim \chi^2(\delta, \kappa(u, t))$ is a non-central chi-squared distributed random variable with degrees of freedom and non-centrality parameters

$$\delta = \frac{4\lambda\mu}{\sigma^2}, \quad \kappa(u, t) = c(u, t)r_u e^{-\lambda(t-u)}, \tag{7.48}$$

respectively.

The CDF and PDF of a non-central chi-squared distributed random variable were given in Definition 7.9 and can be used to write the conditional CDF and PDF for solutions of (7.39).

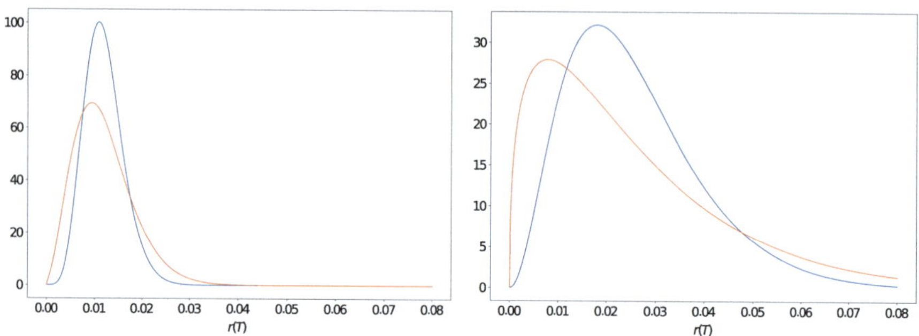

Fig. 7.8 Theoretical density plots of a CIR process for $r_0 = 0.01$, $\lambda = \mu = 0.05$, $\sigma = 0.04$ (blue lines) and $\sigma = 0.06$ (orange lines), $T = 1$ (left plot), $T = 10$ (right plot)

The conditional CDF of $r(t)$ satisfying (7.39) is

$$\mathbb{Q}\left[r(t) \leq x | r(u) = r_u\right] = \mathbb{Q}\left[X_u^2(t) \leq c(u, t) \cdot x \big| r(u) = r_u\right]$$
$$= F_{\chi^2(\delta, \kappa(u,t))}\left(c(u, t) \cdot x\right), \tag{7.49}$$

where $\mathbb{Q}$ is the risk-neutral measure and $F_{\chi^2(\delta, \kappa(t,u))}(\cdot)$ is the CDF defined by (7.41) in Definition 7.9 with δ and $\kappa = \kappa(u, t)$ satisfying (7.48).

The conditional PDF of $r(t)$ satisfying (7.39) is

$$\frac{\partial}{\partial x} F_{\chi^2(\delta, \kappa(t,u))}(c(u, t) \cdot x) = c(u, t) \cdot f_{\chi^2(\delta, \kappa(t,u))}\left(c(u, t) \cdot x\right),$$

where we have applied the chain rule to (7.49), and $f_{\chi^2(\delta, \kappa(t,u))}(\cdot)$ satisfies (7.42) in Definition 7.9, again with δ and $\kappa = \kappa(u, t)$ satisfying (7.48).

Density plots for random variables with this distribution are displayed in Fig. 7.8. For the blue curves, $\sigma = 0.04$, and for the orange curves, $\sigma = 0.06$. The Feller condition is satisfied in both cases and trajectories cannot achieve the reflecting boundary at zero in finite time.

7.4.3.4 The CIR Model: Long-Run Behaviour and Sampling

For $t \in [0, T]$, the mean and variance of solutions of (7.39) are given by

$$\mathbb{E}_{\mathbb{Q}}[r(t)] = (1 - e^{-\lambda t})\mu + r_0 e^{-\lambda t};$$

$$\text{Var}_{\mathbb{Q}}[r(t)] = r_0 \frac{\sigma^2}{\lambda}(e^{-\lambda t} - e^{-2\lambda t}) + \frac{\mu \sigma^2}{2\lambda}(1 - e^{-\lambda t})^2.$$

If we allow for $T = \infty$, then as t grows large we can see that

$$\lim_{t \to \infty} \mathbb{E}_{\mathbb{Q}}[r(t)] = \lim_{t \to \infty} r_0 e^{-\lambda t} + \lim_{t \to \infty} \mu(1 - e^{-\lambda t}) = \mu,$$

and

$$\lim_{t \to \infty} \mathrm{Var}_{\mathbb{Q}}[r(t)] = \lim_{t \to \infty} r_0 \frac{\sigma^2}{\lambda}(e^{-\lambda t} - e^{-2\lambda t}) + \lim_{t \to \infty} \frac{\mu \sigma^2}{2\lambda}(1 - e^{-\lambda t}) = \frac{\mu \sigma^2}{2\lambda}.$$

Just as for the Vasicek model, on an arbitrary mesh $0 = t_0 < t_1 < \cdots < t_N = T$ we can use the exact conditional distribution of the CIR model to produce observations on an ensemble of trajectories and examine its long term behaviour using the relation

$$r(t_{i+1}) = c(t_i, t_{i+1}) X_{t_i}^2(t_{i+1}), \quad i = 0, \ldots, N - 1,$$

where $c(t_i, t_{i+1})$ is given in (7.47) and

$$X_{t_i}^2(t_{i+1}) \sim \chi^2(\delta, \kappa(t_i, t_{i+1})); \quad \delta = \frac{4\lambda \mu}{\sigma^2}, \ \kappa(t_i, t_{i+1}) = c(t_i, t_{i+1}) r(t_i) e^{-\lambda(t_{i+1} - t_i)}.$$

Exercise 7.12 asks you to use this relation to generate trajectories of r, and Fig. 7.9 displays them for parameter values where the Feller condition holds (top), and where the Feller condition does not hold. We see that when the Feller condition does not hold trajectories may spend considerably more time in the vicinity of zero, but nonetheless they do not cross over into negative values.

7.4.3.5 Bond Pricing and Affine Term Structure Under the CIR Model

When r satisfies (7.39), at time $t \in [0, T]$ the price of a zero-coupon bond maturing at time T with face value $P(T, T) = 1$ is given by (7.18) with

$$A(t, T) = \frac{2\lambda \mu}{\sigma^2} \ln \left(\frac{2\gamma e^{(\lambda + \gamma)(T - t)/2}}{(\lambda + \gamma)(e^{\gamma(T - t)} - 1) + 2\gamma} \right);$$

$$B(t, T) = \frac{2(e^{\gamma(T - t)} - 1)}{(\lambda + \gamma)(e^{\gamma(T - t)} - 1) + 2\gamma},$$

where $\gamma := \sqrt{\lambda^2 + 2\sigma^2}$. It is possible to verify by direct computation that A and B satisfy the ODE system (7.23) and (7.24) that arise from the Fundamental Theorem of Bond Pricing when (7.18) holds. A term structure generated by the CIR model is illustrated in Fig. 7.10. Qualitatively it is very similar to the term structure produced by the Vasicek model in Fig. 7.5.

7.4.3.6 CIR Processes and the Heston Model

The Cox-Ingersoll-Ross process also finds a role in a generalisation of the Black-Scholes model where the volatility is allowed to vary stochastically,   and the

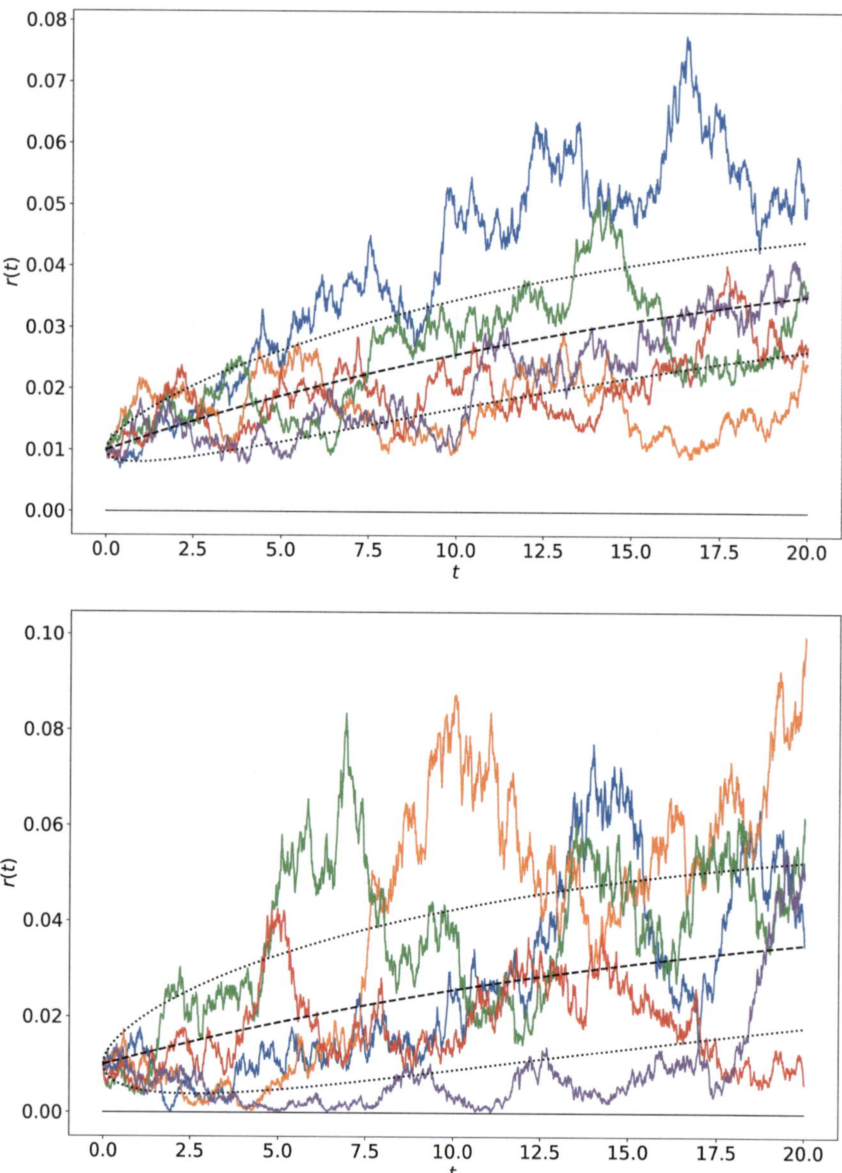

Fig. 7.9 An ensemble of $M = 5$ simulated trajectories of the CIR interest rate model with parameters $r_0 = 0.01$, $\lambda = \mu = 0.05$, $\sigma = 0.04$ (top, Feller condition holds) and $\sigma = 0.075$ (bottom, Feller condition does not hold). Each trajectory is sampled over the interval $[0, 20]$ using a uniform mesh with $N = 2000$ steps. A dashed black line tracks the theoretical mean of the process and dotted lines mark upper and lower values that are one theoretical standard deviation from the mean in each direction

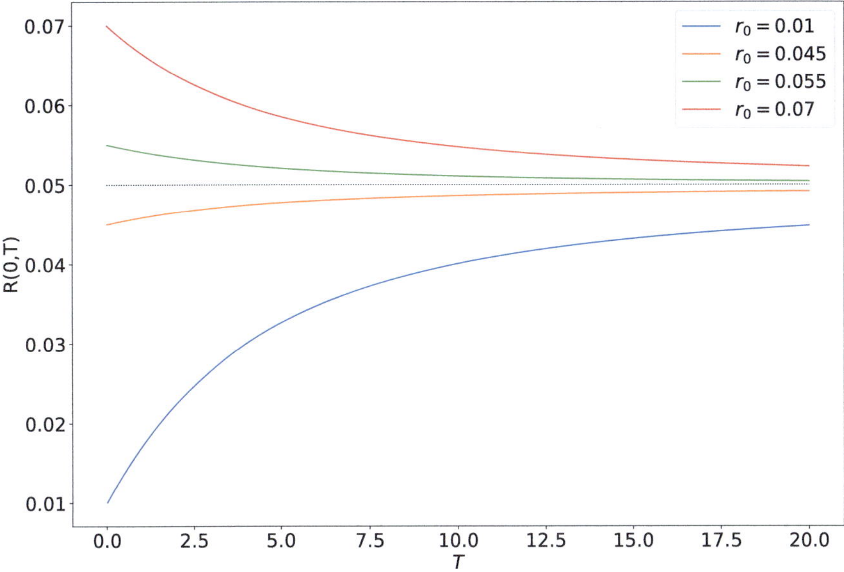

Fig. 7.10 Yield curves, parameterised in r_0, describing a term-structure of interest rates for bonds valued under the CIR model with maturity dates up to $T = 20$, $\lambda = 0.4$, $\mu = 0.05$, and $\sigma = 0.04$. A dotted line marks the value of μ

randomness driving the volatility may be correlated with the randomness driving the asset. Consider the *Heston model*:

$$dS(t) = r S(t)dt + \sqrt{V(t)}S(t)dW_1(t) \tag{7.50}$$

$$dV(t) = \lambda(\mu - V(t))dt + \sigma\sqrt{V(t)}dW_2(t), \tag{7.51}$$

for $t \in [0, T]$, $S(0) = S_0 > 0$, $V(0) = V_0 > 0$, and $W = [W_1, W_2]^T$ is a bivariate standard Brownian motion such that

$$W(t) \sim \mathcal{N}\left(0, \begin{pmatrix} 1 & \rho \\ \rho & 1 \end{pmatrix} t\right); \quad \rho \in (-1, 1), \quad t \in [0, T].$$

Here, the asset price process S is described by an SDE (7.50) that is linear in S, and the volatility process V is described by the SDE governing a CIR process with distribution that is characterised by Proposition 7.11 when not part of a coupled system of SDEs.

Nonetheless, it is not known how to sample from the exact solutions of the system (7.50) and (7.51) for the simple reason that, when sampling from solutions of the CIR process, we do so by sampling directly from the appropriate variant of the non-central chi-squared distribution, as described in Sect. 7.4.3.4. The joint distribution of $W_1(t)$ and $V(t)$, when $W_1(t)$ and $W_2(t)$ are correlated in this way, is unknown.

In order to compute values of S satisfying a Heston model, it is therefore necessary to generate numerical approximations of solutions of the SDE system (7.50) and (7.51), and to understand the nature of the additional discretisation error thus introduced. The numerical solution of systems of SDEs is the subject of our final chapter.

7.5 Sampling for the Monte Carlo Valuation of Interest Rate Derivatives

Consider the case where a stochastic interest rate r remains the risk-free rate for option valuation. Then

$$\beta(t) = e^{\int_0^t r(u)du}, \quad t \in [0, T], \tag{7.52}$$

will again be the value of the numeraire asset at time t. In order to use (7.17) with $t = 0$ and $S = T$ in the statement of Proposition 7.8 to produce a Monte Carlo valuation of an interest rate derivative with payoff $V_{r,T}$ depending on $r(T)$ at the expiry time T, we are interested in sampling not only from the conditional distribution of the risk-neutral form of $r(T)$ but also from

$$\frac{1}{\beta(T)} = e^{-Y(T)}, \quad Y(T) := \int_0^T r(u)du.$$

A possible computational approach to producing a single sample of the latter is to sample values $r(t_0, \omega), r(t_1, \omega), \ldots, r(t_N, \omega)$ along a single trajectory $r(\cdot, \omega)$, $\omega \in \Omega$, over some discretisation of the time set $\{t_0 = 0, t_1, \ldots, t_N = T\}$, and to compute the rectangular approximation

$$Y(T) \approx \sum_{j=0}^{N} r(t_j, \omega) \Delta t_j; \quad \Delta t_j = t_{j+1} - t_j.$$

By approximating the integral as a rectangular sum we introduce discretisation error over each trajectory: an additional source of error over and above that produced by a Monte Carlo approach. The nature of this kind of error in a stochastic setting is explored in Chap. 8.

However, if r is governed by a Gaussian model, the pair $(r(T), Y(T))$ is a bivariate random variable with joint normal distribution and it will be possible to use the techniques introduced in Sect. 6.3.1 of Chap. 6 to simulate without discretisation error, as long as we know the covariance of $r(T)$ and $Y(T)$. We will demonstrate this for the Vasicek model.

7.5.1   Joint Sampling of an Interest Rate Model and Its Discount Factor

Suppose that $r(t)$, $t \in [0, T]$, is described by the Vasicek model (7.29) so that by (7.35), given $r(u) = r_u$, $0 \leq u < t \leq T$, the conditional distribution of $r(t)$ is given by

$$(r(t)|r(u) = r_u) \sim N\left(e^{-\alpha(t-u)}r_u + m_r(u, t), \sigma_r^2(u, t)\right),$$

where

$$m_r(u, t) = b(1 - e^{-\alpha(t-u)}),$$

and

$$\sigma_r^2(u, t) = \frac{\sigma^2}{2\alpha}(1 - e^{-2\alpha(t-u)}).$$

Then it can be shown that $Y(t) = \int_0^t r(v)dv$ has conditional distribution, given $r(u) = r_u$ and $Y(u) = Y_u$

$$(Y(t)|r(u) = r_u, \, Y(u) = Y_u) \sim N\left(Y_u + m_Y(u, t), \sigma_Y^2(u, t)\right),$$

where

$$m_Y(u, t) = \frac{1}{\alpha}(r_u + b)(1 - e^{-\alpha(t-u)}) + b(t - u),$$

and

$$\sigma_Y^2(u, t) = \frac{\sigma^2}{\alpha^2}\left((t - u) + \frac{1}{2\alpha}\left(1 - e^{-2\alpha(t-u)}\right) + \frac{2}{\alpha}\left(e^{-\alpha(t-u)} - 1\right)\right).$$

Moreover the covariance of $r(t)$ and $Y(t)$ given the observed values of $r(u)$ and $Y(u)$ is given by

$$\sigma_{r,Y}(u, t) = \frac{\sigma^2}{2\alpha}\left(1 + e^{-2\alpha(t-u)} - 2e^{-\alpha(t-u)}\right),$$

with corresponding correlation

$$\rho_{r,Y}(u, t) = \frac{\sigma_{r,Y}(u, t)}{\sigma_r(u, t)\sigma_Y(u, t)}.$$

Define the linear correlation matrix

$$\Sigma_{r,Y}(u,t) = \begin{pmatrix} 1 & \rho_{r,Y}(u,t) \\ \rho_{r,Y}(u,t) & 1 \end{pmatrix},$$

so that, for fixed u and t, the Cholesky factor $A_{r,Y}(u,t) \in \mathbb{R}^{2\times2}$

$$A_{r,Y}(u,t)A_{r,Y}(u,t)^T = \Sigma_{r,Y}(u,t)$$

is given by

$$A_{r,Y}(u,t) = \begin{pmatrix} 1 & 0 \\ \rho_{r,Y}(u,t) & \sqrt{1 - \rho_{r,Y}^2(u,t)} \end{pmatrix}.$$

Then by adopting the approach described in Sect. 6.3.1 of Chap. 6 we can sample the bivariate Gaussian pair $(r(t), Y(t))$ given observations on $r(u) = r_u$ and $Y(u) = Y_u$ at times $0 \le u < t \le T$ by

$$\begin{pmatrix} r(t) \\ Y(t) \end{pmatrix} = \begin{pmatrix} e^{-\alpha(t-u)}r_u \\ Y_u \end{pmatrix} + \begin{pmatrix} m_r(u,t) \\ m_Y(u,t) \end{pmatrix} + \begin{pmatrix} \sigma_r(u,t) & 0 \\ 0 & \sigma_Y(u,t) \end{pmatrix} A_{r,Y}(u,t)Z,$$

where $Z = [Z_1, Z_2]^T$ is a pair of independent standard normal random variables. The discount factor can then be recovered as $e^{-Y(t)}$.

Figure 7.11 shows a scatterplot and marginal histograms of a sample of 500 observations on $(r(T), e^{-Y(T)})$ given $r_0 = 0.03$, $Y_0 = 0$, $\alpha = 0.4$, $b = 0.05$, $\sigma = 0.02$, and $T = 2$.

7.5.2 Pricing Interest Rate Derivatives Under the Forward Risk-Neutral Measure

An alternative approach that avoids sampling from a stochastic discount factor entirely, is to express the value of the interest rate derivative as an expectation under a different measure, called the *forward risk-neutral measure*, rather than the risk-neutral measure we have been using up to now.

Let $P(r, t, T)$ denote the price at time $t \in [0, T]$ of a zero-coupon bond with maturity T and face value $P(r, T, T) = 1$, for a given value of the short rate r. Let $V(r, t, T)$ denote the value of an interest rate derivative expiring on the maturity date of the bond (so that $S = T$). Suppose that the short rate dynamics under the risk neutral measure $\mathbb{Q}$ are governed by the SDE

$$dr(t) = m(r,t)dt + \sigma(r,t)dW^{\mathbb{Q}}(t), \quad t \in [0, T].$$

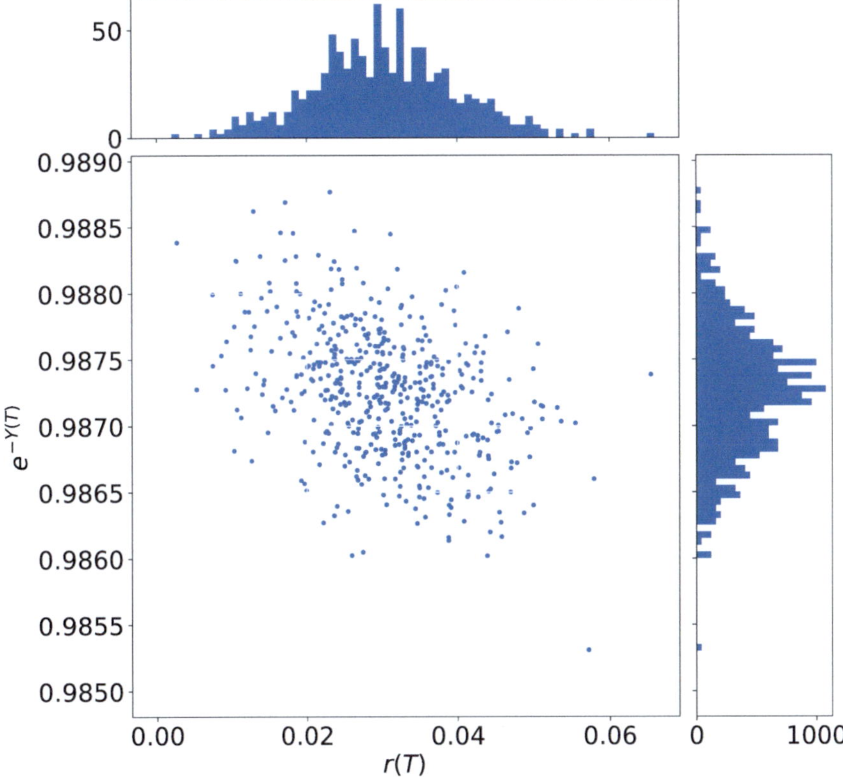

Fig. 7.11 500 observations on the bivariate random pair $\left(r(T), e^{-\int_0^T r(s)ds} \right)$ given $r_0 = 0.03$, $Y_0 = 0$, $\alpha = 0.4$, $b = 0.05$, $\sigma = 0.03$, and $T = 2$

It can be shown by an argument that makes use of Novikov's Theorem (see Theorem 2.6 in Chap. 2) and Girsanov's Theorem (see Theorem 2.7 in Chap. 2) that there exists an equivalent probability measure $\mathbb{Q}^F$ under which the time $t = 0$ value of the interest rate derivative can be expressed as

$$V(r, 0, T) = \mathbb{E}_{\mathbb{Q}} \left[e^{-\int_0^T r(u)du} V_{r,T} \right] = P(r, 0, T)\mathbb{E}_{\mathbb{Q}^F} \left[V_{r,T} \right], \tag{7.53}$$

The measure $\mathbb{Q}^F$ is the forward risk neutral measure, and the dynamics of the short rate under it may be written as the SDE

$$dr(t) = (m(r, t) + \sigma_P(r, t)\sigma(r, t))dt + \sigma(r, t)dW^{\mathbb{Q}^F}(t), \tag{7.54}$$

where $W^{\mathbb{Q}^F}$ is a standard Brownian motion under $\mathbb{Q}^F$ and

$$\sigma_P(r, t) = \frac{1}{P(r, t, T)} P_r(r, t, T) \sigma(r, t).$$

Therefore, as long as we can identify a bond with the same maturity date as the expiry date of the interest rate derivative to be valued, along with a characterisation of the first order partial derivative process $P_r(r, t, T)$, we can apply Monte Carlo methods to the expectation in (7.53) to value this derivative by sampling only from the distribution of r expressed according to the forward risk-neutral dynamics governed by the SDE (7.54). Eq. (7.53) generalises to allow for pricing at time-t as

$$V(r, t, T) = P(r, t, T)\mathbb{E}_{\mathbb{Q}^F}\left[V_{r,T}|\mathcal{F}_t\right], \quad t \in [0, T].$$

Full mathematical details are available in Choe [15].

7.6   Further Reading

We have considered issues arising in Monte Carlo valuation of bonds and interest rate derivatives with payoffs that depend on the short rate. A similar approach to interest rate derivatives with payoffs depending on the instantaneous forward rate (for example (7.15)) would rely upon an appropriate SDE model of forward-rate dynamics in an arbitrage-free setting. This is the basis of the Heath-Jarrow-Morton framework, but it is beyond the scope of this textbook. We point the reader instead to Chapter 3 of Glasserman [25], which also contains a treatment of the interpolation of Brownian paths that includes an additional technique called Principal Components Construction.

In any case, market prices for caplets/floorlets and caps/floors are often determined using Black's model, which gives a closed form solution for the value of European bond options, interest rate caps and floors, and European swaptions. The model was adapted from its original purpose of pricing options on commodity futures, and relies on assumptions that are not always realistic. Such derivatives may be also valued via a discrete no-arbitrage approach, which makes use of binomial tree and other lattice models. Both approaches are well-treated in Hull [36], and we refer interested readers there.

The original article by Cox, Ingersoll, and Ross on their model of the short rate is [18]. For readers interested in understanding the stochastic dynamics of the CIR process in more depth, the characterisation of the reflecting boundary at zero is related to Feller's test for explosions, which may be found in the book of Karatzas and Shreve [39], along with Bessel processes: a class of stochastic process connected to the Lamperti transform of SDEs with square-root diffusion. The NIST Handbook of Mathematical Functions [57] is an excellent reference for special functions such as the gamma and Bessel functions encountered in the characterisation of the non-central chi-squared distibution.

The fact that there are multiple possible SDE models for the short rate raises the question of which best reproduces the empirical density of observed rates, and this was examined by Aït-Sahalia [2], whose analysis led him to propose another, more general and eponymous, model which is presented at the end of this chapter in Exercise 7.11.

In Sect. 7.5 we looked at the joint sampling of a Vasicek interest rate model and the associated numeraire for risk-neutral pricing. For the CIR model joint sampling is not so straightforward and relies upon the bivariate Laplace transform of r. More detail is given in Glasserman [25], with the original work due to Scott [61].

7.6.1 Exercises

7.1 Consider a 1-year bond with face value $P(1, 1) = 100$ paying coupons semi-annually at 4% p.a.

1. What is the value of the bond at time $t = 0$, assuming spot rates as given in Table 7.4?
2. Compute the yield-to-maturity of the bond.

7.2 Consider again Table 7.4.

1. For each maturity date T (measured in years), compute the corresponding prices of a zero-coupon bond with face value $P(T, T) = 100$.
2. Calibrate the risk-neutral form of a Black-Scholes asset model with $S_0 = 1$ and $\sigma = 0.4$, and generate trajectories over the interval $[0, 2]$.
3. Use a Brownian bridge to interpolate these trajectories so that they evolve over a uniform mesh with $N = 24$ steps of length 1 month each.

7.3 Show that (7.3) may be rewritten in the form

$$S(T_{i+1}) = S_0 \frac{P(0, 0)}{P(0, T_{i+1})} e^{-\frac{1}{2} T_{i+1} \sigma^2 + \sigma W^{\mathbb{Q}}(T_{i+1})}, \quad i = 0, \ldots, N - 1,$$

and hence implement the calibration procedure in Sect. 7.1.2 without using a `for` loop.

Table 7.4 US Treasury par yield rates on December 1, 2023. Source of data: https://home.treasury.gov/policy-issues/financing-the-government/interest-rate-statistics

$T \times 12$ (months)	$R(0, T)$ (%)
1	5.55
2	5.53
3	5.43
4	5.45
6	5.33
12	5.05
24	4.56

7.4 Suppose that the short rate r is governed by the Vasicek model with stochastic dynamics given by the SDE (7.29). Use Itô's formula applied to the transformation $Y(r, t) = e^{\alpha t} r(t)$ to provide direct confirmation that r satisfies (7.34).

7.5 Demonstrate that the mean and variance of $r(t)$, where r satisfies (7.29), are as shown in (7.35).

7.6 Verify that A and B satisfy the ODE system (7.24) and (7.23) for bond pricing (with associated boundary conditions $A(T, T) = B(T, T) = 0$) under the Vasicek model: when A and B satisfy (7.36).

7.7 Prove that Eq. (7.37), giving the price of a zero-coupon bond under the Ho-Lee model, holds.

7.8 Consider again Table 7.4.

1. For each maturity date T (measured in years), compute the instantaneous forward rates $f(0, T)$.
2. Calibrate a Ho Lee interest rate model to the forward curve produced by interpolating these rates, and produce an ensemble of $M = 50$ trajectories when $r_0 = 0.01$ and $\sigma = 0.02$, each evolving on the nonuniform mesh induced by the set of maturity dates T.

7.9 In Python, plot yield curves demonstrating the term structure of interest rates under the Vasicek model as shown in Fig. 7.5. You may use the same values for α, b, σ, and r_0. Repeat for the CIR model.

7.10 Consider an SDE of the form

$$dr(t) = \mu(r(t))dt + \sigma(r(t))dW^{\mathbb{Q}}(t); \quad r(0) = r_0,$$

where σ is differentiable and strictly positive. Use Itô's formula to show that the SDE governing the process generated by applying the transformation

$$L(r) = \int^{r} \frac{1}{\sigma(x)} dx,$$

has a diffusion coefficient that is independent of r.

7.11 The Aït-Sahalia model of short rate dynamics is given by the SDE

$$dr(t) = \left(\alpha_{-1} \frac{1}{r(t)} - \alpha_0 + \alpha_1 r(t) - \alpha_2 r(t)^2 \right) dt$$

$$+ \sigma r(t)^{\theta} dW^{\mathbb{Q}}(t), \quad t \in [0, T] \quad (7.55)$$

where $r(0) = r_0 > 0$; $\alpha_{-1}, \alpha_0, \alpha_1, \alpha_2$ are positive constants; and $\theta > 1$. In the appendix of [2], the author sets out additional conditions on these constants to ensure that that unique global solutions of (7.55) exist.

Identify the Lamperti transform $L(r)$ of the SDE (7.55), and write down the SDE that governs its dynamics.

7.12 Consider a CIR process satisfying (7.39) with $\lambda = 0.05$, $\mu = 0.05$, and $\sigma = 0.02$, over the interval $[0, 1]$.

1. Write code that generates an ensemble of $M = 5$ trajectories, each on a uniform mesh with $N = 100$ steps. Use `rng.noncentral_chisquare(df,kappa)` to generate observations from a non-central chi-squared distrtibution, where `df` and `kappa` respectively specify the degrees of freedom and non-centrality parameters, along with Proposition 7.11.
2. For the values of λ and μ given above, identify the range of values of σ for which the Feller condition (7.40) fails to hold.
3. Use your code to plot trajectories for values of σ both where the Feller condition holds, and where it does not hold, and observe the output.

7.13 Under the Vasicek model, the price of a zero-coupon bond with face-value \$1 and maturity S at any time $t \in [0, S]$ is given by

$$P(r, t, S) = e^{A(t,S) - B(t,S)r}$$

where

$$B(t, S) = \frac{1}{\alpha} \left(1 - e^{-\alpha(S-t)} \right);$$

$$A(t, S) = (B(t, S) - (S - t)) \left(b - \frac{\sigma^2}{2\alpha^2} \right) - \frac{\sigma^2}{4\alpha} B(t, S)^2.$$

The price of a call option with strike K and expiry $T < S$ written on this bond is given by

$$\mathbb{E}\left[e^{-\int_0^T r(s)ds} \left(P(r, T, S) - K \right)^+ \right].$$

Using exact sampling of the bivariate random variable $\left(r(T), \int_0^T r(s)ds \right)$, produce a Monte Carlo valuation (with confidence interval) of this option when $r_0 = 0.03$, $K = 1$, $T = 0.5$, $S = 1$.

Simulation III: Numerical Approximation of SDE Models

8

Thus far when using Monte Carlo methods to sample from the distribution of a single asset price, a collection of d assets with correlated noise, or the bivariate distribution of an interest rate and its associated discount factor, we have started by sampling from the Gaussian distribution of one or more independent standard Brownian motions and building up the necessary model from there, allowing us to generate samples from the model distribution exactly. However, only in a few special cases is it possible to sample from the exact distribution of a financial process. To give one example, we saw in Chap. 7 that the Heston model for individual asset prices with stochastic volatility is represented by a 2-dimensional SDE system for which the solution has no closed-form in terms of the noise. Exact sampling simply isn't possible in such cases, and we turn to numerical approximation methods for SDE systems.

A critical notion of error in this context is known as *weak discretisation error*. Let P denote the discounted payoff of a European-style derivative $P(s) = e^{-rT} \Lambda(s)$. We know by now that if S describes the risk-neutral price evolution of the underlying asset, $\mathbb{E}[P(S(T))]$ is the time $t = 0$ value of the derivative. However, since the mean of the distribution of $P(S(T))$ is not known in general, we approximate $\mathbb{E}[P(S(T))]$ in two stages, first numerically solving for $S(T)$ on each trajectory ω over a mesh with N uniform steps by generating a discrete-time approximation $S_N(\omega) \approx S(T, \omega)$, and second using Monte Carlo sampling to approximate the expectation by averaging over the sampled trajectories:

$$\mathbb{E}[P(S(T))] \approx \mathbb{E}[P(S_N)] \approx \mu_M := \frac{1}{M} \sum_{i=1}^{M} P(S_N(\omega_i))$$

© The Author(s), under exclusive license to Springer Nature Switzerland AG 2024
C. Kelly, *Computation and Simulation for Finance*,
Springer Undergraduate Texts in Mathematics and Technology,
https://doi.org/10.1007/978-3-031-60575-8_8

The overall (signed) error of the approximation can be decomposed into *bias* and *statistical error* as follows:

$$\underbrace{\mathbb{E}\left[P(S(T))\right] - \mu_M}_{\text{Overall error}} = \underbrace{\mathbb{E}\left[P(S(T))\right] - \mathbb{E}\left[P(S_N)\right]}_{\text{Bias}} + \underbrace{\mathbb{E}\left[P(S_N)\right] - \mu_M}_{\text{Statistical error}}.$$

The bias is a form of error arising from the use of a numerical discretisation method to approximate $S(T)$, and its absolute value is known as *weak discretisation error*. The statistical error arises from the use of a Monte-Carlo approach with finite sample size M to approximate the expectation $\mathbb{E}[P(S_N)]$: see Chap. 4.

We are also concerned with *strong discretisation error*, which measures the average error between true and approximate trajectories, and which is important for path-dependent applications. Weak and strong discretisation error will be defined and explored in Sect. 8.3.

Our goals for this final chapter are to describe approximation methods for SDE systems, define and analyse appropriate notions of convergence and stability for such methods, and illustrate their application for Monte Carlo sampling. We will finish by describing a variance reduction technique known as *Multilevel Monte Carlo (MLMC)* which can in some circumstances mitigate the additional error introduced by the use of approximate sampling due to the numerical method, and which relies on understanding of both its weak and strong convergence properties.

Python Setup To follow the Python code in this chapter, you will need to import the NumPy and PyPlot packages, and the `optimize.fsolve` module from the SciPy package.

```
In []: import numpy as np
        import matplotlib.pyplot as plt
        from scipy.optimize import fsolve
```

For the final time, instantiate a local RNG:

```
In []: rng=np.random.default_rng()
```

We will also make use of the user-defined function from Chap. 7 listed in Table 8.1.

Table 8.1 User-defined function `bbridge()` required in Chap. 8

Section	Function name	Function output	Parameter	Parameter meaning
7.1.4	`bbridge()`	$W(t, \omega)$	t1	t_1
			t	t
			t2	t_2
			x1	$W(t_1, \omega) = x_1$
			x2	$W(t_2, \omega) = x_2$
			rng	local RNG

8.1   Strong Approximation Methods for Itô-Type SDEs

Let $W = [W_1, \ldots, W_m]^T$ be an m-dimensional vector of independent standard Brownian motions defined on a complete filtered probability space $(\Omega, \mathcal{F}, (\mathcal{F}_t)_{t \in [0,T]}, \mathbb{P})$ adapted to the filtration $(\mathcal{F}_t)_{t \in [0,T]}$. Consider the d-dimensional Itô-type stochastic differential equation with time-homogeneous coefficients given by

$$dX(t) = f(X(t))dt + G(X(t))dW(t), \quad t \in [0, T], \tag{8.1}$$

where $X(0) = X_0$, $X \in \mathbb{R}^d$, and $f : \mathbb{R}^d \to \mathbb{R}^d$, and $G : \mathbb{R}^d \to \mathbb{R}^{d \times m}$ satisfy appropriate integrability conditions. Eq. (8.1) has integral form

$$X(t) = X_0 + \int_0^t f(X(s))ds + \sum_{i=1}^m \int_0^t G_i(X(s))dW_i(s), \quad t \in [0, T]. \tag{8.2}$$

where G_i represents the ith column of the matrix-valued function G.

This form for an SDE system is extremely general, and includes all of the SDE models that we have encountered in the book thus far. For example:

1. The Black-Scholes asset price model under the risk-neutral measure

$$dS(t) = rS(t)dt + \sigma S(t)dW(t), \quad t \in [0, T], \tag{8.3}$$

   corresponds to (8.1) with $d = m = 1$, $X(t) = S(t)$, $f(s) = rs$, and $G(s) = \sigma s$.
2. The Vasicek model of the short rate

$$dr(t) = \alpha(b - r(t))dt + \sigma dW(t), \quad t \in [0, T], \tag{8.4}$$

   corresponds to (8.1) with $d = m = 1$, $X(t) = r(t)$, $f(r) = \alpha(b - r)$, and $G(r) \equiv \sigma$.
3. The Heston model introduced as the SDE system (7.50) and (7.51) in Chap. 7 may be written in this setting as

$$dS(t) = rS(t)dt + \sqrt{V(t)}S(t)\left(\rho dW_1(t) + \sqrt{1 - \rho^2}dW_2(t)\right);$$

$$dV(t) = \lambda(\mu - V(t))dt + \sigma\sqrt{V(t)}dW_2(t), \quad t \in [0, T],$$

   where we have used Cholesky factorisation to incorporate the correlated diffusions. The SDE system (7.50)-(7.51) corresponds to (8.1) if we set $d = m = 2$, the state vector $X(t) = [S(t), V(t)]^T$, and

$$f(s, v) - \begin{pmatrix} rs \\ \lambda(\mu - v)] \end{pmatrix}; \quad G(s, v) - \begin{pmatrix} \sqrt{v}s\rho & s\sqrt{v(1 - \rho^2)} \\ 0 & \sigma\sqrt{v} \end{pmatrix}.$$

Therefore we will develop our framework for numerical discretisation in terms of the general SDE (8.1), before applying it to specific models. We wish to generate approximate trajectories for systems of the form (8.1) by discretisation over the interval $[0, T]$ using a uniform mesh of N steps $0 = t_0 < t_1 < \cdots < t_N = T$ with fixed step h, so that $Nh = T$. Since we are able to sample from trajectories of W exactly, we will use the same trajectories for the noise in the discretisation as appear in the original SDE. Such approximations are called *strong numerical approximations*. By contrast, numerical discretisation methods that do not sample from the trajectories of W but which instead seek to reproduce only the mean and variance of each increment of W are referred to as *weak numerical approximations*: see Exercise 8.5.

8.1.1 The θ-Maruyama Class of Numerical Methods

Let $X(t_n)$ be given and consider the integral form (8.2) over a single step of length h over the interval $[nh, (n+1)h]$ for some $n = 0, \ldots, N - 1$. On each trajectory we will approximate the ith stochastic integral on the RHS using the rectangular approximation

$$\int_{nh}^{(n+1)h} G_i(X(s, \omega)) dW_i(s, \omega) \approx G_i(X(t_n), \omega) \int_{nh}^{(n+1)h} dW_i(s, \omega)$$

$$= G_i(X(t_n, \omega)) \underbrace{(W_i((n+1)h, \omega) - W_i(nh, \omega))}_{=: \Delta W_{n+1}(\omega)},$$

for each $i = 1, \ldots, m$, and $\omega \in \Omega$. As usual we will suppress the dependence on ω in the notation unless we wish to specifically emphasise an individual trajectory ω.

The θ-Maruyama class of numerical discretisation methods for (8.1) is given by the sequence $\{X_n^h\}_{n=0}^N$ where

$$X_{n+1}^h = X_n^h + (1 - \theta)hf(X_n^h) + \theta h f(X_{n+1}^h) + \sum_{i=1}^m G_i(X_n^h)\Delta W_{n+1}^i, \qquad (8.5)$$

for $n = 0, \ldots, N$, and any $\theta \in [0, 1]$. The choice of θ determines the type of method.

- If $\theta = 0$, then we have the Euler-Maruyama, or explicit Euler, method for (8.1), given by

$$X_{n+1}^h = X_n^h + f(X_n^h)h + \sum_{i=1}^m G_i(X_n^h)\Delta W_{n+1}^i, \quad n = 0, \ldots, N - 1.$$

- If $\theta = 1$, then we have the implicit Euler method for (8.1), given by

$$X_{n+1}^h = X_n^h + f(X_{n+1}^h)h + \sum_{i=1}^{m} G_i(X_n^h)\Delta W_{n+1}^i, \quad n = 0, \ldots, N-1.$$

- If $\theta = 0.5$, we have a stochastic variant of the trapezoid rule applied to the drift of (8.1), resulting in a method of the form

$$X_{n+1}^h = X_n^h + \frac{1}{2}h\left(f(X_n^h) + f(X_{n+1}^h)\right) + \sum_{i=1}^{m} G_i(X_n^h)\Delta W_{n+1}^i, \quad n = 0, \ldots, N-1.$$

8.1.2   Pathwise Error for Euler Schemes Applied to a Linear SDE Model

We will first examine the error introduced on individual trajectories by these numerical schemes. Since an exact solution for the Black-Scholes asset model (8.3) is available in terms of the standard Brownian motion W, we can use it as a test case. The θ-Maruyama discretisation of (8.3) is given by the random sequence $\{S_n^h\}_{n=0}^N$, where

$$S_{n+1}^h = S_n^h + (1-\theta)hr S_n^h + \theta hr S_{n+1}^h + \sigma S_n^h \Delta W_{n+1}, \quad n = 0, \ldots, N-1. \qquad (8.6)$$

Since this equation is linear, we can rewrite by gathering the S_{n+1} terms together on the LHS of the equation and dividing through by $1 - \theta hr$ to get

$$S_{n+1}^h = \frac{S_n^h}{1 - \theta hr}[1 + (1-\theta)rh + \sigma\Delta W_{n+1}], \quad n = 0, \ldots, N-1.$$

Now set $\theta = 0$ to construct the explicit Euler method:

$$S_{n+1}^h = S_n^h[1 + hr + \sigma\Delta W_{n+1}], \quad n = 0, \ldots, N-1; \qquad (8.7)$$

and $\theta = 1$ to construct the implicit Euler method:

$$S_{n+1}^h = \frac{S_n^h}{1 - hr}[1 + hr + \sigma\Delta W_{n+1}], \quad n = 0, \ldots, N-1. \qquad (8.8)$$

In Python, the following code generates an exact trajectory of (8.3) over a mesh with $N = 10$ steps, contained in the array variable `exactS`, and an approximate trajectory using the θ-Maruyama method, contained in the array variable `approxS`:

```
In []: # Model and discretisation parameters
        S0=1; T=1; r=0.01; sig=0.3
        theta = 0; N=10; dt=T/N

        # Generate Brownian path and mesh
        inc = rng.normal(0,np.sqrt(dt),N)
        W=np.insert(np.cumsum(inc),0,0)
        t=np.arange(0,T+dt,dt)

        # Generate exact trajectory
        exactS = S0*np.exp((r-0.5*sig**2)*t+sig*W)

        # Precompute constant terms
        dampFac=1/(1-theta*dt*r)
        drift=(1-theta)*dt*r

        # Generate approximate trajectory
        approxS=S0*np.ones(N+1)

        for i in range(N):
            approxS[i+1]=approxS[i]*dampFac*(1+drift+sig*inc[i])
```

There are two ways to view error on an individual trajectory. The (unsigned) *pathwise error at the final time* T is given for any $\omega \in \Omega$ by

$$\varepsilon_\omega^{\text{final}}(h) := \left| S(T, \omega) - S_N^h(\omega) \right|.$$

This can be computed as

```
In []: errorFinal=abs(exactS[-1]-approxS[-1])
```

The *maximum (or supremum) pathwise error* is given for any $\omega \in \Omega$ by

$$\varepsilon_\omega^{\text{sup}}(h) := \sup_{i=1,\dots,N} \left| S(t_i, \omega) - S_i^h(\omega) \right|.$$

This can be computed as

```
In []: errorMax=np.max(abs(exactS-approxS))
```

Final-time errors are of interest when valuing European-style options with a payoff that depends only on the value of the underlying asset at time T. Supremum errors are more relevant for path-dependent problems such as the valuation of barrier-style options. Notice that in both cases we have chosen to emphasise the dependence on the stepsize h, and the specific trajectory ω.

Figure 8.1 compares an exact trajectory, $S(t, \omega)$ for $t \in [0, 1]$, of the Black-Scholes asset model (8.3) to the explicit Euler and implicit Euler approximation methods given by (8.7) and (8.8) respectively, at two different resolutions. The top plot shows exact (solid line) and approximate (dashed lines) trajectories computed over a mesh with uniform stepsize $h = 0.1$. The bottom plot uses a Brownian bridge (see Sect. 7.1.3 in Chap. 7) to interpolate the same asset price trajectory over a mesh with smaller uniform stepsize $h = 0.05$.

Table 8.2 shows error computations for two trajectories. On each trajectory we compute the final-time and supremum pathwise errors at two different stepsizes for each of the explicit Euler and implicit Euler schemes. We conclude that these errors change with ω, i.e. that they are random, and for these specific trajectories at least, reducing the stepsize h leads to a corresponding reduction in pathwise error.

The θ-Maruyama discretisation for the Vasicek model of the short rate (8.4) is given by the random sequence $\{r_n^h\}_{n=0}^{N}$, where

$$r_{n+1}^h = r_n^h + (1 - \theta)h\alpha(b - r_n^h) + \theta h\alpha(b - r_{n+1}^h) + \sigma \Delta W_{n+1}, \quad n = 0, \ldots, N - 1. \tag{8.9}$$

Exercise 8.1 asks you to implement this numerical scheme.

8.1.2.1 The CIR Model and Dynamical Consistency

The Black-Scholes and Vasicek models are both linear SDEs. However we sound a note of caution when applying θ-Maruyama methods directly to nonlinear models. To see this, consider the Cox-Ingersoll-Ross interest rate model given by the SDE

$$dr(t) = \lambda(\mu - r(t))dt + \sigma\sqrt{r(t)}dW(t), \quad t \in [0, T], \tag{8.10}$$

with $r(0) = r_0 > 0$. Once again, we draw a correspondence with (8.1), setting $d = m = 1$, and coefficients

$$f(r) = \lambda(\mu - r); \quad G(r) = \sigma\sqrt{r}.$$

Then applying (8.5) with $\theta = 0$ we construct an explicit Euler scheme as the random sequence $\{r_n\}_{n=0}^{N}$, where

$$r_{n+1}^h = r_n^h + h\lambda(\mu - r_n^h) + \sigma\sqrt{r_n^h}\,\Delta W_{n+1}, \quad n = 0, \ldots, N - 1. \tag{8.11}$$

However, the stochastic difference equation (8.11) is not well defined, since at any t_n there is a non-zero probability that the solution will become negative after the

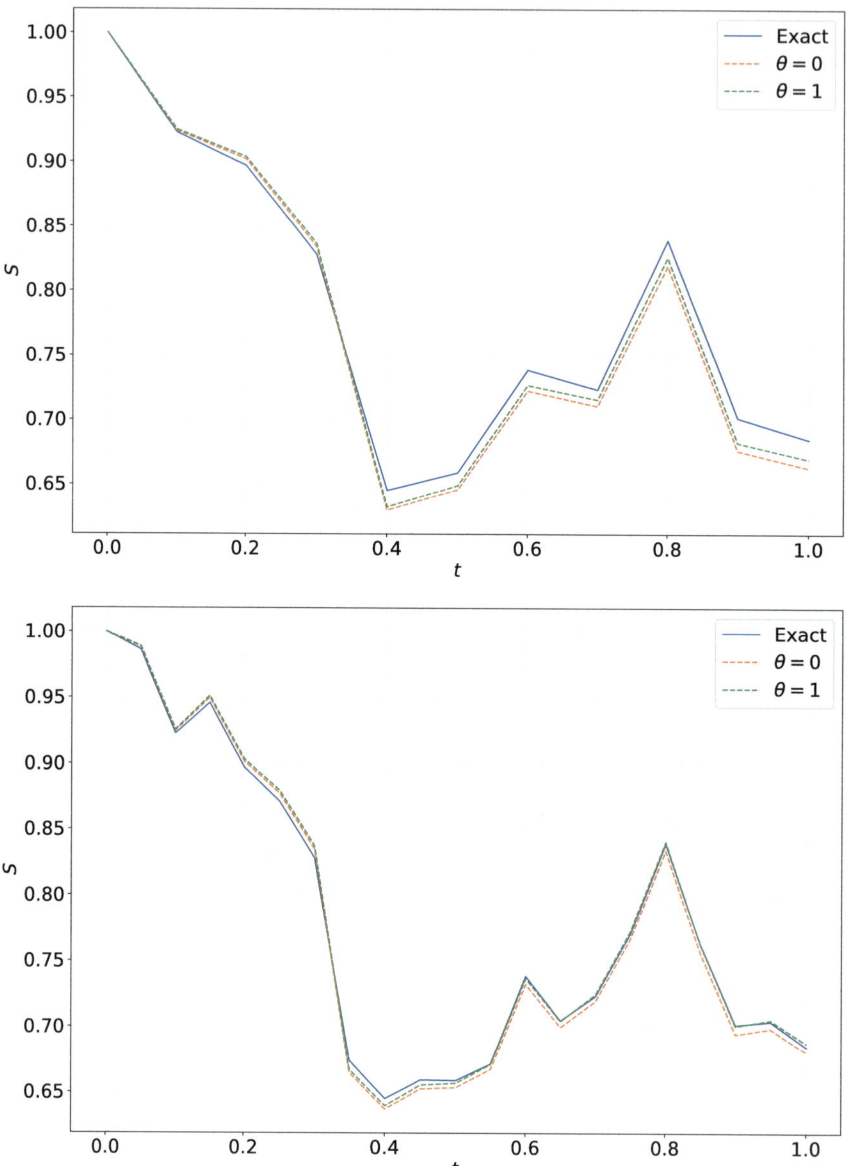

Fig. 8.1 An exact trajectory of the Black-Scholes asset model (8.3) compared to approximate trajectories produced by the explicit Euler method (8.7) ($\theta = 0$), implicit Euler method (8.8) ($\theta = 1$). Here, $S_0 = 1$, $r = 0.01$, $\sigma = 0.3$, $T = 1$, $N = 10$ (top), and $N = 20$ (bottom)

Table 8.2 Tabulated errors for Euler schemes applied to two trajectories $S(t, \omega_1)$ (left) and $S(t, \omega_2)$ (right) of the Black-Scholes SDE (8.3) with $S_0 = 1$, $r = 0.01$, $\sigma = 0.3$, $T = 1$. On each trajectory final-time and supremum errors are shown for both explicit ($\theta = 0$) and implicit ($\theta = 1$) Euler schemes, in each case at two different mesh sizes

Scheme	h	$\varepsilon_{\omega_1}^{\mathrm{final}}(h)$	$\varepsilon_{\omega_1}^{\sup}(h)$
$\theta = 0$	0.1	0.0219	0.0254
	0.05	0.0041	0.0092
$\theta = 1$	0.1	0.0152	0.0192
	0.05	0.0028	0.0092
Scheme	h	$\varepsilon_{\omega_2}^{\mathrm{final}}(h)$	$\varepsilon_{\omega_2}^{\sup}(h)$
$\theta = 0$	0.1	0.0039	0.0085
	0.05	0.0013	0.0070
$\theta = 1$	0.1	0.0109	0.0120
	0.05	0.0083	0.0083

next step. To see this, suppose that $r_n = r > 0$. Then

$$
\mathbb{P}\left[r_{n+1}^h < 0 \mid r_n^h = r > 0 \right] = \mathbb{P}\left[r_n^h + h\lambda(\mu - r_n^h) + \sigma\sqrt{r_n^h}\,\Delta W_{n+1} < 0 \mid r_n = r > 0 \right]
$$

$$
= \mathbb{P}\left[\Delta W_{n+1} < \frac{-r_n^h - h\lambda(\mu - r_n^h)}{\sigma\sqrt{r_n^h}} \,\middle|\, r_n^h = r > 0 \right]
$$

$$
= \mathbb{P}\left[Z < \frac{-r - h\lambda(\mu - r)}{\sigma\sqrt{r}\sqrt{h}} \right]
$$

$$
= \Phi\left(\frac{-r - h\lambda(\mu - r)}{\sigma\sqrt{r}\sqrt{h}} \right),
$$

where $Z \sim \mathcal{N}(0, 1)$ and $\Phi(\cdot)$ is the CDF of a standard normal random variable. It is clear that this probability will be nonzero for any fixed values of $r, h, \lambda, \mu > 0$. Once we observe a negative value in the sequence defined by (8.11) we must stop, since the square-root term will be undefined at the next step. Figure 8.2 displays probability surfaces representing the probability of a negative value over a single step when the Feller condition (7.40) in Chap. 7 holds (left) and when it does not (right). Note that probabilities of a spurious negative value, shown on the vertical axes, are up to an order of magnitude higher where the Feller condition does not hold.

We can handle this issue by either truncating negative values at zero or by taking the absolute value. For example, Bossy and Diop [9] presented the *symmetrised Euler scheme*

$$
r_{n+1} = \left| r_n + \lambda(\mu - r_n)h + \sigma\sqrt{r_n}\,\Delta W_{n+1} \right|, \quad n = 0, \ldots, N - 1,
$$

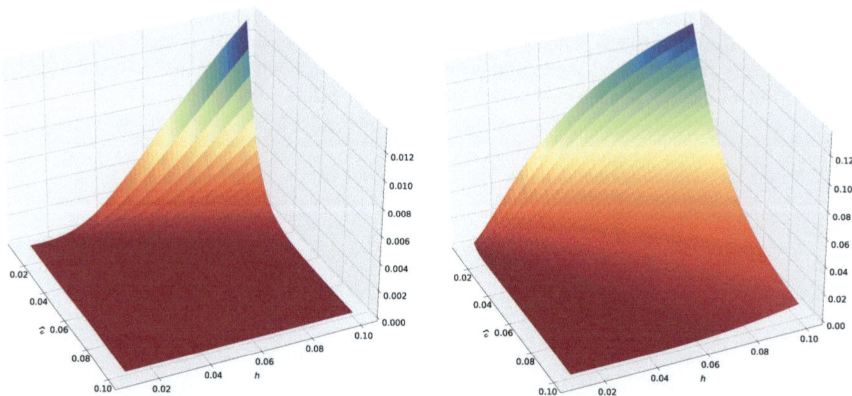

Fig. 8.2 The probability of a negative value produced after a single step of the explicit Euler discretisation of the CIR model (7.39) with $\lambda = 1$ and $\mu = 0.05$. On the left, $\sigma = 0.2$ so that $2\lambda\mu > \sigma^2$, and on the right, $\sigma = 0.4$ so that $2\lambda\mu < \sigma^2$

and demonstrated weak convergence to the true solution for this scheme, but there are many other possible approaches. In fact the development of effective numerical approximation methods for solutions of the Cox-Ingersoll-Ross model is an active and ongoing area of research and we will discuss this in more detail later in the chapter, in Sect. 8.5.

8.1.2.2 Nonlinear Stochastic and Local Models of Volatility

Let $F(t), t \in [0, T]$ describe the forward price of an asset on which we have written an option expiring at time T. The process F might represent a forward interest rate, the forward yield on a bond, or the forward price of an equity asset. The *stochastic α-β-ρ (SABR) model* describes the stochastic dynamics of the forward price according to the 2-dimensional system of SDEs

$$dF(t) = \sigma(t)F(t)^\beta \, dW_1(t); \tag{8.12}$$

$$d\sigma(t) = \alpha\sigma(t) \, dW_2(t), \quad t \in [0, T], \tag{8.13}$$

with initial values $F(0) = F_0 > 0$, $\sigma(0) = \sigma_0 > 0$. Here, we suppose that pricing is to take place using a suitably chosen numeraire associated with a probability measure under which F is a martingale. Therefore, under this measure, $W = [W_1, W_2]$ is a 2-dimensional standard Brownian motion with distribution

$$\mathcal{N}(0, \Sigma); \quad \Sigma = \begin{pmatrix} 1 & \rho \\ \rho & 1 \end{pmatrix} t.$$

The process σ represents the stochastically varying volatility of F, and $\alpha \geq 0$ is the volatility of σ (sometimes referred to as the *vol-vol* or *volga*). Normally we take $\beta \in [0, 1]$. If $\alpha = 0$ then the model reduces to a scalar SDE

$$dF(t) = \sigma_0 F(t)^\beta \, dW_1(t), \quad t \in [0, T].$$

If we allow $\beta \geq 0$, this can be viewed as the martingale form of a *constant elasticity of variance (CEV)* model, a local volatility model previously described in Sect. 2.3.2.3 of Chap. 2.

Returning to the more general SABR model, notice that the SDE for the forward price process F given by (8.12) depends both on F itself and on the volatility process σ, but the SDE for σ, given by (8.13), is linear and depends only on σ. Therefore it has exact solution

$$\sigma(t) = \sigma_0 \exp\left(-\frac{\alpha^2}{2}t + \alpha W_2(t)\right). \tag{8.14}$$

If we wish to simulate trajectories of this system we can start by generating exact trajectories of σ over the mesh $\{t_i\}_{i=0}^N$ where $t_i = ih$: see Exercise 8.4.

This done, we need to generate values for the forward price $F(t_i), i = 0, \ldots, N$. Even if we already know the value of σ at each of the meshpoints t_i the SDE

$$dF(t) = \sigma(t)F(t)^\beta dW_1(t) \tag{8.15}$$

has a known exact solution in terms of the Brownian trajectory W_1 only in the special cases where $\beta = 0$ or 1. If (say) $\beta = 3/4$ then we need to numerically approximate F. Moreover, if $\beta = 1/2$ the same issues with dynamical consistency identified for the CIR model in the previous section apply. A suitable variant of the θ-Maruyama discretisation of (8.12) is given by the sequence $\{F_n\}_{n=0}^N$, where

$$F_{n+1} = F_n + \sigma(t_n)|F_n|^\beta \Delta W_{1,n+1}, \quad n = 0, \ldots, N-1, \tag{8.16}$$

where $\Delta W_{1,n+1} := W_1((n+1)h) - W_1(nh)$. Notice that, since there is no drift coefficient, the choice of θ is immaterial.

8.2   Stochastic Notions of Stability

8.2.1   Asymptotic Stability in Mean-Square

Recall from Chap. 5 that the stability of a finite difference scheme for solving a PDE describes the way that errors generated locally at a mesh point are propagated by that scheme through subsequent iterations. Similar considerations hold if we numerically solve an SDE. Certainly if we are interested in using approximate numerical solutions to value long-term derivatives we should understand the long run stability of the scheme that we choose.

Higham [30] provided an analysis by which to judge the stability in mean-square of a numerical scheme by applying it to a class of linear test SDEs with an equilibrium solution at zero. This test SDE can be viewed as a more general form of the Black-Scholes asset model with complex parameters.

$$dX(t) = \lambda X(t)dt + \sigma X(t)dW(t), \quad t \in [0, \infty); \quad X(0) = X_0 \in \mathbb{R}, \qquad (8.17)$$

where $\lambda, \sigma \in \mathbb{C}$, evolving on the infinite time set $[0, \infty)$. Let's review the relevant notions of stability in the SDE setting.

Definition 8.1 The equilibrium solution $X(t) \equiv 0$ of (8.17) is *pth-moment stable* if and only if, for each $\epsilon > 0$, there exists a $\delta > 0$ such that

$$\mathbb{E}\left[|X(t)|^p\right] < \epsilon, \quad t \in [0, \infty),$$

whenever $|X_0| < \delta$. It is *globally pth-moment asymptotically stable* if and only if it is *pth*-moment stable, and for all $X_0 \in \mathbb{R}$,

$$\lim_{t \to \infty} \mathbb{E}\left[|X(t)|^p\right] = 0.$$

If $p = 2$, the equilibrium is said to be *globally asymptotically stable in mean-square*.

It is known (see for example Saito and Mitsui [60]) that the equilibrium solution of (8.17) is globally asymptotically stable in mean-square if and only if

$$2\Re(\lambda) + |\sigma|^2 < 0,$$

where, for any $z \in \mathbb{C}$, $\Re(z)$ and $|z|$ denote the real part and modulus of z, respectively.

The θ-Maruyama method applied to the test equation (8.17) is a sequence $\left\{X_n^h\right\}_{n \geq 0}$ satisfying

$$X_{n+1}^h = X_n^h + (1 - \theta)h\lambda X_n^h + \theta h\lambda X_{n+1}^h + \sigma X_n^h \Delta W_{n+1}, \quad n \geq 0, \qquad (8.18)$$

with $X_0^h = X_0$, and equivalent definitions of stability for the numerical scheme (8.18) are as follows.

Definition 8.2   The equilibrium solution $X_n^h \equiv 0$ of (8.18) is *pth-moment stable* if and only if, for each $\epsilon > 0$, there exists a $\delta > 0$ such that

$$\mathbb{E}\left[\left|X_n^h\right|^p\right] < \epsilon, \quad n \geq 0,$$

whenever $\left|X_0^h\right| < \delta$. It is *globally pth-moment asymptotically stable* if and only if it is *pth*-moment stable, and for all $X_0^h \in \mathbb{R}$,

$$\lim_{n \to \infty} \mathbb{E}\left[\left|X_n^h\right|^p\right] = 0.$$

If $p = 2$, the equilibrium is said to be *globally asymptotically stable in mean-square.*

If these definitions are not satisfied, then the relevant equilibrium is termed *pth-mean unstable.*

The question we wish to answer is: when the equilibrium solution of (8.17) is mean-square asymptotically stable, for what range of parameters (including the stepsize h and the parameter θ) will the equilibrium solution of the numerical approximation (8.18) also be globally mean-square asymptotically stable?

Let us start with the following result, which gives exact criteria for global mean-square asymptotic stability of the equilibrium solution of (8.18) without making assumptions on the stability of the test SDE (8.17).

Theorem 8.3   *Let* $\{X_n^h\}_{n=0}^N$ *be a solution of (8.18), the θ-Maruyama method applied to the test equation (8.17), and suppose that we choose h and θ to ensure that $1 - \theta h\lambda \neq 0$. The equilibrium solution $X_n^h \equiv 0$ of (8.18) is globally asymptotically stable in mean-square if and only if h and θ are such that*

$$\frac{|1 + (1 - \theta)h\lambda|^2}{|1 - \theta h\lambda|^2} + \frac{|\sigma|^2 h}{|1 - \theta h\lambda|^2} < 1. \tag{8.19}$$

Proof   Fix $n \in \mathbb{N}$. From (8.18),

$$X_{n+1}^h = X_n^h + (1 - \theta)h\lambda X_n^h + \theta h\lambda X_{n+1}^h + \sigma X_n^h \Delta W_{n+1}.$$

Hence, by gathering the X_{n+1}^h terms on the LHS, we have

$$X_{n+1}^h = \left(\frac{1 + (1 - \theta)h\lambda}{1 - \theta h\lambda} + \frac{\sigma}{1 - \theta h\lambda}.\Delta W_{n+1}\right) X_n.$$

Then, using the fact that each ΔW_{n+1} is independent of X_n^h, $\mathbb{E}\left[\Delta W_{n+1}\right] = 0$, and $\mathbb{E}\left[(\Delta W_{n+1})^2\right] = h$, we get

$$\mathbb{E}\left[\left|X_{n+1}^h\right|^2\right] = \left(\frac{|1 + (1 - \theta)h\lambda|^2}{|1 - \theta h\lambda|^2} + \frac{|\sigma|^2 h}{|1 - \theta h\lambda|^2}\right)\mathbb{E}\left[\left|X_n^h\right|^2\right]$$

$$\vdots$$

$$= \left(\frac{|1 + (1 - \theta)h\lambda|^2}{|1 - \theta h\lambda|^2} + \frac{|\sigma|^2 h}{|1 - \theta h\lambda|^2}\right)^{n+1}\left|X_0^h\right|^2,$$

since $\mathbb{E}\left[\left|X_0^h\right|^2\right] = \left|X_0^h\right|^2$. It follows that

$$\lim_{n \to \infty} \mathbb{E}\left[\left|X_n^h\right|^2\right] = 0 \quad \text{if and only if} \quad \frac{|1 + (1 - \theta)h\lambda|^2}{|1 - \theta h\lambda|^2} + \frac{|\sigma|^2 h}{|1 - \theta h\lambda|^2} < 1,$$

as required.
$\square$

8.2.2 Regions of Stability and A-Stability in Mean-Square

Suppose $\lambda, \sigma \in \mathbb{R}$, and set $\theta = 0$ in (8.18), so that we have chosen the explicit Euler method to discretise (8.17). Suppose also that $2\lambda + \sigma^2 < 0$, so that the equilibrium solution of (8.17) is globally mean-square asymptotically stable. Then the condition (8.19) becomes

$$(1 + h\lambda)^2 + \sigma^2 h < 1$$

or equivalently

$$h < -\frac{1}{\lambda^2}(2\lambda + \sigma^2). \tag{8.20}$$

In this case therefore, the explicit Euler method requires an additional constraint on the stepsize as given by (8.20) in order to reproduce the mean-square asymptotic stability of the test equilibrium. Notice that the RHS of the inequality (8.20) is positive if and only if $2\lambda + \sigma^2 < 0$.

One must take care when drawing a connection from this analysis back to a financial model. If r represents a (real, positive) risk-free rate of interest and σ a (real, positive) volatility then in fact $2r + \sigma^2 > 0$, and the SDE model of the asset price process has a mean-square unstable equilibrium at zero. We do not necessarily expect our chosen numerical method to damp out mean-square errors asymptotically

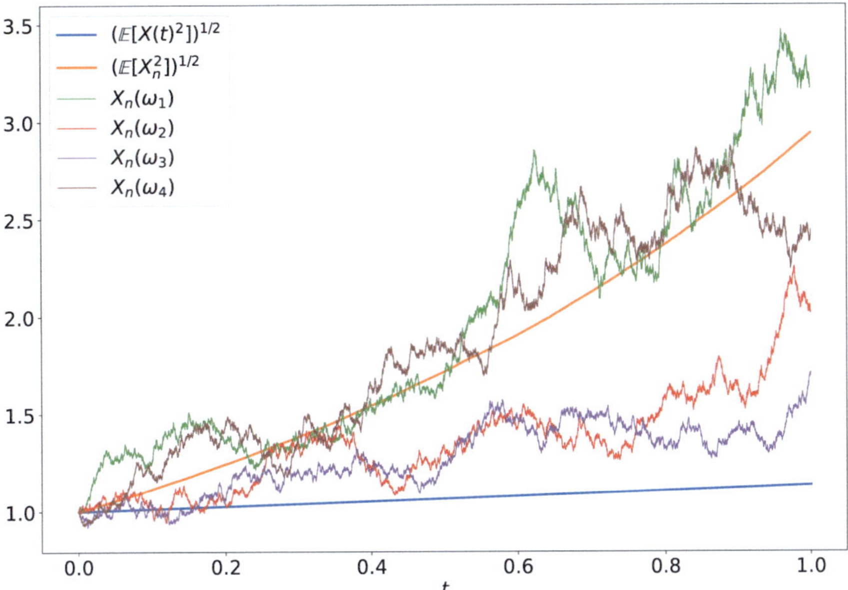

Fig. 8.3 A Monte Carlo estimate of the root-mean-square $\left(\mathbb{E}[X_n^2]\right)^{1/2}$ (orange line) for the θ-Maruyama method with $\theta = 1/2$ applied to the Black-Scholes asset model (8.3) with $r = 0.045$ and $\sigma = 0.4$. The root-mean-square of the true solution, denoted $\left(\mathbb{E}[X(t)^2]\right)^{1/2}$ is shown as the blue line, and we also display 4 trajectories from the ensemble of size $M = 10^4$ used to estimate $\left(\mathbb{E}[X_n^2]\right)^{1/2}$

since the underlying SDE has an underlying dynamic that grows in that norm. This is illustrated in Fig. 8.3, where we have applied the θ-Maruyama method with $\theta = 1/2$ and $N = 10^4$ to (8.3) with $r = 0.045$, $\sigma = 0.4$, and $T = 1$. Here, $2r + \sigma^2 = 0.25 > 0$, the equilibrium solution of the underlying SDE is mean-square unstable, and as we will see later in this section, this is perfectly reflected in our numerical method. Nonetheless we see that a Monte Carlo estimate of the root-mean-square (RMS) of the numerical solution represented by an orange line and given by

$$\left(\mathbb{E}\left[\left(X_n^h\right)^2\right]\right)^{1/2} \approx \left(\frac{1}{M}\sum_{i-1}^{M} X_n^h(\omega_i)^2\right)^{1/2}, \quad n = 0, \ldots, N,$$

deviates considerably from the RMS of the true solution evaluated at the same times, given by $\left(\mathbb{E}\left[X(nh)^2\right]\right)^{1/2}$ and plotted as a blue line, due to the accumulation of error over time.

However since we can always sample from the exact distribution of the Black-Scholes asset model we would not choose to discretise the SDE in this case. Rather the analysis is meant to guide us on the anticipated stability properties of an explicit method applied to more complex SDE models where an equivalent stability analysis may not be tractable. In such situation we may seek to use a method that is guaranteed to be mean-square asymptotically stable without stepsize constraints when applied to the test SDE (8.17) as long as the equilibrium solution of the underlying SDE is stable. Such methods are called *A-stable in mean-square*, and can be thought of as a stochastic analogue of the unconditionally stable finite difference methods encountered in Chap. 5.

This notion of *A*-stability in mean-square was proposed in [30], and we follow their approach by describing and then comparing the parameter ranges where mean-square stability is observed for the test SDE (8.17) and the numerical discretisation (8.18).

Definition 8.4 The *mean-square stability region* of (8.17) is given by

$$S_{\mathrm{SDE}}(\lambda, \sigma) = \left\{ \lambda, \sigma \in \mathbb{C} \,:\, 2\Re(\lambda) + |\sigma|^2 < 0 \right\}, \tag{8.21}$$

and represents the set of parameter values for which the equilibrium solution of (8.17) is mean-square asymptotically stable in the particular sense of Definition 8.1 with $p = 2$.

The mean-square stability region of (8.18) is given by

$$S_{\mathrm{STM}}(h, \theta; \lambda, \sigma) = \{ h > 0, \, \theta \in [0, 1]; \, \lambda, \sigma \in \mathbb{C} \,:\, (8.19) \text{ holds.} \}, \tag{8.22}$$

and represents the set of parameter values for which the equilibrium solution of (8.18) is stable in the particular sense of Definition 8.2 with $p = 2$.

Definition 8.5 The stochastic θ-method described by (8.18) is A-stable in mean-square if and only if, for all $h > 0$,

$$S_{\mathrm{SDE}}(\lambda, \sigma) \subseteq S_{\mathrm{STM}}(h, \theta; \lambda, \sigma).$$

The following theorem from [30] summarises the results of its mean-square stability analysis.

Theorem 8.6 *For all $h > 0$,*

$$S_{STM}(\theta, h; \lambda, \sigma) \subset S_{SDE}(\lambda, \sigma), \quad for \quad 0 \le \theta < 1/2;$$

$$S_{STM}(\theta, h; \lambda, \sigma) \equiv S_{SDE}(\lambda, \sigma), \quad for \quad \theta = 1/2;$$

$$S_{STM}(\theta, h; \lambda, \sigma) \supset S_{SDE}(\lambda, \sigma), \quad for \quad \theta > 1/2.$$

For $0 \leq \theta < 1/2$, given $(\lambda, \sigma) \in S_{SDE}(\lambda, \sigma)$, the equilibrium solution of Eq. (8.18) is globally asymptotically stable in mean-square if and only if

$$h < \frac{2\Re(\lambda) + |\sigma|^2}{|\lambda|^2(2\theta - 1)}.$$

Hence, the stochastic θ-Maruymama method is A-stable in mean-square if and only if $\theta \in [1/2, 1]$.

Theorem 8.6 demonstrates that the explicit Euler-Maruyama method, which corresponds to $\theta = 0$, is not A-stable in the sense of Definition 8.5, since it is necessary to impose a constraint on the stepsize h to ensure mean-square stability. However both the implicit Euler ($\theta = 1$), and trapezoid rule ($\theta = 1/2$) methods are A-stable in mean-square.

Following [30], we can visually compare mean-square stability for the test SDE and the θ-Maruyama method in the case where $\lambda, \sigma \in \mathbb{R}$, providing a useful illustration of the statement of Theorem 8.6. The mean-square stability condition for (8.17) can be written (multiplying though the inequality in (8.21) by the stepsize h)

$$2h\lambda + h\sigma^2 < 0$$

and the equivalent condition for the θ-Maruyama method from (8.22) is

$$2h\lambda + h\sigma^2 + (1 - 2\theta)h^2\lambda^2 < 0.$$

Choose new coordinates $(x, y) := (h\lambda, h\sigma^2)$ and rewrite these conditions as

$$x + \frac{1}{2}y < 0; \quad 2\left(x + \frac{1}{2}y\right) + (1 - 2\theta)x^2 < 0,$$

respectively. Figure 8.4 illustrates Theorem 8.6 by overlaying the mean-square stability regions of the test equation and the θ-Maruyama method for $\theta = 0, 0.5, 1$ in this coordinate system. Given a pair of coordinates (x, y) representing a particular set of values for h, μ, and σ, the choice of coordinate system allows us to observe the effect of changes in h alone on the stability of the discretisation by moving along lines of slope -1 if $x < 0$, and along lines of slope 1 if $x > 0$.

A discussion of the suitability of (8.17) as a test equation for general SDE systems and an extension of this framework to incorporate higher-dimensional test SDEs with real coefficients may be found in [11] and [12].

8.2.3   Almost Sure Asymptotic Stability

Let us return again to the case of the Black-Scholes asset price model (8.3) with θ-Maruyama discretisation (8.6), and allow each to evolve over the time set $[0, \infty)$

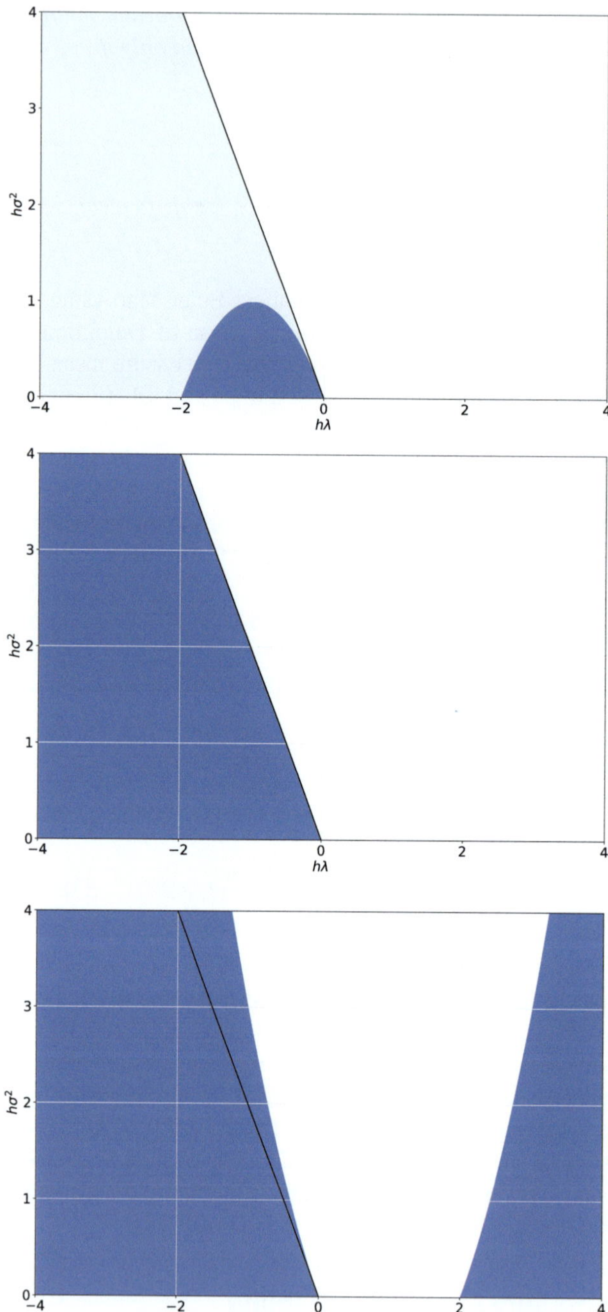

Fig. 8.4 The area to the left of the black line, shaded light blue in the first plot, represents S_{SDE}, the mean-square asymptotic stability region for the linear test equation (8.17). The dark blue shaded areas represent $S_{\mathrm{STM}}(\theta, h)$, mean-square asymptotic stability regions for the explicit Euler ($\theta = 0$, top), trapezoid rule ($\theta = 1/2$, middle), and implicit Euler ($\theta = 1$, bottom) schemes, respectively

(in the case of (8.3)) and associated uniform discretisation with stepsize h (in the case of (8.6)). Recall from Sect. 3.1.4 in Chap. 3 that

$$\lim_{t\to\infty} \frac{1}{t} \ln(S(t)) = r - \frac{1}{2}\sigma^2, \quad a.s,$$

and therefore $\lim_{t\to\infty} S(t) = 0$ a.s. if and only if $2r - \sigma^2 < 0$. This describes an alternative notion of stability where we are concerned with the long-run behaviour of solutions on almost all trajectories. The relevant definitions for stability here are as follows.

Definition 8.7 The equilibrium solution $X(t) \equiv 0$ of (8.17) is *a.s. stable* if and only if, for each $\epsilon > 0$, there exists a $\delta > 0$ such that

$$|X(t)| < \epsilon, \quad t \in [0, \infty), \quad a.s,$$

whenever $|X(0)| < \delta$. The equilibrium is *globally a.s. asymptotically stable* if and only if it is a.s. stable and, for all $X_0 \in \mathbb{R}$,

$$\lim_{t\to\infty} X(t) = 0 \quad a.s.$$

Definition 8.8 The equilibrium solution $X_n^h \equiv 0$ of (8.18) is *a.s. stable* if and only if, for each $\epsilon > 0$, there exists a $\delta > 0$ such that

$$|X_n| < \epsilon, \quad n \geq 0, \quad a.s,$$

whenever $|X_0| < \delta$. The equilibrium is *globally a.s. asymptotically stable* if and only if it is a.s. stable and, for all $X_0 \in \mathbb{R}$,

$$\lim_{n\to\infty} X_n = 0 \quad a.s.$$

If these definitions are not satisfied, then the relevant equilibrium is termed *a.s. unstable*. Notice that it is possible for (8.3) to have an equilibrium that is globally a.s. asymptotically stable but mean-square unstable, if $2r - \sigma^2 < 0$ but $2r + \sigma^2 > 0$.

The article [30] showed that A-stability in the a.s. sense is impossible for the scalar test SDE (8.17), in the sense that for each $\theta \geq \frac{1}{2}$, there exist $\lambda, \sigma \in \mathbb{R}$ and $h > 0$ such that $2\lambda - \sigma^2 < 0$ but the θ-Maruyama method is not a.s. asymptotically stable. The article [11] identifies certain classes of linear stochastic feedback types in higher dimensional test SDEs for which methods that are A-stable in the a.s. sense may be possible.

8.3   Stochastic Notions of Convergence

We turn now to the question of whether or not the error induced by the numerical approximation scheme over a bounded interval of simulation $[0, T]$ can be made arbitrarily small by reducing the stepsize h, and if so at what rate this occurs. We will distinguish between weak convergence, which is concerned with the error of an estimated expected value, and strong convergence, which is concerned with the average error on individual trajectories.

Definition 8.9 Let $(X(t))_{t \in [0,T]}$ be the true solution of the SDE (8.1), and suppose $\{X_n\}_{n=0}^{N}$ is a numerical approximation over a uniform mesh $0 = t_0 < t_1 < \cdots < t_N = T$ where $t_n = nh$ for some fixed stepsize $h = T/N$.

1. $\{X_n\}_{n=0}^{N}$ *converges weakly* with order γ to the true solution of the SDE (8.1) if there exist constants $\bar{h}, C > 0$ such that

$$|\mathbb{E}[P(X_N)] - \mathbb{E}[P(X(T))]| \leq Ch^{\gamma}, \quad \text{for all } h \leq \bar{h}.$$

   where C may depend on T but not on h, and P is some real-valued test function satisfying appropriate conditions.
2. Fix $p \geq 1$. $\{X_n\}_{n=0}^{N}$ *converges strongly* in pth moment with order γ to the true solution of the SDE (8.1) if there exist constants $\bar{h}, C > 0$ such that

$$\left(\mathbb{E}\left[|X_N - X(T)|^p\right]\right)^{1/p} \leq Ch^{\gamma}, \quad \text{for all } h \leq \bar{h},$$

   where C may depend on T but not on h.

As we saw in the discussion at the start of the chapter, the role of the test function P in the definition of weak convergence can be motivated in financial applications by thinking of it as a discounted payoff function in the valuation of European-style financial derivatives. When determining bounds on the order of weak convergence of a numerical method, we will assume that P is a continuous polynomial function of $X(T)$ with constant coefficients. The quantity $|\mathbb{E}[P(X_N)] - \mathbb{E}[P(X(T))]|$ is often referred to as the (unsigned) *weak discretisation error*.

For some applications, such as path-dependent derivatives, we may be interested in the simulation of entire trajectories and in ensuring that these approximate trajectories remain close on average to their continuous-time counterparts. In this case we seek numerical methods that are strongly convergent. Such methods are also necessary in order to guarantee the effectiveness of an advanced variance reduction technique known as multi-level-Monte Carlo, which is treated in Sect. 8.6. If $p = 2$, this type of convergence is referred to as *root-mean-square strong convergence of order γ*. The quantity $\left(\mathbb{E}\left[|X_N - X(T)|^2\right]\right)^{1/2}$ is referred to as the *strong root-mean-square error*.

8.3.1   Theoretical and Numerical Demonstrations of Convergence

In this section, we prove the strong and weak convergence of the Euler-Maruyama numerical approximation method to the true solutions of a diffusion-only SDE.

8.3.1.1 The SDE Model and Its Properties

Motivated by the martingale form of the SABR model SDE (8.12) of a forward price in Sect. 8.1.2.2, we will demonstrate the weak and strong convergence of the explicit Euler method to the true solutions of the SDE

$$dF(t) = G(F(t))dW(t), \quad t \in [0, T].$$ (8.23)

where $F(0) = F_0 > 0$, W is a scalar standard Brownian motion, and the diffusion coefficient G satisfies the following conditions:

1. G satisfies a *linear growth bound*: for all $f \in \mathbb{R}$ there exists a constant $0 < K_1 < \infty$ such that

$$G(f)^2 \leq K_1 \left(1 + f^2\right);$$ (8.24)

2. $G : \mathbb{R} \to \mathbb{R}$ is a *globally Lipschitz continuous function*, so that for all $f_1, f_2 \in \mathbb{R}$, there exists a contant $0 < K_2 < \infty$ such that

$$|G(f_1) - G(f_2)| \leq K_2|f_1 - f_2|.$$ (8.25)

In fact the existence of a linear growth bound is a consequence of global Lipschitz continuity. However since we will use the linear growth bound by itself, it is convenient to state it as a separate condition.

In order to prove our main convergence results, we will make use of the following properties of solutions of (8.23). First, moments of all positive orders are uniformly bounded from above.

Proposition 8.10 *Let F be a solution of (8.23) and suppose that G satisfies the linear growth condition (8.24). Then for any $0 < p < \infty$ there exists $0 < K_3 < \infty$ such that*

$$\mathbb{E}\left[\sup_{r \in [0,T]} |F(r)|^p\right] \leq K_3.$$ (8.26)

This is proved as Theorem 2.4.4 in Mao [53]. Solutions of (8.23) also satisfy the following *Hölder regularity* condition, which characterises how rapidly solutions change over time in mean-square:

Lemma 8.11 *Let F be a solution of (8.23), and suppose that G G satisfies the linear growth condition (8.24). Then there exists $0 < K_4 < \infty$ such that*

$$\mathbb{E}\left[|F(t) - F(s)|^2\right] \leq K_4|t - s|, \quad 0 \leq s < t \leq T. \tag{8.27}$$

Proof We use Itô's isometry (Part 2 of Proposition 1.9 in Chap. 1), followed by the linear growth bound (8.24) to estimate from above

$$\mathbb{E}\left[|F(t) - F(s)|^2\right] = \mathbb{E}\left[\left|\int_s^t G(F(u))dW(u)\right|^2\right]$$

$$\leq \int_s^t \mathbb{E}\left[G^2(F(u))\right] du$$

$$\leq \int_s^t \mathbb{E}\left[K_1(1 + F^2(u))\right] du,$$

by (8.24). Now apply the moment bound (8.26) in the statement of Proposition 8.10, to get

$$\mathbb{E}\left[|F(t) - F(s)|^2\right] \leq K_1 \int_s^t \mathbb{E}\left[1 + \sup_{r \in (s,t)} |F(r)|^2\right] du$$

$$\leq K_4 \int_s^t du = K_4|t - s|,$$

with $K_4 := K_1(1 + K_3)$, as required.

8.3.1.2 The Discretisation and Its Properties

The explicit Euler-Maruyama discretisation of (8.23) over the mesh

$$0 = t_0 < \cdots < t_N = T$$

with $t_n = nh$ and $h = T/N$ is given by

$$F_{n+1} = F_n + G(F_n)\Delta W_{n+1}, \quad n = 0, \ldots, N - 1, \tag{8.28}$$

For all $t \in [0, T]$, let $\bar{F}_h(t)$ be a continuous-time interpolant of (8.28) given over each individual step $[t_n, t_{n+1})$ as

$$d\bar{F}_h(t) = G\left(\bar{F}_h(t_n)\right) dW(t), \quad t \in [t_n, t_{n+1}), \tag{8.29}$$

where $\bar{F}_h(t_n) = F_n$. We can write (8.29) for any $t \in [t_n, t_{n+1}]$ as

$$\bar{F}_h(t) = F_h(t_n) + G\left(\bar{F}_h(t_n)\right)\left(W(t) - W(t_n)\right).$$

By construction, $\bar{F}_h(t_n) = F_n$ and $\bar{F}_h(t_{n+1}) = F_{n+1}$, and intermediate values are determined by an interpolation of the Brownian path.

Finally, we will also make use of the fact that the continuous-time interpolant of the explicit Euler-Maruyama scheme has moments of all orders that are uniformly bounded from above.

Proposition 8.12 *Let $\bar{F}_h$ be a solution of* (8.29) *and suppose that G satisfies the linear growth condition* (8.24). *For any $t \in [0, T]$ and nonzero $0 < p < \infty$ there exists a constant $K_5 < \infty$, independent of the stepsize h, such that*

$$\mathbb{E}\left[|\bar{F}_h(t)|^p\right] \le K_5. \tag{8.30}$$

The proof is similar to that of Theorem 2.4.4 in Mao [53], the statement of which theorem is given as Proposition 8.10 here.

8.3.2 Order of Weak Convergence for the Explicit Euler-Maruyama Scheme

We will use the following result, which is closely related to the theorem of Feynman-Kac, previously encountered as Theorem 2.10 in Chap. 2. It provides a PDE that governs the quantity $\mathbb{E}[P(F(t))]$ when it is considered as a bivariate function of time t and the initial value F_0 of the process solving (8.23).

Proposition 8.13 (Kolmogorov Backward Equation) *Let F be the solution of* (8.23) *with $F(0) = F_0$. Let $P : \mathbb{R} \to \mathbb{R}$ be a polynomial function, and set*

$$V(t, F_0) := \mathbb{E}[P(F(t))]. \tag{8.31}$$

Then the function $V \in C^{1,2}([0, T] \times \mathbb{R})$ satisfies the PDE

$$V_t(t, f) = \frac{1}{2}G(f)^2 V_{ff}(t, f), \quad t \in [0, T],\ f \in \mathbb{R}; \tag{8.32}$$

$$V(0, f) = P(f).$$

This result is given, with proof, as Proposition 8.43 in Lord et al. [50]. That text also provides a more general version of the following weak convergence theorem, which we state and prove as a special case here.

Theorem 8.14 *The explicit Euler-Maruyama method given by (8.29) converges weakly with order $\gamma = 1$ to the true solution of the SDE (8.23).*

Proof Let $V \in C^{1,2}([0, T] \times \mathbb{R})$. For each $n = 0, \ldots, N - 1$ set $\bar{t} = t_n$ when $t \in [t_n, t_{n+1})$. Apply Itô's formula, given as Theorem 1.7 in Chap. 1 to (8.29) to get

$$dV(t, \bar{F}_h(t)) = \left(V_t(t, \bar{F}_h(t)) + \frac{1}{2}G^2(\bar{F}_h(\bar{t}))V_{ff}(t, \bar{F}_h(t))) \right) dt$$

$$+ G(\bar{F}_h(\bar{t}))V_f(t, \bar{F}_h(t))dW(t), \quad t \in [0, T]. \tag{8.33}$$

Now suppose that V is defined according to (8.31), so that it also satisfies the PDE (8.32). We have, since $F_0 = \bar{F}_h(0)$,

$$\mathbb{E}\left[P(F(T)) - P(\bar{F}_h(T)) \right] = V(T, F_0) - V(0, \bar{F}_h(T))$$

$$= V(T, \bar{F}_h(0)) - V(0, \bar{F}_h(T)). \tag{8.34}$$

Substitute (8.33) to the RHS of (8.34) and take the expectation again. Since the Itô integral has zero expectation (see Proposition 1.9 in Chap. 1), we get

$$\mathbb{E}\left[P(F(T)) - P(\bar{F}_h(T)) \right]$$

$$= \mathbb{E}\left[V(T, \bar{F}_h(0)) - V(0, \bar{F}_h(T)) \right]$$

$$= \mathbb{E}\left[\int_0^T V_t(T - s, \bar{F}_h(s)) + \frac{1}{2}G^2(\bar{F}_h(\bar{s}))V_{ff}(T - s, \bar{F}_h(s))ds \right]$$

$$= \mathbb{E}\left[\int_0^T \left[-\frac{1}{2}G^2(\bar{F}_h(s))V_{ff}(T - s, \bar{F}_h(s)) + \frac{1}{2}G^2(\bar{F}_h(\bar{s}))V_{ff}(T - s, \bar{F}_h(s)) \right] ds \right].$$

At the second-to-last step we used the backward Kolmogorov equation (8.32) with substitution $t \mapsto T - t$, so that $-V_t(T - t, f) = V_{ff}(T - t, f)$, and this allowed us to eliminate the time derivative of V. Now decompose the term

$$- G^2(\bar{F}_h(s))V_{ff}(T - s, \bar{F}_h(s)) + G(\bar{F}_h(\bar{s}))V_{ff}(T - s, \bar{F}_h(s))$$

$$= Q^{(1)}(s, \bar{F}_h(s)) - Q^{(1)}(\bar{s}, \bar{F}_h(\bar{s})) + Q^{(2)}(\bar{s}, \bar{F}_h(\bar{s})) - Q^{(2)}(s, \bar{F}_h(s)),$$

where we have defined the new polynomial functions

$$Q^{(1)}(s, f) := G^2(\bar{F}_h(\bar{s}))V_{ff}(T - s, f);$$

$$Q^{(2)}(s, f) := G^2(f)V_{ff}(T - s, f).$$

For each of $Q^{(1)}$ and $Q^{(2)}$, an application of Itô's formula followed by taking expectation gives, for $i = 1, 2$ and $s \in [0, T]$,

$$\mathbb{E}\left[Q^{(i)}(s, \bar{F}_h(s))\right] = \mathbb{E}\left[Q^{(i)}(\bar{s}, \bar{F}_h(\bar{s}))\right]$$
$$+ \int_{\bar{s}}^{s} \mathbb{E}\left[Q_r^{(i)}(r, \bar{F}_h(r)) + \frac{1}{2}G^2(\bar{F}_h(\bar{r}))Q_{ff}^{(i)}(r, \bar{F}_h(r))\right] dr.$$

Now define

$$\widetilde{Q}^{(i)}(r, f) := Q_r^{(i)}(r, f) + \frac{1}{2}G^2(\bar{F}_h(\bar{r}))Q_{ff}^{(i)}(r, f). \tag{8.35}$$

Exchanging the order of the partial derivatives and applying the backward Kolmogorov equation (8.32) again, we have

$$\frac{\partial}{\partial s}V_{ff}(T - s, f) = -\frac{1}{2}G^2(f)V_{ffff}(T - s, f).$$

Each $Q^{(i)}$ is a polynomial function in f and therefore by (8.24) and (8.35), when $f = \bar{F}_h(r)$, each $\widetilde{Q}^{(i)}$ is bounded above by a polynomial function in the same argument. So there exists some constant $0 < K_6 < \infty$ and integer $p \geq 2$ such that

$$\mathbb{E}\left[\widetilde{Q}^{(i)}(r, \bar{F}_h(r))\right] \leq K_6\mathbb{E}\left[1 + |\bar{F}_h(r)|^p\right].$$

Finally we have, using the bound (8.30) from the statement of Proposition 8.12

$$\mathbb{E}\left[P(F(T)) - P(\bar{F}_h(T))\right]$$
$$= \int_0^T \int_{\bar{s}}^{s} \frac{1}{2}\left(\mathbb{E}\left[\widetilde{Q}^{(1)}(r, \bar{F}_h(r))\right] + \mathbb{E}\left[\widetilde{Q}^{(2)}(r, \bar{F}_h(r))\right]\right) dr\, ds$$
$$\leq K_6 \int_0^T \int_{\bar{s}}^{s} \mathbb{E}\left[1 + |\bar{F}_h(r)|^p\right] dr\, ds$$
$$\leq K_6 T(1 + K_5)h, \tag{8.36}$$

where in the last step we have bounded $s - \bar{s}$ from above by h. By comparing (8.36) with Part 1 of Definition 8.9 we see that the statement of the Theorem is proved.

An order of convergence of $\gamma = 1$ means that if we wish to decrease the weak error by a factor of 10, we should reduce our stepsize by a factor of 10. We will demonstrate this numerically. Let X be a solution of the linear test SDE (8.17) with $X_0 = 1$, $\lambda = -0.5$, $\sigma = 0.3$, so that $\lambda + \sigma^2/2 = -0.455 < 0$ and the equilibrium solution $X(t) \equiv 0$ is mean-square asymptotically stable. Let $\{X_n^h\}_{n=0}^{N}$ be the explicit Euler-Maruyama approximation of (8.17) over a uniform mesh with N steps each of length $h = T/N$. Using the identity $P(x) = x$ as test function, denote the weak discretisation error at time $T = 1$ as

$$\mathcal{E}_{\text{weak}}(h) := \left| \mathbb{E}[X(T)] - \mathbb{E}\left[X_N^h \right] \right|$$

By Theorem 8.14, there exist constants $0 < C < \infty$ and $0 < \bar{h} < \infty$ such that, for all $h < \bar{h}$, $\mathcal{E}_{\text{weak}}(h) \leq Ch^\gamma$, where $\gamma = 1$. Take logs of both sides of this inequality to get

$$\log(\mathcal{E}_{\text{weak}}(h)) \leq \log(C) + \gamma \log(h).$$

If we numerically approximate $\mathcal{E}_{\text{weak}}(h)$ using a Monte Carlo estimation of $\mathbb{E}\left[X_N^h \right]$ with M samples, for a set of decreasing values of h (for example $h_l = 2^{-l}$, for $l = 3, 4, \ldots, 7$), and plot the estimator $\widehat{\mathcal{E}}_{\text{weak}}^M(h_l)$ against h_l on a log-log scale, we should observe a linear relationship, and the slope of the regression line can be taken as a numerical estimate of the order of weak convergence.

The code for this is as follows, and a log-log plot of the weak error as a function of stepsize is displayed in Fig. 8.5, showing a numerical approximation of the order of weak convergence that is consistent with the statement of Theorem 8.14.

```
In []: # Model and discretisation parameters
       X0=1; T=1; lam=-0.5; sig=0.3
       M=10**6


       # Compute exact mean of X(T)
       muX=X0*np.exp(r*T)


       # Set up an array containing 5 values of N and
       # corresponding stepsizes
       NVals=2**np.arange(3,8); dtVals=T/NVals


       # Create an array to contain the weak error estimates
       # for each N
       weakError=np.zeros(len(NVals))


       # At each level corresponding to a value of N, compute
       # mean of explicit Euler approximation at time T
       for k in range(len(NVals)):
           N=NVals[k]; dt=dtVals[k]


           # Explicit Euler discretisation across M
           # trajectories at this level
           inc=rng.normal(0,np.sqrt(dt),(M,N))
           W=np.c_[np.zeros(M),np.cumsum(inc,axis=1)]
```

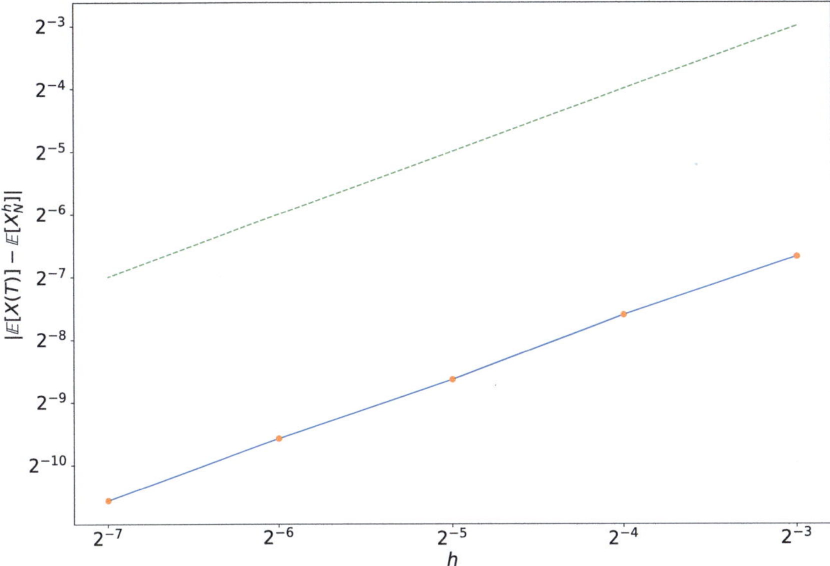

Fig. 8.5 Numerical demonstration of weak convergence with order $\gamma = 1$ of the explicit Euler-Maruyama method applied to the SDE (8.17) with $\lambda = -0.5$, $\sigma = 0.3$, $T = 1$, and $X_0 = 1$. Monte Carlo estimates of the weak error (orange dots), using $M = 10^6$ simulated trajectories, are plotted against h on a log-log-scale. A reference line (green, dashed) with slope 1 is provided for comparison

```
drift=dt*lam

EEMS=X0*np.ones((M,N+1))

for i in range(N):
    EEMS[:,i+1]=EEMS[:,i]*(1+drift+sig*inc[:,i])

    # Compute weak error at this level
    weakError[k]=abs(np.mean(EEMS[:,-1])-muS)

# Output weak errors, stepsizes, and slope of the
# regression line
print(weakError)
print(dtVals)
print(np.polyfit(np.log(dtVals),np.log(weakError),1)[0])
```

```
Out[]: [0.00976105 0.00513551 0.00250763 0.0013095  0.00065957]
       [0.125      0.0625     0.03125    0.015625   0.0078125]
       0.9746379143235475
```

8.3.3   Order of Strong Convergence for the Explicit Euler-Maruyama Scheme

In this section we show that the order of root-mean-square strong convergence of the explicit Euler method is $\gamma = 1/2$. This is lower than the order of the bound determined for weak convergence given in Theorem 8.14. It predicts that, in order to reduce the strong root-mean-square error by a factor of 10, we should reduce the stepsize by a factor of 100. This difference is visible in practice: see Fig. 8.6.

By contrast with the proof of weak convergence order, which computed an estimate of the weak error at time T by integrating over the entire interval of simulation from the beginning, our approach for strong convergence will be to, first, determine a bound for the local mean-square strong error produced over an individual step, and then to show the cumulative effect of these errors as they propagate over the interval of simulation. For the latter step we will need the following result, which can be found, for example, in Stuart and Humphries [64].

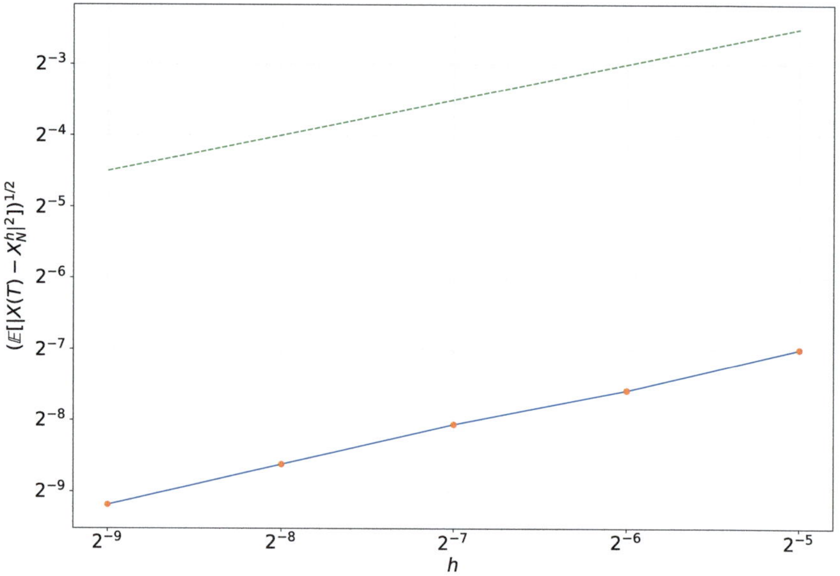

Fig. 8.6 Numerical demonstration of strong root-mean-square convergence with order $\gamma = 1/2$ of the explicit Euler-Maruyama method applied to the SDE (8.17) with $\lambda = -0.5, \sigma = 0.3, T = 1$, and $X_0 = 1$. Monte Carlo estimates of the strong error (orange dots), using $M = 10^3$ simulated trajectories, are plotted against h on a log-log-scale. A reference line (green, dashed) with slope $1/2$ is provided for comparison

Proposition 8.15 (Discrete Gronwall Inequality)   *Let $(u_n)_{n=1}^{\infty}$ and $(v_n)_{n=1}^{\infty}$ be non-negative sequences satisfying, for some $\alpha > 0$*

$$u_n \le \alpha + \sum_{k=0}^{n-1} u_k v_k, \quad n \in \mathbb{N} \setminus \{0\}.$$

Then

$$u_n \le \alpha \exp\left(\sum_{k=0}^{n-1} v_k \right), \quad n \in \mathbb{N} \setminus \{0\}.$$

Now we can state and prove our main strong convergence theorem.

Theorem 8.16 *The explicit Euler-Maruyama method given by (8.29) displays root-mean-square strong convergence with order $\gamma = 1/2$ to the true solution of the SDE (8.23).*

Proof Fix $n = 0, \ldots, N - 1$. For any $t \in [t_n, t_{n+1}]$ the signed pathwise error at time t is given by

$$E(t) := F(t) - \bar{F}_h(t), \tag{8.37}$$

and may be written in stochastic integral equation form for any $t \in [t_n, t_{n+1}]$ as

$$E(t) = E(t_n) + \int_{t_n}^{t} \left(G(F(s)) - G(\bar{F}_h(t_n)) \right) dW(s), \quad a.s. \tag{8.38}$$

Apply Itô's formula to (8.38) with the transformation $V(E) = E^2$ and take expectations on both sides, noting that the Itô integral has zero expectation:

$$\mathbb{E}\left[E(t)^2 \right]$$

$$= \mathbb{E}\left[E(t_n)^2 \right] + \int_{t_n}^{t} \mathbb{E}\left[\left(G(F(s)) - G(\bar{F}_h(t_n)) \right)^2 \right] ds$$

$$= \mathbb{E}\left[E(t_n)^2 \right] + \int_{t_n}^{t} \mathbb{E}\left[\left(G(F(t_n)) - G(\bar{F}_h(t_n)) + G(F(s)) - G(F(t_n)) \right)^2 \right] ds.$$

Now set $t = t_{n+1}$ and apply the elementary inequality $(a + b)^2 \leq 2a^2 + 2b^2$ to the RHS to get

$$\mathbb{E}\left[E(t_{n+1})^2\right] \leq \mathbb{E}\left[E(t_n)^2\right] + 2\int_{t_n}^{t_{n+1}} \mathbb{E}\left[\left(G(F(t_n)) - G(\bar{F}_h(t_n))\right)^2\right]ds$$

$$+ 2\int_{t_n}^{t_{n+1}} \mathbb{E}\left[\left(G(F(s)) - G(F(t_n))\right)^2\right]ds.$$

Next, we make use of the global Lipschitz bound (8.25) to get

$$\mathbb{E}\left[E(t_{n+1})^2\right] \leq \mathbb{E}\left[E(t_n)^2\right] + 2K_2^2\int_{t_n}^{t_{n+1}} \mathbb{E}\left[E(t_n)^2\right]ds$$

$$+ 2K_2^2\int_{t_n}^{t_{n+1}} \mathbb{E}\left[|F(s) - F(t_n)|^2\right]ds,$$

where we used the fact that $E(t_n) = F(t_n) - \bar{F}_h(t_n)$, by (8.37). Resolve the first integral on the RHS, since the integrand is constant over the interval of integration, and apply the Hölder regularity bound given by (8.27) in the statement of Lemma 8.11. This yields the following bound on the local (one-step) error of the scheme in mean-square:

$$\mathbb{E}\left[E(t_{n+1})^2\right] \leq \mathbb{E}\left[E(t_n)^2\right] + 2K_2^2 h\mathbb{E}\left[E(t_n)^2\right] + 2K_2^2\int_{t_n}^{t_{n+1}} |s - t_n|ds$$

$$\leq \mathbb{E}\left[E(t_n)^2\right] + 2K_2^2 h\mathbb{E}\left[E(t_n)^2\right] + 2K_2^2 h^2.$$

This can be rewritten

$$\mathbb{E}\left[E(t_{n-1})^2\right] - \mathbb{E}\left[E(t_n)^2\right] \leq 2K_2^2 h\mathbb{E}\left[E(t_n)^2\right] + 2K_2^2 h^2. \tag{8.39}$$

Now we compute the cumulative effect of all the local errors propagating over the interval $[0, T]$. For any $n = 1, \ldots, N$, sum both sides of (8.39) over $k = 0, \ldots, n - 1$ and use the fact that $nh \leq T$:

$$\sum_{k=0}^{n-1}\left(\mathbb{E}\left[E(t_{k+1})^2\right] - \mathbb{E}\left[E(t_k)^2\right]\right) \leq \sum_{k=0}^{n-1} 2K_2^2 h\mathbb{E}\left[E(t_k)^2\right] + 2K_2^2 Th. \tag{8.40}$$

The LHS of (8.40) is a telescoping sum, and when written out in full one can see that all terms except for the first and last will cancel:

$$\sum_{k=0}^{n-1} \left(\mathbb{E}\left[E(t_{k+1})^2 \right] - \mathbb{E}\left[E(t_k)^2 \right] \right) = \mathbb{E}\left[E(t_n)^2 \right] - \mathbb{E}\left[E(t_0)^2 \right].$$

Since $E(t_0)^2 = 0$ we get

$$\mathbb{E}\left[E(t_n)^2 \right] \leq K_7 h + \sum_{k=0}^{n-1} \mathbb{E}\left[E(t_k)^2 \right] K_8 h,$$

where $K_7 := 2K_2^2 T$ and $K_8 := K_7 T$ are new constants. Now apply Gronwall's inequality (see Proposition 8.15) with $u_n = \mathbb{E}\left[E(t_n)^2 \right]$, $v_n \equiv K_8 h$, and $\alpha = K_7 h$, giving

$$\mathbb{E}\left[E(t_N)^2 \right] \leq K_7 h \exp\left(\sum_{n=0}^{N-1} h K_8 \right) = K_7 h e^{T K_8} = K_9 h,$$

with $K_9 := K_7 e^{T K_8}$. Taking square roots on both sides and recalling that $t_N = T$ gives the result, by reference to Part 2 of Definition 8.9.

Just as in Sect. 8.3.2, we can provide a numerical demonstration of the order of strong root-mean-square convergence of the explicit Euler-Maruyama method using the test equation (8.17) with the same parameter values. Here we denote the strong error as

$$\mathcal{E}_{\text{strong}}(h) := \left(\mathbb{E}\left[\left| X(T) - X_N^h \right|^2 \right] \right)^{1/2}.$$

By Theorem 8.16, there exist constants $0 < C < \infty$ and $0 < \bar{h} < \infty$ such that, for all $h < \bar{h}$, $\mathcal{E}_{\text{strong}}(h) \leq C h^\gamma$, where $\gamma = 1/2$. Figure 8.6 displays values of the Monte Carlo estimator $\widehat{\mathcal{E}}_{\text{strong}}(h_l)$ for $h_l = 2^{-l}$, $l = 5, \ldots, 9$, on a log-log plot. The observed slope of the linear regression line is approximately 0.5, as expected.

The Python code necessary to produce these plots is as follows. Notice that since strong error can be viewed as an average pathwise error, we use the same ensemble of Brownian trajectories at all levels of the mesh to ensure that convergence is observed, and NumPy array slicing is very useful for this. It was not necessary to preserve the ensemble between levels in order to demonstrate weak convergence.

```
In []: # Model and discretisation parameters
       X0=1; T=1; r=-0.5; sig=0.3
       M=10**3

       # Set up an array containing 5 values of N and
       # corresponding stepsizes
       NVals=2**np.arange(5,10); dtVals=T/NVals

       # Create an array to contain the strong error estimates
       # for each N
       strongErr=np.zeros(len(NVals))

       # Generate M Brownian trajectories at the smallest
       # stepsize
       inc = rng.normal(0,np.sqrt(dtVals[-1]),(M,NVals[-1]))
       W=np.c_[np.zeros(M),np.cumsum(inc,axis=1)]
       t=np.arange(0,T+dtVals[-1],dtVals[-1])

       # Generate X(T) on M trajectories
       exactX = X0*np.exp((r-0.5*sig**2)*t[-1]+sig*W[:,-1])

       # Reverse the order of N and dt
       NFlip=np.flip(NVals); dtFlip=np.flip(dtVals)

       for k in range(len(NVals)):
           N=NFlip[k]; dt=dtFlip[k]

           # Use array slicing to step through W and
           # select values for current level
           Wk=W[:,::2**k]

           # Explicit Euler discretisation across the same M
           # trajectories
           EEMS=X0*np.ones((M,N+1))

           for i in range(N):
               EEMS[:,i+1]=EEMS[:,i]
                           *(1+dt*r+sig*(Wk[:,i+1]-Wk[:,i]))

           # Compute strong error at this level
           strongErr[k]=np.sqrt(np.mean((EEMS[:,-1]-exactS)**2))

       print(strongErr)
       print(dtFlip)
       print(np.polyfit(np.log(dtFlip),np.log(strongErr),1)[0])
```

```
Out[]:  [0.0017173   0.00254686 0.00375632 0.0052277   0.00781713]
        [0.00195312 0.00390625 0.0078125  0.015625    0.03125   ]
        0.5410461543178742
```

8.4   The Milstein Scheme

Let F be a solution of the SDE forward price model (8.23). We have just seen that
the order of weak convergence of the explicit Euler-Maruyama scheme is $\gamma = 1$,
but that the order of strong root-mean-square convergence is $\gamma = 1/2$. It is possible
to construct a numerical method that converges strongly with order of convergence
1 rather than 1/2 by changing the way that we approximate the Itô integral on the
RHS.

8.4.1   Motivation in the 1-Dimensional Diffusion-Only Case

Consider the SDE (8.23). Suppose that instead of approximating G at the left hand
point of the interval of integration we use a Taylor expansion around the point $F(t_n)$
that has been truncated at first order

$$G(F(t)) \approx G(F(t_n)) + G'(F(t_n))(F(t) - F(t_n)), \quad t \in [t_n, T]. \tag{8.41}$$

We can also approximate

$$F(t) - F(t_n) \approx G(F(t_n))(W(t) - W(t_n)), \quad t \in [t_n, T]. \tag{8.42}$$

Substituting (8.41) and (8.42) with $t = s$ into the RHS of the integral form of (8.23)
yields the approximation

$$\int_{t_n}^{t_{n+1}} G(F(s))dW(s) \approx \int_{t_n}^{t_{n+1}} G(F(t_n))dW(s)$$

$$+ G'(F(t_n))G(F(t_n)) \int_{t_n}^{t_{n+1}} [W(s) - W(t_n)]dW(s). \tag{8.43}$$

Exercise 8.7 asks you to show that

$$\int_{t_n}^{t_{n+1}} G(F(s))dW(s) \approx G(F(t_n))\Delta W_{n+1}$$

$$+ \frac{1}{2} G'(F(t_n))G(F(t_n)) \left[\Delta W_{n+1}^2 - h \right]. \tag{8.44}$$

This yields the difference form of the Milstein scheme for (8.23), given by

$$F_{n+1} = F_n + G(F(t_n))\Delta W_{n+1} + \frac{1}{2}G'(F(t_n))G(F(t_n))\left[\Delta W_{n+1}^2 - h\right],$$

for $n = 0, \ldots, N - 1$. Notice that if $G(f) \equiv c$ for some constant $c \in \mathbb{R}$, the Euler-Maruyama and Milstein schemes coincide, since in this case $G'(f) \equiv 0$: see Exercise 8.3.

8.4.2　The General Form of the Milstein Scheme

The Milstein scheme can be extended to d-dimensional systems of SDEs with nonzero drift coefficient and m independent noise perturbations given by (8.1) with integral form (8.2).

For $n \in \mathbb{N}$, $s \in [t_n, t_{n+1}]$ and given $\bar{X}_h(t_n)$, the Milstein scheme for (8.1), interpolated over the interval $[t_n, t_{n+1}]$, is given by

$$\bar{X}_h(s) := \bar{X}_h(t_n) + f\big(\bar{X}_h(t_n)\big)|s - t_n| + \sum_{i=1}^{m} G_i\big(\bar{X}_h(t_n)\big)I_i^{t_n,s}$$

$$+ \sum_{i,j=1}^{m} G_i'\big(\bar{X}_h(t_n)\big)G_j\big(\bar{X}_h(t_n)\big)I_{j,i}^{t_n,s}, \qquad (8.45)$$

where, denoting $G_{i,j}(x)$ the jth entry of the $\mathbb{R}^d$-valued function $G_i(x)$, we define

$$G_i'(x) := \begin{pmatrix} \frac{\partial}{\partial x_1}G_{i,1}(x) & \cdots & \frac{\partial}{\partial x_d}G_{i,1}(x) \\ \vdots & \ddots & \vdots \\ \frac{\partial}{\partial x_1}G_{i,d}(x) & \cdots & \frac{\partial}{\partial x_d}G_{i,d}(x) \end{pmatrix}$$

to be the Jacobian matrix of $G_i(x)$ for any $x \in \mathbb{R}^d$ and, following [6, 67], the stochastic integral and the iterated stochastic integral are defined as

$$I_i^{t_n,s} := \int_{t_n}^{s} dW_i(r), \qquad I_{j,i}^{t_n,s} := \int_{t_n}^{s}\int_{t_n}^{r} dW_j(p)dW_i(r).$$

Making use of the relations

$$I_{i,i}^{t_n,s} = \frac{1}{2}\left((I_i^{t_n,s})^2 - |s - t_n|\right); \qquad I_{i,j}^{t_n,s} + I_{j,i}^{t_n,s} = I_i^{t_n,s}I_j^{t_n,s}.$$

we can expand the last term in (8.45) to get

$$\sum_{i,j=1}^{m} G_i'\big(\bar{X}_h(t_n)\big)G_j\big(\bar{X}_h(t_n)\big)I_{j,i}^{t_n,s}$$

$$= \frac{1}{2}\sum_{i=1}^{m} G_i'\big(\bar{X}_h(t_n)\big)G_i\big(\bar{X}_h(t_n)\big)\left(\big(I_i^{t_n,s}\big)^2 - |s - t_n|\right)$$

$$+ \frac{1}{2}\sum_{\substack{i,j=1\\i<j}}^{m}\left(G_i'\big(\bar{X}_h(t_n)\big)G_j\big(\bar{X}_h(t_n)\big) + G_j'\big(\bar{X}_h(t_n)\big)G_i\big(\bar{X}_h(t_n)\big)\right)I_i^{t_n,s}I_j^{t_n,s}$$

$$+ \sum_{\substack{i,j=1\\i<j}}^{m}\left(G_i'\big(\bar{X}_h(t_n)\big)G_j\big(\bar{X}_h(t_n)\big) - G_j'\big(\bar{X}_h(t_n)\big)G_i\big(\bar{X}_h(t_n)\big)\right)A_{ij}^{t_n,s}. \tag{8.46}$$

On the last line the terms $A_{ij}^{t_n,s}$ are called *Lévy areas* (see for example [46]), and are defined by

$$A_{ij}^{t_n,s} := \frac{1}{2}\left(I_{i,j}^{t_n,s} - I_{j,i}^{t_n,s}\right).$$

In general, full implementation of the Milstein scheme requires that we sample from these Lévy areas. See Malham and Wiese [52] for an efficient approach to solving this problem. If a commutativity condition of the form

$$G_i'(x)G_j(x) = G_j'(x)G_i(x)$$

holds for all $i, j = 1, \ldots, m$ and $x \in \mathbb{R}^d$, the last term in (8.46) is zero, and we can avoid this step. This is always the case when $d = 1$.

Figure 8.7 demonstrates weak and strong convergence plots for the Milstein method applied to the scalar linear SDE (8.17). The slope of the linear regression line for the weak error plot (top) is computed to be 1.0416462848555017, and the slope for the strong error plot (bottom) is computed to be 0.9588811215119359. Both display order of convergence close to 1.

8.4.3 Mean-Square Asymptotic Stability of the Milstein Scheme

Consider the following variant of the Milstein scheme, proposed by Saito and Mitsui [60], which uses the same semi-implicit discretisation of the drift coefficient as the θ-Maruyama method, applied to the scalar test SDE (8.17) with real parameters $\lambda, \sigma \in \mathbb{R}$.

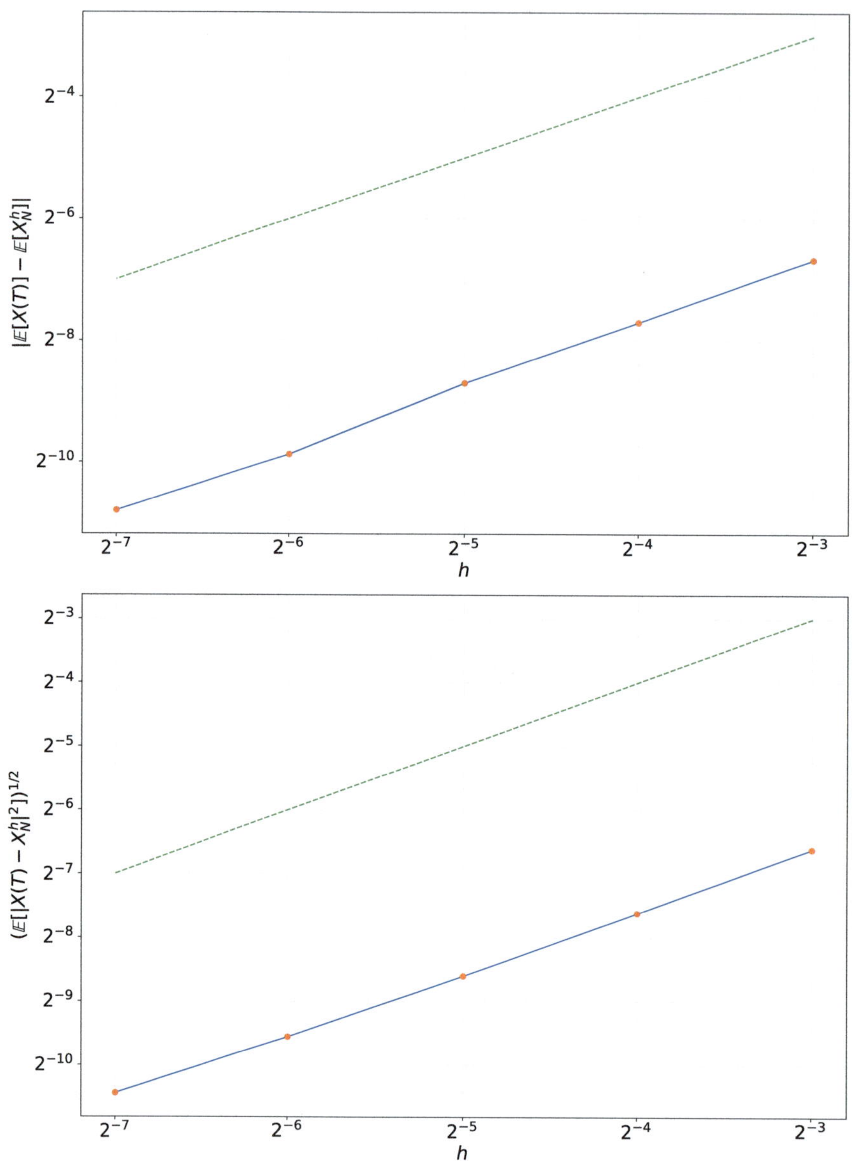

Fig. 8.7 Numerical demonstration of convergence of the Milstein method applied to the SDE (8.17) with $\lambda = -0.5$, $\sigma = 0.1$, $T = 1$, and $X_0 = 1$. Monte Carlo estimates of the weak error (top) using $M = 10^3$ simulated trajectories, and strong root-mean-square error (bottom) using $M = 10^3$ simulated trajectories, are plotted against h on a log-log-scale. A reference line (green, dashed) with slope 1 is provided for comparison in both cases

$$X_{n+1}^h = X_n^h + (1-\theta)h\lambda X_n^h + \theta h\lambda X_{n+1}^h$$

$$+ \sigma X_n^h \Delta W_{n+1} + \frac{1}{2}\sigma^2 X_n^h \left(\Delta W_{n+1}^2 - h\right), \quad n = 0, \ldots, N-1. \qquad (8.47)$$

Exercise 8.8 asks you to show that the equilibrium solution $X_n^h \equiv 0$ of (8.47) is mean-square asymptotically stable if and only if h and θ are such that

$$\lambda + \frac{1}{2}\sigma^2 + \frac{1}{2}h\left((1-2\theta)\lambda^2 + \frac{1}{2}\sigma^4\right) < 0, \qquad (8.48)$$

and this condition determines the mean-square stability region of the scheme.

As in Sect. 8.2.2, if we define coordinates $(x, y) = (h\lambda, h\sigma^2)$, we can visually compare mean-square stability regions for the test SDE (8.17) and those of the semi-implicit Milstein scheme (8.47), which are characterised by the inequalities

$$x + \frac{1}{2}y < 0; \quad 2\left(x + \frac{1}{2}y\right) + (1-2\theta)x^2 + \frac{1}{2}y^2 < 0,$$

respectively. Figure 8.8 shows this for the explicit ($\theta = 0$), trapezoid rule ($\theta = 1/2$), and implicit ($\theta = 1$) versions of the method. We can see that, by contrast with the θ-Maruyama method, the semi-implicit Milstein method is not A-stable in mean-square for values of $\theta \in [0, 1]$, in spite of its faster rate of root-mean-square strong convergence. It was shown in [29] that A-stability in mean square could be recovered for a method of this kind by choosing $\theta \geq 3/2$, though outside of a scalar linear test case it may be difficult to predict what other qualitative distortions are induced by choosing θ in this way.

8.5   A Review of Advanced Numerical Methods for Nonlinear Financial Models

The analysis of weak and strong convergence of the Euler-Maruyama scheme applied to the diffusion-only SDE (8.23) presented in Sect. 8.3 makes the assumption that the diffusion coefficient G is globally Lipschitz continuous (see (8.25)) and satisfies a linear growth bound (see (8.24)). These assumptions are not generally satisfied by the coefficients of the SABR, CEV, and CIR models, and their convergence needs to be examined separately.

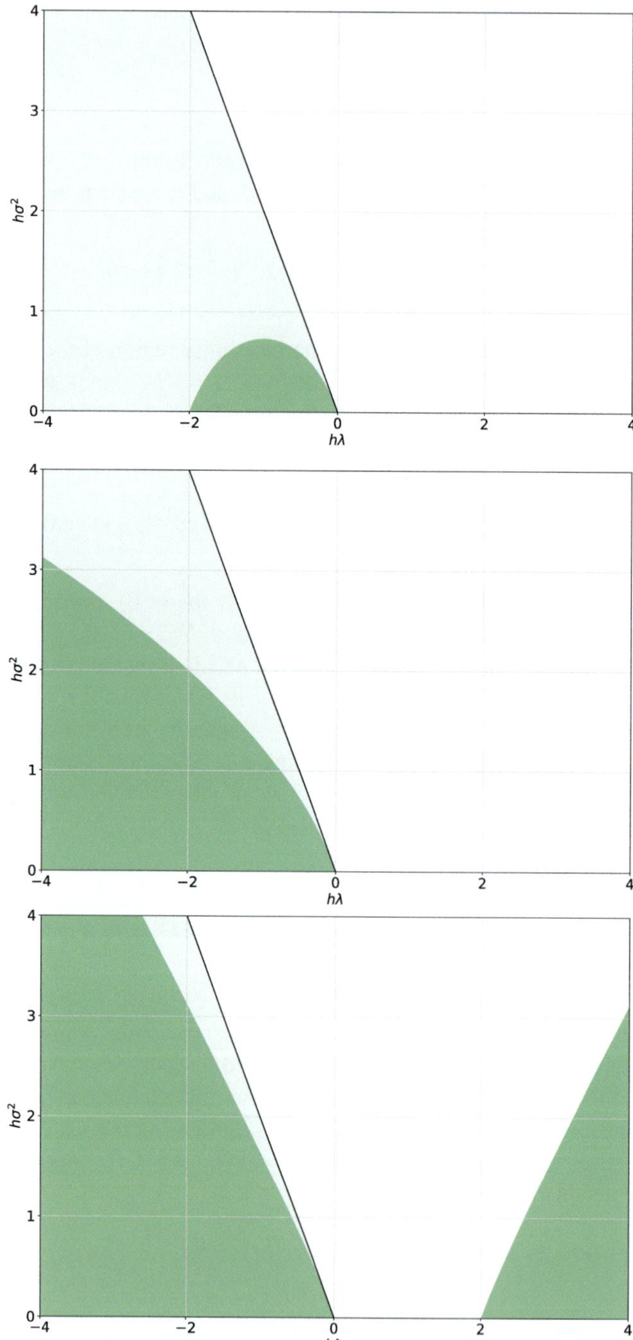

Fig. 8.8 The area to the left of the black line, shaded light green in the first plot, represents S_{SDE}, the mean-square asymptotic stability region for the linear test equation (8.17). The dark green shaded areas represent $S_{\mathrm{Mil}}(\theta, h)$, mean-square asymptotic stability regions for the explicit ($\theta = 0$, top), trapezoid rule ($\theta = 1/2$, middle), and implicit ($\theta = 1$, bottom) Milstein schemes, respectively

8.5.1   Weakly Convergent Numerical Methods

Consider a class of SDE models given by

$$dX(t) = f(X(t))dt + \sigma|X(t)|^{\beta}dW(t), \quad t \in [0, T], \tag{8.49}$$

where $X(0) = X_0 \in \mathbb{R}$, f is a globally Lipschitz continuous function, and $\beta \in [1/2, 1)$.

This class of SDEs includes the CEV model: set $f(x) = \lambda(\mu - x)$; the Cox-Ingersoll-Ross model: set $f(x) = \lambda(\mu - x)$ and $\beta = 1/2$; and the forward price process in the SABR model: set $f(x) \equiv 0$. The function $G(x) = \sigma|x|^{\beta}$ does not satisfy a global Lipschitz condition. In particular, when $\beta = 1/2$, G is the diffusion coefficient for the CIR model (8.10), does not satisfy even a local Lipschitz condition, and has an unbounded gradient near zero. The development of efficient and effective numerical methods for such models is an active area of research.

Bossy and Diop [9] analysed the symmetrised Euler scheme for the SDE (8.49), defined over the usual mesh as

$$X_{n+1} = \left| X_n + f(X_n)h + \sigma X_n^{\beta} \Delta W_{n+1} \right|, \quad n = 0, \ldots, N - 1,$$

and demonstrated weak convergence of order 1 when f is sufficiently smooth, $\alpha = 1/2$ and $b(0) > \sigma^2$. For the Cox-Ingersoll-Ross model, the latter condition is equivalent to $\lambda\mu > \sigma^2$. In the case where $\beta \in (1/2, 1)$, which includes the CEV and SABR models, weak convergence of order 1 is ensured without constraints on f.

8.5.2   Strongly Convergent Numerical Methods

Let us now turn to strongly convergent schemes, with a specific focus on the CIR model (8.10). There are broadly three approaches to its strong numerical approximation: discretise (8.10) directly using an Euler-type method, discretise the Lamperti transform of (8.10) using an Euler-type method, and the truncated Milstein method, which merits its own category. In this section we will describe a representative selection of methods and summarise their known convergence properties.

8.5.2.1 Direct Approximation

As we saw in Sect. 8.1.2.1, in order for a direct Euler discretisation of (8.10) to be well-defined, it is necessary to handle solution values that become negative by truncation or symmetrisation. We review two of the most commonly used of these in practice here.

The first method is the *fully truncated Euler scheme*, which produces an approximate solution $\{r_n^h\}_{n=0}^N$ characterised by the two-stage relation

$$\tilde{r}_{n+1}^h = \tilde{r}_n^h + h\lambda\left(\mu - \max\{\tilde{r}_n^h, 0\}\right) + \sigma\sqrt{\max\left\{\tilde{r}_n^h, 0\right\}}\,\Delta W_{n+1};$$

$$r_{n+1}^h = \max\left\{\tilde{r}_{n+1}^h, 0\right\}, \quad n = 0, \ldots, N-1,$$

where $\tilde{r}_0 = r_0$. This scheme was proposed by Lord, Koekkoek and van Dijk [49]. They showed that it converges strongly in first moment and that in comparative empirical testing it appeared to minimise the bias introduced by truncation of the solution at zero. Cozma and Reisinger [19] demonstrated order $\gamma = 1/2$ strong convergence in pth mean for $p \in [2, 2\kappa\mu/\sigma^2 - 1)$ when $\kappa\mu > 3\sigma^2/2$. With $p = 2$, this gives strong root-mean-square convergence with order $\gamma = 1/2$ for $2\kappa\mu > 3\sigma^2$.

The second method is a *partially truncated Euler scheme*, given by

$$r_{n+1} = r_n + \lambda(\mu - r_n) + \sigma\sqrt{|r_n|}\,\Delta W_{n+1}, \quad n = 0, \ldots, N-1.$$

Higham and Mao [34] demonstrated strong convergence (without specifying a polynomial order) in pth mean for $p = 1$ and $p = 2$.

8.5.2.2 Approximation of the Lamperti Transform

Recall next from Chap. 7 that if we set $s(t) = \sqrt{r(t)}$, where r is the solution of (8.10), then

$$ds(t) = (\alpha s^{-1}(t) - \beta s(t))dt + \bar{\gamma}dW(t), \quad t \in [0, T]. \tag{8.50}$$

In (8.50), $\alpha := (4\lambda\mu - \sigma^2)/8$, $\beta := \lambda/2$, and $\bar{\gamma} := \sigma/2$. The SDE (8.50) is well-defined as long as Feller's condition $2\lambda\mu > \sigma^2$ holds. Since there is no square-root dependence on the state variable s in the diffusion coefficient we do not need to worry about potential negative values causing the approximation to be undefined, though we must exclude the possibility of solutions taking a zero value due to the nonlinear term $1/s$ in the drift.

We review two methods that approximate solutions of (8.10) by discretising (8.50) and reversing the transform. The first approach applies an implicit Euler scheme to (8.50) to get

$$s_{n+1} = s_n + h\left[\frac{\alpha}{s_{n+1}} + \beta s_{n+1}\right] + \bar{\gamma}\Delta W_{n+1}, \quad n = 0, \ldots, N-1. \tag{8.51}$$

Multiplying both sides of (8.51) by s_{n+1} and rearranging yields

$$(1 - h\beta)s_{n+1}^2 - (s_n + \bar{\gamma}\Delta W_{n+1})s_{n+1} - h\alpha = 0, \quad n = 0, \ldots, N-1. \tag{8.52}$$

an equation which, given s_n and ΔW_{n+1} as well as $\alpha, \beta, \bar\gamma, h$, is quadratic in the unknown quantity s_{n+1}. The positive root of (8.52) is

$$s_{n+1} = \frac{s_n + \bar\gamma \Delta W_{n+1}}{2(1 - \beta h)} + \sqrt{\frac{(s_n + \bar\gamma \Delta W_{n+1})^2}{4(1 - \beta h)^2} + \frac{\alpha h}{1 - \beta h}}, \quad n = 0, \ldots, N - 1.$$

(8.53)

Using this relation to generate successive observations on $\{s_i\}_{i=0}^{N}$ and then reversing the transform via $r_n = s_n^2$, we can recover simulated trajectories of (7.39) that are positive a.s. This scheme is known as the *drift-implicit square-root Euler scheme*. It was proposed by Alphonsi in [3], and strong convergence in pth mean for $p \in [1, 4\kappa\mu/2\sigma^2)$ is known to be $\gamma = 1$ when $\kappa\mu > (1 \vee 3p/4)\sigma^2$ (see [4, 20]).

The second approach uses an explicit Euler discretisation of (8.50) and modifies solution values that get too close to zero by projecting them upwards. It is known as the *projected Euler scheme*, due to Chassagneux et al. [13] and is given by

$$Y_{n+1} = \widehat{Y}_n + \left(\frac{\alpha}{\widehat{Y}_n}\right) h + \bar\gamma \Delta W_{n+1}; \quad \widehat{Y}_n = \max(h^\delta, Y_n), \quad n = 0, \ldots, N - 1,$$

for some user-selected parameter $\delta \in (0, 1)$. Orders of strong convergence in first moment ($p = 1$) are known for the following parameter ranges:

- If $\kappa\mu > \frac{5}{2}\sigma^2$ then $\gamma = 1$;
- If $\frac{5}{2}\sigma^2 > \kappa\mu > \frac{3}{2}\sigma^2$ then $\gamma = 1/2$;
- If $\frac{3}{2}\sigma^2 > \kappa\mu > \sigma^2$ then $\gamma \in \left(1/6, 1/2 - \sigma^2/(2\kappa\mu + \sigma^2)\right)$.

Alternative approaches that use adaptive timestepping and a splitting of the drift coefficient to construct explicit discretisations of the auxiliary SDE (8.53) may be found in [40, 41], and we direct the reader to the first of these articles for a more comprehensive literature review and comparative numerical investigation of strong approximation methods for (8.10).

8.5.2.3 The Truncated Milstein Method

This method, proposed and analysed by Hefter and Hurzwurm in [28], is the only method we are aware of where a positive polynomial order of strong convergence is known for all parameter values. The method itself is characterised by the relation

$$R_1 = \max\left\{\frac{\sigma}{2}\sqrt{h}, \sqrt{\max\left\{\frac{\sigma^2}{4}h, X_n\right\}} + \frac{\sigma}{2}\Delta W_{n+1}\right\};$$

$$X_{n+1} = \max\left\{R_1^2 + h\left(\kappa(\mu - X_n) - \frac{\sigma^2}{4}\right), 0\right\}, \quad n = 0, \ldots, N - 1,$$

and the order of strong convergence in pth mean is

$$\gamma = \min\left\{\frac{1}{2p}, \frac{2\kappa\mu}{p\sigma^2} - \epsilon\right\},$$

for all $\epsilon > 0$. When $p = 2$ this corresponds to strong root-mean-square convergence of order $\gamma = 1/4$ when $4\kappa\mu > \sigma^2$ and $\gamma = \kappa\mu/\sigma^2 < 1/4$ otherwise.

A clear understanding of weak and strong orders of convergence is necessary to characterise the effectiveness of a variance reduction technique known as multilevel Monte Carlo sampling, and our final subsection explores this.

8.6   Multilevel Monte Carlo (MLMC) Approximation

Let X be a solution of the SDE (8.1) with $d = m = 1$, where the drift coefficient f and the diffusion coefficient G both satisfy a global Lipschitz condition.

Suppose we wish to produce a MC estimation of $\mathbb{E}[P(X(T))]$, where P is a globally Lipschitz test function and the value of $X(T)$ must itself be approximated as X_N^h, either using an Euler-Maruyama or a Milstein scheme over a mesh with fixed stepsize h. Recall from the outset of the chapter that if we let $Y = P(X(T))$, the brute-force MC estimator $\widehat{Y}$ using M samples is given by

$$\widehat{Y} = \frac{1}{M}\sum_{i=1}^{M} P(X_N(\omega_i)).$$

When an SDE numerical method has been used to approximate $P(X(T))$, the mean-square error (MSE) of the brute-force MC estimator can be decomposed into a sum of the variance $\mathbb{E}\left[\left(\widehat{Y} - \mathbb{E}\left[\widehat{Y}\right]\right)^2\right]$ and the square of the bias $\left(\mathbb{E}\left[\widehat{Y}\right] - \mathbb{E}[Y]\right)^2$. As we saw in Chap. 4, the first of these terms is $O(1/M)$. By Theorem 8.14 in Sect. 8.3.2, which provides the order of weak convergence $\gamma = 1$ of the explicit Euler scheme, the square of the bias is $O(h^2)$, as $h \to 0$. The same is true for the Milstein scheme. When combined, the MSE of the brute-force MC estimate can be expressed in big-O notation as

$$\mathbb{E}\left[\left(\widehat{Y} - \mathbb{E}[Y]\right)^2\right] = O(1/M) + O(h^2).$$

If we wish to achieve a RMS error within a tolerance ϵ (equivalent to a MSE within a tolerance ϵ^2), $1/\sqrt{M}$ and h must both be $O(\epsilon)$. Equivalently, we must have $M = O(\epsilon^{-2})$ and $h = O(\epsilon)$.

To determine the computational cost to achieve the desired MSE ϵ^2, first consider the cost of simulating a single trajectory, taken as the number of timesteps used. For both explicit Euler and Milstein schemes, the cost of a trajectory is $O(1/h)$. Hence M independent trajectories have an overall computational cost of

$$O(M/h) = O(\epsilon^{-2}/\epsilon) = O(\epsilon^{-3}).$$

8.6.1   Motivating the MLMC Estimator

Consider numerical simulations of the same trajectory $X(\cdot, \omega)$ over $L + 1$ different levels of resolution, such that each level uses $N_l = 2^l$ steps, for $l = 0, \ldots, L$. Then associated with each level is a timestep $h_l = T/N^l$. On the coarsest level, corresponding to $l = 0$, simulated trajectories use only a single step, while on the finest level, corresponding to $l = L$, they use 2^L steps. This notion of generating the same trajectories at different levels of resolution is illustrated using a Brownian bridge in Figs. 7.2 and 7.3 in Chap. 7.

Let

$$\widehat{P}_l := P\left(X_N^{h_l}\right), \quad l = 0, \ldots, L,$$

denote each numerical approximation of $P(X(T))$ using a numerical discretisation $P\left(X_N^{h_l}\right)$ with timestep h_l. Then

$$\mathbb{E}\left[\widehat{P}_L\right] = \underbrace{\mathbb{E}\left[\widehat{P}_0\right]}_{=:A} + \sum_{l=1}^{L} \underbrace{\mathbb{E}\left[\widehat{P}_l - \widehat{P}_{l-1}\right]}_{=:B_l} \tag{8.54}$$

holds, since the RHS of (8.54) is a telescoping sum. This allows us to express the expectation computed using the finest resolution at level L, given by $\mathbb{E}\left[\widehat{P}_L\right]$, in terms of the expectation computed using the coarsest resolution, $\mathbb{E}\left[\widehat{P}_0\right]$, plus a sum of corrections at intermediate levels of resolution. The idea of MLMC is to independently estimate each of the expectations on the RHS of (8.54) in a way that minimises the overall variance for a given computational cost.

Let $\widehat{Y}_0$ be an estimator for A using M_0 samples, and let $\widehat{Y}_l$ be an estimator for B_l using M_l samples. Then we can estimate $\mathbb{E}\left[\widehat{P}_L\right]$ by summing over all these estimators:

$$\mathbb{E}\left[\widehat{P}_L\right] \approx \widehat{Y} := \sum_{l=0}^{L} \widehat{Y}_l = \sum_{l=0}^{L} \left[\frac{1}{M_l} \sum_{i=1}^{M_l} \left(\widehat{P}_l(\omega_i) - \widehat{P}_{l-1}(\omega_i)\right)\right]; \tag{8.55}$$

$$\widehat{P}_{-1}(\omega_i) := 0, \; \omega_i \in \Omega.$$

The key point here is that each $\widehat{P}_l(\omega_i) - \widehat{P}_{l-1}(\omega_i)$ is the difference of two discrete approximations of the same trajectory but with different stepsizes: see Fig. 8.1.

The advantage of MLMC over brute force MC is that the algorithm is designed so that most of the samples are taken on the coarser levels with relatively large stepsizes, and only a few samples are taken with high resolution, leading to a more efficient sampling method for a given bound on the MSE.

8.6.2 A Python Demonstration of Variance Reduction by MLMC Sampling

We will apply brute-force MC sampling and MLMC sampling to compute an estimate for $\mathbb{E}[X(T)]$ (choosing the identity $P(x) = x$ as our test function), where X satisfies the SDE

$$dX(t) = -X(t)dt + \frac{1}{4}X(t)dW(t), \quad t \in [0, 1] \tag{8.56}$$

with $X(0) = X_0 = 1$. We will sample trajectories of (8.56) approximately using the Milstein scheme (8.47) with $\lambda = -1$, $\sigma = 0.25$, and $\theta = 0$.

Before we start, check the mean-square stability of both the SDE and Milstein scheme. For these parameter values, we can confirm that $2\lambda + \sigma^2 < 0$ with $\lambda = -1$ and $\sigma = 0.25$, and therefore that the equilibrium solution $X(t) \equiv 0$ of (8.56) is globally-mean-square asymptotically stable. By further setting $h = 1$ and $\theta = 0$ in the condition (8.48), we can confirm that the equilibrium solution at zero of the Milstein scheme applied to (8.56) is also mean-square asymptotically stable for all stepsizes used.

Our goal is to demonstrate that if both the brute-force MC and MLMC sampling are carried out with the same computational cost, then the sample variance of the MLMC approach is smaller. For the brute force MC approach, we will compute $M = 10^4$ samples using $N = 2$ uniform steps over the interval of simulation $[0, 1]$. The computational cost of this is the total number of steps taken to generate the ensemble,

$$C = M \times N = 2 \times 10^4,$$

which corresponds to the total number of observations on a normal random variable generated by the RNG during the execution of the code. For the MLMC approach, we will compute $M_0 = 10^4$ trajectories using $N_0 = 1$ step over $[0, 1]$ and, by applying a Brownian bridge, recompute $M_1 = 5 \times 10^3$ of those trajectories using $N_1 = 2$ steps over $[0, 1]$. The computational cost associated with this is the same:

$$C = M_0 \times N_0 + M_1 \times N_1 = 2 \times 10^4.$$

First, set up model parameters, along with samples sizes $M = M_0, M_1$ and corresponding timesteps h_0, h_1.

```
In []: # SDE parameters
       r=-1; sig=0.25; S0=1; T=1


       # Steps and stepsizes for levels 0 and 1
       N0=2**0; N1=2**1; dt0=T/N0; dt1=T/N1
```

```
# Sample sizes for brute force MC and each level of MLMC
M=10**4; M0=int(M*1); M1=int(M*0.5)
```

Confirm that computational costs are the same

```
In []:  C=M*N1
        Cm=M0*N0+M1*N1
        print([C,Cm])
```

```
Out[]:  [20000, 20000]
```

For the brute-force MC approach, generate $M = 10^4$ samples at level $L = 1$ using the Milstein method. We define a function `milstein()` to apply the map associated with the scheme at every step of the iteration.

```
In []:  S=S0*np.ones((M,N1+1))
        Z=rng.normal(0,dt1,(M,N1))

        def milstein(Si,dt,dW,r,sig):
            return(Si*(1+r*dt+sig*dW+0.5*sig**2*(dW**2-dt)))

        for i in range(N1):
            S[:,i+1]=milstein(S[:,i],dt1,Z[:,i],r,sig)
```

For the MLMC approach, start by generating $M_1 = 10^4$ samples at level $l = 0$.

```
In []:  S10=S0*np.ones((M0,N0+1))
        Z10=rng.normal(0,dt0,(M0,N0))
        W0=np.c_[np.zeros(M0),np.cumsum(Z10,axis=1)]

        for i in range(N0):
            S10[:,i+1]=milstein(S10[:,i],dt0,Z[:,i],r,sig)
```

We now interpolate a subset of these trajectories to move from level $l = 0$ to level $l = 1$ using a Brownian bridge. To prepare for the application of the `bbridge()` function constructed in Sect. 7.1.4 of Chap. 7, use array slicing to insert columns of zeros at every second column index, and create a time set.

```
In []:  Wb1=np.zeros((M1,2*(np.shape(W0)[1])-1))
        Wb1[:M1,::2]=W0[:M1,::]
        t1=np.arange(0,T+dt1,dt1)
```

Now apply the Brownian bridge to the first $M_1 = 5000$ trajectories of the ensemble.

```
In []: for i in range(1,N1+1,2):
           Wb1[i]=b_bridge(t1[i-1],t1[i],t1[i+1],
                           Wb1[i-1],Wb1[i+1],rng)
```

Use these interpolated trajectories to generate $M_1 = 5000$ samples at level $l = 1$.

```
In []: Sl1=S0*np.ones((M1,N1+1))

       for i in range(N1):
           dW1=Wb1[:,i+1]-Wb1[:,i]
           Sl1[:,i+1]=milstein(Sl1[:,i],dt1,dW1,r,sig)
```

Compute the telescoping sum (8.54) (without taking expectations, so that we can examine the variance of the resulting samples) with $L = 1$.

```
In []: P0=Sl0[:,-1];
       P1=Sl1[:,-1]-Sl0[:M1,-1]
```

Finally, compute the standard deviation of the brute force and MLMC estimators respectively

```
In []: print(np.std(S))
       print(np.std(np.concatenate((P0,P1))))

Out[]: 0.32772228731404607
       0.20172668562020776
```

We see that there has been a reduction of approximately 38% in the sample standard deviation due to the use of MLMC sampling here.

8.6.3 Algorithm, and the Roles of Weak and Strong Convergence

We present a more general version of the MLMC algorithm here that avoids the use of a Brownian bridge, and which dynamically estimates the optimal number of levels L and sample sizes M_l for each level $l = 0, \ldots, L$. Let V_0 and V_l denote the variances of $\widehat{Y}_0$ and $\widehat{Y}_l$ respectively, $l = 1, \ldots, L$, and C_0 and C_l to be their respective computational cost. In estimating $\mathbb{E}[P(X(T))]$ the total computational cost is $C = \sum_{l=0}^{L} M_l C_l$.

Although we don't give a detailed theoretical justification, we do wish to highlight the roles of both weak and strong order of convergence in the effectiveness of the method. Let α be the order of weak convergence of the numerical method used to discretise (8.1) (recall that $\alpha = 1$ for both the Euler-Maruyama and Milstein schemes). Let β be the square of the order of root-mean-square strong convergence of the numerical method (so that $\beta = 1$ for the Euler-Maruyama method and $\beta = 2$ for the Milstein method).

The MLMC algorithm may be expressed as follows (see for example Giles [24]).

1. Choose a tolerance ϵ and set $L = 0$;
2. Estimate $\widehat{Y}$ and V_L using an initial $M_L = 10^4$ samples;
3. Determine the optimal number of samples at each level $l = 0, \ldots, L$ to be

$$M_l = 2\epsilon^{-2}\sqrt{V_l/C_l} \sum_{i=0}^{L} \sqrt{V_i C_i}.$$

4. Evaluate extra samples at each level as needed for each new M_l;
5. If $L \geq 2$, test for convergence using the criterion

$$\left| \mathbb{E}\left[\widehat{P}_L - \widehat{P}_{L-1} \right] \right| < \frac{\epsilon}{2}(2^\alpha - 1)$$

If the method has converged, stop, and compute the final estimator according to (8.55).
6. If $L < 2$ or the method has not converged, set $L := L + 1$ and go back to step 2.

We can see that the order of weak convergence α plays a direct role in the convergence test used to determine the optimal value for L given in Step 5. This is intuitively because the order of weak convergence controls the rate at which the bias of the method decreases as the stepsize h_l decreases, and the faster this occurs, the fewer levels are required. Giles showed that, for this algorithm, the MLMC estimator (8.55) has MSE that satisfies the bound

$$\mathbb{E}\left[\left(\widehat{Y} - \mathbb{E}[P(X(T))] \right)^2 \right] < \epsilon^2,$$

and an overall computational cost that satisfies the bound

$$C = \begin{cases} O(\epsilon^{-2}), & \beta = 2, \ \text{(e.g. using the Milstein scheme);} \\ O(\epsilon^{-2}(\log \epsilon)^2), & \beta = 1, \ \text{(e.g. using the Euler-Maruyama scheme).} \end{cases}$$

The order of strong convergence also plays a role in the final computational cost through the parameter β. We can see that if we use the Milstein scheme to discretise (8.1), the computational cost of the MLMC method required to bound the MSE above by ϵ^2 is the same as the computational complexity of brute-force MC in the case where $P(X(T))$ can be sampled exactly. In effect the MLMC approach eliminates the additional cost associated with the use of the Milstein method.

More comprehensive demonstration code for MLMC computations using Python has been made available online under a GNU General Public Licence at [23].

8.7   Further Reading

For readers who made it this far and would like to step further into the world of numerical methods for SDEs, we recommend the excellent tutorial article by Higham [31] which has by now introduced several generations of students to this area. Sample code in MATLAB, though not in Python, is provided. A more recent textbook treatment is in Higham and Kloeden [33].

Other texts of interest include that of Kloeden and Platen [43], representing an early codification of an emerging field, Milstein and Tretyakov [54], who give an in-depth treatment of numerical techniques for stochastic modelling problems arising in physics. Numerical methods for stochastic PDEs are covered in Lord, Powell, and Shardlow [50].

An interesting recent application of importance sampling to demonstrate the relationship of a.s. and mean-square stability for linear SDEs and Monte Carlo sampling is given in Ableidinger, Buckwar, and Thalhammer [1]. Finally, there is a considerable literature on MLMC methods: see Giles [24] and the references therein.

8.7.1   Exercises

8.1 For the Vasicek interest rate model with $T = 10$, $r_0 = 0.05$, $\lambda = 1$, $\mu = 0.03$, and $\sigma = 0.015$, implement the θ-Maruyama discretisation method given by (8.9) corresponding to $\theta = 0, 1, 0.5$ (i.e. the explicit Euler-Maruyama scheme, the trapezoid rule, and the implicit Euler-Maruyama scheme), and generate an ensemble of $M = 10$ trajectories with $N = 10^3$ uniform steps over the interval $[0, T]$ in each case.

8.2 Suppose that $\lambda, \sigma \in \mathbb{R}$ in the linear test equation (8.18). For the implicit Euler-Maruyama method, prove that

$$S_{\text{SDE}} \subseteq S_{\text{STM}}(h, 1),$$

where S_{SDE} and $S_{\text{STM}}(h, \theta)$ are as defined in (8.21) and (8.22) of Definition 8.4 respectively.

8.3 Explain why

1. the Euler-Maruyama and Milstein discretisations for the Vasicek model SDE (8.4) are the same;
2. the Euler-Maruyama scheme converges strongly in root-mean-square to solutions of (8.4) with order $\gamma = 1$.

8.4 In Python, implement a numerical approximation of the SABR model (8.13) over a uniform mesh by solving the SDE for the volatility process σ exactly and solving the forward price process F using an

1. Euler-Maruyama scheme;
2. Milstein scheme.

You may take $T = 1$, $N = 100$, $\rho = -0.4$, $\alpha = 0.01$, $\beta = 0.75$, $\sigma_0 = 0.2$.

8.5 In Sect. 8.3.2 we examined the weak convergence of the explicit Euler scheme (8.28). Since this scheme samples from the increments of the standard Brownian motion W it is known as a strong approximation. A weak approximation, by contrast, replaces the random variable ΔW_{n+1} in (8.28) with a two- or three- point random variable with mean zero and variance h. Such schemes are intended to preserve the statistical properties of the SDE while allowing for more computationally efficient sampling.

1. Show that if we define

$$V^h_{n+1} = \begin{cases} \sqrt{h}, & \text{with probability } 1/2; \\ -\sqrt{h}, & \text{with probability } 1/2, \end{cases}$$

then

$$\mathbb{E}\left[V^h_{n+1}\right] = 0 \quad \text{and} \quad \mathbb{E}\left[\left(V^h_{n+1}\right)^2\right] = h.$$

2. Modify the code in Sect. 8.3.2 to produce a weak convergence plot of the weak Euler approximation of (8.17) given by

$$X_{n+1} = X_n \left(1 + \lambda h + \sigma V^h_{n+1}\right), \quad n \in 0, \ldots, N - 1,$$

with $X_0 = 1$, $\lambda = -0.5$, and $\sigma = 0.3$.

8.6 All of the models examined in this chapter have a drift coefficient that is either linear or zero. In the case where the drift coefficient is nonlinear, the implementation of the implicit Euler scheme requires the solution of a fixed-point problem at each step. Consider the following general example:

$$dX(t) = f(X(t))dt + \sigma X(t)dW(t), \quad t \in [0, T], \tag{8.57}$$

$X(0) = X_0 > 0$. An implicit Euler-Maruyama scheme takes the form

$$X_{n+1} = X_n + f(X_{n+1})h + \sigma X_n \Delta W_{n+1}, \quad n = 0, \dots, N-1.$$

At time t_n, when we will have computed values for $X_n = x$ and $\Delta W_{n+1} = w$, the value of the approximation at time t_n is given by the relation

$$X_{n+1} = x + f(X_{n+1})h + \sigma x w,$$

and so solving for X_{n+1} is equivalent to solving the fixed point problem $H(y) = y$ where $H(y) = x + f(y)h + \sigma x w$.

Use the `fsolve()` call to implement an implicit Euler-Maruyama method to solve the SDE (8.57) with $\sigma = 0.1$ and $X_0 = 10$ over the interval $[0, 1]$ when $f(x) = \sin(x)$, and plot $M = 10$ trajectories.

8.7 Show that (8.44) follows directly from (8.43).

8.8 Let $\{X_n^h\}_{n=0}^N$ be the solution of (8.47), the Milstein discretisation of (8.17) with real coefficients $\lambda, \sigma \in \mathbb{R}$. Prove that the equilibrium solution $X_n^h \equiv 0$ is mean-square asymptotically stable if and only if h and θ are such that (8.48) holds.

8.9 Consider the one-dimensional SDE model given by

$$dF(t) = \sigma F(t)^\beta dW(t), \quad t \in [0, T], \tag{8.58}$$

where $\sigma > 0$, $\beta \in [1/2, 1)$ and W is a standard Brownian motion.

1. Use Itô's formula to show that the SDE governing the transformed process

$$Y(t) = \frac{1}{1 - \beta} F(t)^{1 - \beta}, \quad t \in [0, T],$$

is given by

$$dY(t) = \frac{\gamma}{Y(t)} dt + \sigma B(t), \quad t \in [0, T], \tag{8.59}$$

where

$$\gamma = -\frac{\sigma^2 \beta}{2(1 - \beta)},$$

for those trajectories of F that never achieve a value of 0.

2. Derive a numerical scheme for Eq. (8.58) by applying a drift-implicit Euler-Maruyama method to (8.59). Is the scheme well-defined for all $\sigma > 0$ and $\beta \in [1/2, 1)$?

8.10 Using the code given in Sect. 8.6.2, explore the effect of varying the volatility parameter σ on the extent of the variance reduction provided by two-level MLMC when the computational cost is held constant. What happens if either the SDE (8.56) or the Milstein discretisation of (8.56) have a mean-square unstable equilibrium solution at zero?

Correction to: Computation and Simulation for Finance

Correction to:
C. Kelly, *Computation and Simulation for Finance*,
Springer Undergraduate Texts in Mathematics and Technology,
https://doi.org/10.1007/978-3-031-60575-8

The original version of the book was inadvertently published with ESM files. The removed online material is now available at sn.pub/lecturer-material

---

The updated version of this book can be found at
https://doi.org/10.1007/978-3-031-60575-8

References

1. M. Ableidinger, E. Buckwar, A. Thalhammer, An importance sampling technique in Monte Carlo methods for SDEs with a.s. stable and mean-square unstable equilibrium. J. Comput. Appl. Math. **316**, 3–14 (2017)
2. Y. Aït-Sahalia, Testing continuous-time models of the spot interest rate. Rev. Financ. Stud. **9**(2), 385–426 (1996)
3. A. Alfonsi, On the discretization schemes for the CIR (and Bessel squared) processes. Monte Carlo Methods Appl. **11**, 355–384 (2005)
4. A. Alfonsi, Strong order one convergence of a drift implicit Euler scheme: application to the CIR process. Stat. Probab. Lett. **83**, 602–607 (2013)
5. R. Bellman, *Dynamic Programming* (Dover Publications, 1957)
6. W.-J. Beyn, E. Isaak, R. Kruse, Stochastic C-stability and B-consistency of explicit and implicit Milstein-type schemes. J. Sci. Comput. **70**(3), 1042–1077 (2017)
7. T. Björk, *Arbitrage Theory in Continuous Time*, second, reprint edition (Oxford University Press, Oxford, 2005)
8. F. Black, M. Scholes, The pricing of options and corporate liabilities. J. Polit. Econ. **81**(3), 637–654 (1973)
9. M. Bossy, A. Diop, An efficient discretization scheme for one dimensional SDEs with a diffusion coefficient function of the form $|x|^{\alpha}$, $\alpha \in [\frac{1}{2}, 1]$. Technical Report 5396, INRIA working paper, 2004
10. P.P. Boyle, S.H. Lau, Bumping up against the barrier with the binomial method. J. Derivatives **1**(4), 6–14 (1994)
11. E. Buckwar, C. Kelly, Towards a systematic linear stability analysis of numerical methods for systems of stochastic differential equations. SIAM J. Numer. Anal. **48**(1), 298–321 (2010)
12. E. Buckwar, C. Kelly, Non-normal drift structures and linear stability analysis of numerical methods for systems of stochastic differential equations. Comput. Math. Appl. **64**(7), 2282–2293 (2012). Recent Developments in Difference Equations
13. J-F. Chassagneux, A. Jacquier, I. Mihaylov, An explicit Euler scheme with strong rate of convergence for financial SDEs with non-Lipschitz coefficients. SIAM J. Financ. Math. **7**(1), 993–1021 (2016)
14. U. Cherubini, E. Luciano, W. Vecchiato, *Copula Methods in Finance*. Wiley Finance Series (Wiley, 2013)
15. G.H. Choe, *Stochastic Analysis for Finance with Simulations*. Universitext (Springer International Publishing, 2016)
16. M. Costabile, A discrete-time algorithm for pricing double barrier options. Decis. Econ. Finance **24**, 49–58 (2001)
17. J.C. Cox, S.A. Ross, M. Rubinstein, Option pricing: A simplified approach. J. Financ. Econ. **7**(3), 229–263 (1979)

© The Author(s), under exclusive license to Springer Nature Switzerland AG 2024

C. Kelly, *Computation and Simulation for Finance*,

Springer Undergraduate Texts in Mathematics and Technology,

https://doi.org/10.1007/978-3-031-60575-8

18. J.C. Cox, J.E. Ingersoll Jr., S.A. Ross, A theory of the term structure of interest rates. Econometrica **53**(2), 385–407 (1985)
19. A. Cozma, C. Reisinger, Strong order 1/2 convergence of full truncation Euler approximations to the Cox-Ingersoll-Ross process. IMA J. Numer. Anal. **40**, 358–376 (2020)
20. S. Dereich, A. Neuenkirch, L. Szpruch, An Euler-type method for the strong approximation of the Cox-Ingersoll-Ross process. Proc. R. Soc. Lond. A Math. Phys. Eng. Sci. **468**(2140), 1105–1115 (2012)
21. A. Di Clemente, C. Romano, Calibrating and simulating copula functions in financial applications. Front. Appl. Math. Stat. **7**, 11 (2021)
22. M.F. Dixon, I. Halperin, P. Bilokon, *Machine Learning in Finance: From Theory to Practice* (Springer International Publishing, 2020)
23. P.E. Farrell, M.B. Giles, M. Croci, T. Roy, C. Beentjes, A Python implementation of MLMC. https://bitbucket.org/pefarrell/pymlmc/ (2020). Date of last access: March 5, 2024
24. M.B. Giles, Multilevel Monte Carlo methods. Acta Numer. **24**, 259–328 (2015)
25. P. Glasserman, *Monte Carlo Methods in Financial Engineering*, volume 53 of Applications of Mathematics (New York) (Springer, New York, 2004). Stochastic Modelling and Applied Probability
26. G.R. Grimmett, D.R. Stirzaker, *Probability and Random Processes*. Probability and Random Processes (Oxford University Press, Oxford, 2001)
27. E.G. Haug, *The Complete Guide to Option Pricing Formulas*, second, illustrated edition (McGraw-Hill Education, 2007)
28. M. Hefter, A. Herzwurm, Strong convergence rates for Cox-Ingersoll-Ross processes — full parameter range. J. Math. Anal. Appl. **459**, 1079–1101 (2018)
29. D.J. Higham, A-stability and stochastic mean-square stability. BIT **40**(2), 404–409 (2000)
30. D.J. Higham, Mean-square and asymptotic stability of numerical methods for stochastic ordinary differential equations. SIAM J. Numer. Anal. **38**(3), 753–769 (2000)
31. D.J. Higham, An algorithmic introduction to numerical simulation of stochastic differential equations. SIAM Rev. **43**(3), 525–546 (2001)
32. D.J. Higham, *An Introduction to Financial Option Valuation: Mathematics, Stochastics and Computation* (Cambridge University Press, 2004)
33. D.J. Higham, P.E. Kloeden, *An Introduction to the Numerical Simulation of Stochastic Differential Equations*. Other Titles in Applied Mathematics (Society for Industrial and Applied Mathematics, 2021)
34. D.J. Higham, X. Mao, Convergence of Monte Carlo simulations involving the mean-reverting square root process. J. Comput. Finance **8**(3), 35–61 (2005)
35. R.A. Horn, C.R. Johnson, *Matrix Analysis*, 2nd edn. (Cambridge University Press, 2012)
36. J.C. Hull, *Options, Futures, and Other Derivatives*, sixth, Pearson international edition (Pearson Prentice Hall, Upper Saddle River, NJ, 2006)
37. P. Jaeckel, *Monte Carlo Methods in Finance* (Wiley, 2002)
38. S. Jain, C.W. Oosterlee, The stochastic grid bundling method: efficient pricing of Bermudan options and their Greeks. Appl. Math. Comput. **269**(C), 412–431 (2015)
39. I. Karatzas, S. Shreve, *Brownian Motion and Stochastic Calculus*. Graduate Texts in Mathematics (Springer New York, 2014)
40. C. Kelly, G.J. Lord, An adaptive splitting method for the Cox-Ingersoll-Ross process. Appl. Numer. Math. **186**, 252–273 (2023)
41. C. Kelly, G.J. Lord, M. Maulana, The role of adaptivity in a numerical method for the Cox-Ingersoll-Ross model. J. Comput. Appl. Math. **410**, 114208 (2022)
42. J.F.C. Kingman, S.J. Taylor, *Introduction to Measure and Probability* (Cambridge University Press, 1966)
43. P.E. Kloeden, E. Platen, *Numerical Solution of Stochastic Differential Equations*. Stochastic Modelling and Applied Probability (Springer Berlin Heidelberg, 2011)
44. M. Lauko, D. Ševčovič, Comparison of numerical and analytical approximations of the early exercise boundary of American put options. ANZIAM J. **51**, 430–448 (2010)

45. D.P.J. Leisen, M. Reimer, Binomial models for option valuation - examining and improving convergence. Appl. Math. Finance **3**(4), 319–346 (1996)
46. P. Lévy, Wiener's random function, and other Laplacian random functions, in *Proceedings of the Second Berkeley Symposium on Mathematical Statistics and Probability* (The Regents of the University of California, 1951)
47. F. Lindskog, A. McNeil, U. Schmock, Kendall's tau for elliptical distributions, in *Credit Risk: Measurement, Evaluation and Management.* Contributions to Economics (Physica-Verlag HD, Heidelberg, 2003)
48. F. Longstaff, E. Schwartz, Valuing American options by simulation: A simple least-squares approach. Rev. Financ. Stud. **14**, 113–47 (2001)
49. R. Lord, R. Koekkoek, D. Van Dijk, A comparison of biased simulation schemes for stochastic volatility models. Quant. Finance **10**(2), 177–194 (2010)
50. G.J. Lord, C.E. Powell, T. Shardlow, *An Introduction to Computational Stochastic PDEs.* Cambridge Texts in Applied Mathematics (Cambridge University Press, 2014)
51. D. MacKenzie, T. Spears, The formula that killed Wall street: The Gaussian copula and modelling practices in investment banking. Soc. Stud. Sci. **44**(3), 393–417 (2014)
52. S.J.A. Malham, A. Wiese, Efficient almost-exact Lévy area sampling. Stat. Probab. Lett. **88**, 50–55 (2014)
53. X. Mao, *Stochastic Differential Equations and Applications* (Elsevier Science, 2007)
54. G.N. Milstein, M.V. Tretyakov, *Stochastic Numerics for Mathematical Physics.* Scientific Computation (Springer Berlin Heidelberg, 2013)
55. R.B. Nelson, *An Introduction to Copulas*, 2nd edn. (Springer, New York, 2006)
56. B. Øksendal, *Stochastic Differential Equations: An Introduction with Applications.* Universitext (Springer Berlin Heidelberg, 2010)
57. F. Olver, D. Lozier, R. Boisvert, C. Clark, *The NIST Handbook of Mathematical Functions* (Cambridge University Press, New York, 2010)
58. C.W. Oosterlee, L.A. Grzelak, *Mathematical Modeling and Computation in Finance: with Exercises and Python and Matlab Computer Codes* (World Scientific Publishing Company, 2019)
59. L.C.G. Rogers, E.J. Stapleton, Fast accurate binomial pricing. Finance Stoch. **2**, 3–17 (1997)
60. Y. Saito, T. Mitsui, Stability analysis of numerical schemes for stochastic differential equations. SIAM J. Numer. Anal. **33**(6), 2254–2267 (1996)
61. L.O. Scott, Simulating a multi-factor term structure model over relatively long discrete time periods, in *Proceedings of the IAFE First Annual Computational Finance Conference* (1996)
62. S.E. Shreve, *Stochastic Calculus for Finance II: Continuous-Time Models*, volume 11 of Springer Finance Textbooks (Springer, 2004)
63. J.C. Strikwerda, *Finite Difference Schemes and Partial Differential Equations, Second Edition* (Society for Industrial and Applied Mathematics, 2004)
64. A. Stuart, A.R. Humphries, *Dynamical Systems and Numerical Analysis.* Number v. 8 in Cambridge Monographs on Applied and Computational Mathematics (Cambridge University Press, 1998)
65. J.N. Tsitsiklis, B. van Roy, Optimal stopping of Markov processes: Hilbert space theory, approximation algorithms, and an application to pricing high-dimensional financial derivatives. IEEE Trans. Autom. Control **44**(10), 1840–1851 (1999)
66. O. Vasicek, An equilibrium characterization of the term structure. J. Financ. Econ. **5**(2), 177–188 (1977)
67. X. Wang, S. Gan, The tamed Milstein method for commutative stochastic differential equations with non-globally Lipschitz continuous coefficients. J. Difference Equations Appl. **19**(3), 466–490 (2013)
68. M. Wiese, R. Knobloch, R. Korn, P. Kretschmer, Quant GANs: deep generation of financial time series. Quant. Finance **20**(9), 1419–1440 (2020)
69. P. Wilmott, S. Howison, J. Dewynne, *The Mathematics of Financial Derivatives: A Student Introduction* (Cambridge University Press, 1995)

Index

A

Absolute error, 91
  binomial tree method, 91
American option, 93, 170
  compact linearity form, 172
  free boundary formulation, 171
Arithmetic-geometric mean inequality, 119
Asian option, 117–120
  arithmetic average, 117
  geometric average, 118
a.s. stability and instability, 289
A-stability
  in mean-square, 286, 307, 318
  in the a.s. sense, 289

B

Barrier option, 11, 96–100, 115–117, 167, 185
Basket option, 185, 222
Bellman equation, 94
Bermudan option, 93, 179
Bessel process, 266
Binary option, 108
Binomial algorithm, 84–85
Black-Scholes asset price model, 22, 80, 273
  calibration, 228, 267
  correlated assets, 193, 221
  log-returns, 22
  moments, 31, 80
  probability density function, 31
Black-Scholes formula, 55
  European call option, 55
  European put option, 56
Black-Scholes-Merton PDE, 39, 54, 138, 177
  down-and-out barrier call boundary
      conditions, 140
  European call boundary conditions, 54, 139
  European put boundary conditions, 70, 139,
      178
  general boundary conditions, 139
  up-and-out barrier call boundary
      conditions, 140
Black's model, 266
Bond
  coupon-bearing, 237
  coupon dates, 226
  European option on, 242
  face value, 226
  issue price, 226
  issue rate, 226
  maturity date, 226
  par yield, 226
  yield-to-maturity, 237
  zero-coupon, 15, 236, 242
Brownian bridge, 232–234
Brute-force Monte Carlo, 107
Bump-and-revalue, 130

C

Caplet, cap, 242
Cash-or-nothing option, 134, 168, 179
Central limit theorem, 107
CEV model, 281, 309
Cholesky factorisation, 189, 220, 264, 273
Complete market, 42
Conditional stability, 151, 154
Conditioning formula, 232
Continuation value, 94, 113
Copula
  Archimedean, 208, 222
  Clayton, 223
  definition, 207
  elliptical, 208